Clinical Cultural Neuroscience

National Academy of Neuropsychology Series on Evidence-Based Practices

SERIES EDITOR

L. Stephen Miller

SERIES CONSULTING EDITORS

Glenn J. Larrabee
Martin L. Rohling

Civil Capacities in Clinical Neuropsychology
Research Findings and Practical Applications
Edited By George J. Demakis

Secondary Influences on Neuropsychological Test Performance
Edited By Peter Arnett

Neuropsychological Aspects of Substance Use Disorders
Evidence-Based Perspectives
Edited By Daniel N. Allen and Steven Paul Woods

Neuropsychological Assessment in the Age of Evidence-Based Practice
Diagnostic and Treatment Evaluations
Edited By Stephen C. Bowden

Clinical Cultural Neuroscience
An Integrative Approach to Cross-Cultural Neuropsychology
Edited By Otto Pedraza

Clinical Cultural Neuroscience

An Integrative Approach to Cross-Cultural Neuropsychology

EDITED BY OTTO PEDRAZA

Oxford University Press is a department of the University of Oxford. It furthers
the University's objective of excellence in research, scholarship, and education
by publishing worldwide. Oxford is a registered trade mark of Oxford University
Press in the UK and certain other countries.

Published in the United States of America by Oxford University Press
198 Madison Avenue, New York, NY 10016, United States of America.

© Oxford University Press 2020

All rights reserved. No part of this publication may be reproduced, stored in
a retrieval system, or transmitted, in any form or by any means, without the
prior permission in writing of Oxford University Press, or as expressly permitted
by law, by license, or under terms agreed with the appropriate reproduction
rights organization. Inquiries concerning reproduction outside the scope of the
above should be sent to the Rights Department, Oxford University Press, at the
address above.

You must not circulate this work in any other form
and you must impose this same condition on any acquirer.

Library of Congress Control Number: 2019949627
ISBN 978-0-19-061930-5

To my wife Wendy, who makes me a better person each and every day.

Contents

Preface to the Fifth Volume in the *National Academy of Neuropsychology Series on Evidence-Based Practices* ix

Acknowledgments xi

List of Contributors xiii

1. Introduction to Clinical Cultural Neuroscience 3
   Otto Pedraza

2. Vygotsky, Luria, and the Cultural-Historical Approach to Cognitive Psychology and Neuropsychology: Experimental and Theoretical Beginnings 32
   David E. Tupper

3. Challenges in the Neuropsychological Assessment of Ethnic Minorities 55
   Laura A. Rabin, Donald L. Brodale, Milushka M. Elbulok-Charcape, and William B. Barr

4. Culture and Memory 81
   Lixia Yang, Brenda Wong, and Lingqian Li

5. Assessment of Mood Disorders in Ethnic Minorities 100
   Vonetta M. Dotson, Shellie-Anne Levy, Deirdre M. O'Shea, Molly E. McLaren, and Sarah M. Szymkowicz

6. Visual Cognition and Culture 124
   Joshua O. S. Goh, Chun-Yih Li, Yu-Zhen Tu, and Caroline Dallaire-Théroux

7. Cognitive Reserve, Bilingualism, and the Aging Brain 151
   Brian T. Gold

8. Neuropsychological Assessment of Non-English Speakers 169
   Octavio A. Santos, Daryl E. M. Fujii, and Otto Pedraza

9. Neurocognitive Development of Semantics in Chinese- and English-Speaking Children With and Without Autism 200
   Tai-Li Chou and James Booth

10. Culture and Language Diversity in Pediatric Neuropsychology 215
    Veronica Bordes Edgar and Regilda Anne Romero

11. Racial Disparities in Alzheimer's Disease: Biological, Social, and Methodological Considerations 233
    Megan Zuelsdorff, Lisa L. Barnes, and Ozioma C. Okonkwo

12. Bias, Equivalence, and Fairness 252
    Otto Pedraza and Fons J. R. van de Vijver

Index 277

Preface to the Fifth Volume in the *National Academy of Neuropsychology Series on Evidence-Based Practices*

We are excited to provide this latest edited volume for the National Academy of Neuropsychology (NAN) series focusing on an integrative approach to cross-cultural neuropsychology from a clinical cultural perspective. This volume has been a long time coming for the field of clinical neuropsychology. It is unique in that it approaches the area of cultural influences on neuropsychological assessment using as its emphasis the interplay of culture and the neural basis of behavior. As such, each chapter not only focuses on key cultural variables that influence our assessment methods; it does so from a clear and thoughtful understanding of the neural underpinnings of behavior. As a field, clinical neuropsychology has been strongly associated with the bidirectional interplay of brain function and behavior. This volume expands that general perspective to explain and use as a backdrop the neuroscientific evidence for cultural influences on our assessment methods.

Here, in this fifth volume of the National Academy of Neuropsychology's Series on Evidence-Based Practices—*Clinical Cultural Neuroscience: An Integrative Approach to Cross-Cultural Neuropsychology*—Otto Pedraza has brought together a selection of cultural and neuroscience experts to address this important topic area. I hope that it is clear from this esteemed group that the issues of cultural factors within a neuroscience framework as they related to clinical neuropsychological assessment provides a combination of best practice approaches that are clearly tied to our contemporary understanding of brain–behavior relationships borne from current scientific evidence. It is presented in a way that teaches and informs, yet remains accessible to the clinician and applicable to the real world of the practicing clinical neuropsychologist.

This important volume provides the best current evidence across an array of clinically significant cultural factors that include race, ethnicity, age, and language and presents the major challenges as well as potential solutions to appropriately and accurately measure cognitive processes in an age of increasing diversity and difference. Pedraza begins with an introduction and succinct definition of clinical cultural neuroscience in his introductory chapter, followed by invited contributions across the gamut of most important cultural variables. Authors use evidence-driven findings to argue the relative influence of these many factors and to offer where possible solutions for best practices in attending and incorporating cultural differences into our understanding and interpretation of our assessment data. It is hoped that clinicians can and should adopt at the individual patient level. This will inform researchers and practitioners alike and make available the latest science examining these relationships.

Pedraza is Associate Professor of Psychology in the Department of Psychiatry and Psychology at the Mayo Clinic in Jacksonville, Florida. He is a prolific researcher focusing on cognitive endophenotypes associated with the major neurodegenerative diseases, as well as publishing extensively on psychometrics and test validation of some of our most used testing instruments. He is an established clinical neuropsychologist and an American Board of Professional Psychology diplomate in clinical neuropsychology and a longtime and active NAN member. He is a perfect choice for providing guidance for evidence-based methods of bringing the application of the field of clinical cultural neuroscience to the practice of clinical neuropsychology.

As with all of the previous volumes in the NAN Series, this volume is aimed primarily at neuropsychologists but should also be of use to a broad range of professionals interested in understanding cultural influences from a neuroscience perspective.

It is my hope that this volume provides an appropriate beginning for increasing our understanding of the way culture is intimately tied to our functioning and our abilities as clinical professionals to assess and interpret those ties.

Sincerely,
L. Stephen Miller
Editor-in-Chief
National Academy of Neuropsychology Book
Series on Evidence-Based Practices
January 2019

Acknowledgments

I would like to express my gratitude and appreciation to L. Stephen Miller, Series Editor, for his steady guidance, encouragement, and support throughout the development and production of this volume. I also would like to convey my gratitude to the Board of Directors of the National Academy of Neuropsychology for their consideration of this topic within the Evidence-Based Practices book series. I am particularly indebted to the wonderful group of chapter authors, whose willingness to donate their time and expertise enabled this volume to be realized. The contents were further enhanced by the keen eyes and thoughtful comments from Glenn Larrabee, Martin Rohling, and Steve Miller. At Oxford University Press, Phil Velinov and Lynnee Argabright provided invaluable assistance, and Joan Bossert shepherded the volume through all stages of production with kind words and a knowledgeable hand. To my colleagues at Mayo Clinic in Florida, thank you for your enthusiasm and encouragement. To my parents, I offer the deepest gratitude for your love, courage, and sacrifice, which have paved the way for so much in my life. And most of all, I extend heartfelt thanks to my wife Wendy, whose patience, understanding, and support during long weeknight and weekend hours made it all possible.

Contributors

Lisa L. Barnes, PhD
Department of Neurological Sciences,
  Department of Behavioral Sciences,
  Rush Alzheimer's Disease Center,
  Rush University Medical Center
Chicago, IL, USA

William B. Barr, PhD
Department of Neurology, NYU
  School of Medicine
New York, NY, USA

James Booth, PhD
Department of Psychology and
  Human Development, Vanderbilt
  University
Nashville, TN, USA

Donald L. Brodale, MA
Department of Sociology, Brooklyn
  College of the City University of
  New York
Brooklyn, NY, USA

Tai-Li Chou, PhD
Department of Psychology, National
  Taiwan University
Taipei, Taiwan

Caroline Dallaire-Théroux, MD, MSc
Centre de recherche de l'Institut
  universitaire en santé mentale
  de Québec; Faculty of Medicine,
  Université Laval
Quebec City, QC, Canada

Vonetta M. Dotson, PhD
Department of Psychology, Georgia
  State University
Atlanta, GA, USA

Veronica Bordes Edgar, PhD
Departments of Psychiatry and
  Pediatrics, University of Texas
  Southwestern Medical Center
Dallas, TX, USA

Milushka M. Elbulok-Charcape, MPhil
Department of Educational
  Psychology, The Graduate Center
  of the City University of New York
New York, NY, USA

Daryl E. M. Fujii, PhD
Veterans Affairs Pacific Islands Health
  Care System
Honolulu, HI, USA

Joshua O. S. Goh, PhD
Graduate Institute of Brain and Mind
  Sciences, College of Medicine;
  Psychology Department; and
  Neurobiological and Cognitive
  Science Center, National Taiwan
  University
Taipei, Taiwan

Brian T. Gold, PhD
Department of Neuroscience,
  College of Medicine, University of
  Kentucky
Lexington, KY, USA

Shellie-Anne Levy, PhD
Department of Clinical and
  Health Psychology,
  University of Florida
Gainesville, FL, USA

Chun-Yih Li, MS
Graduate Institute of Brain and Mind
  Sciences, College of Medicine,
  National Taiwan University
Taipei, Taiwan

Lingqian Li, MA
Department of Psychology, Ryerson
  University
Toronto, ON, Canada

Molly E. McLaren, PhD
Department of Rehabilitation
  Medicine, Emory University School
  of Medicine
Atlanta, GA, USA

Deirdre M. O'Shea, MS
Department of Clinical and
  Health Psychology,
  University of Florida
Gainesville, FL, USA

Ozioma C. Okonkwo, PhD
Department of Medicine, Wisconsin
  Alzheimer's Disease Research
  Center, and Wisconsin Alzheimer's
  Institute; University of Wisconsin
  School of Medicine and
  Public Health
Madison, WI, USA

Otto Pedraza, PhD
Department of Psychiatry and
  Psychology, Mayo Clinic
Jacksonville, FL, USA

Laura A. Rabin, PhD
Department of Psychology,
  Brooklyn College of CUNY and
  The Graduate Center of the City
  University of New York
Brooklyn, NY, USA

Regilda Anne Romero, PhD
Keystone Behavioral Pediatrics
Jacksonville, FL, USA

Octavio A. Santos, PhD
Department of Psychiatry and
  Psychology, Mayo Clinic
Jacksonville, FL, USA

Sarah M. Szymkowicz, PhD
Department of Neurological
  Sciences, University of Nebraska
  Medical Center
Omaha, NE, USA

Yu-Zhen Tu, MS
Graduate Institute of Brain and Mind
  Sciences, College of Medicine,
  National Taiwan University
Taipei, Taiwan

David E. Tupper, PhD
Neuropsychology Section (G8),
  Hennepin Healthcare; Neurology
  Department, University of
  Minnesota Medical School
Minneapolis, MN, USA

Fons J. R. van de Vijver, PhD
Department of Culture Studies,
  Tilburg University
Tilburg, Germany

Brenda Wong, PhD
Kinark Child and Family Services
Markham, ON, Canada

Lixia Yang, PhD
Department of Psychology, Ryerson
  University
Toronto, ON, Canada

Megan Zuelsdorff, PhD
Wisconsin Alzheimer's Disease
  Research Center and Department
  of Obstetrics and Gynecology,
  University of Wisconsin School of
  Medicine and Public Health
Madison, WI, USA

1

Introduction to Clinical Cultural Neuroscience

OTTO PEDRAZA

A dialectical tension courses throughout the history of psychology between the relative contribution of nature and nurture. The phrase, first used in the modern sense by Galton (1874), became synonymous with the philosophical and scientific debate that regards human phenomena as either innate or acquired. For nearly a century following Darwin's proposition that descent with modification (evolution) proceeds through natural selection, biologists and social anthropologists advanced opposing views about the importance of nature or nurture in shaping human diversity. In the psychological tradition, an inclination toward nurture and empiricist accounts of social learning and behavior dominated the 20th century (Plomin, Shakeshaft, McMillan, & Trzaskowski, 2014). This inclination was understandable for a discipline that emphasizes the capacity for environmentally induced plasticity in brain structure and aims to provide clinical interventions for neurobehavioral disorders (Meany, 2010). The contemporary and prevailing scientific consensus has cast aside this dichotomy and instead posits that complex human behavioral traits (e.g., cognition, personality, emotion) are under substantial genetic and environmental pressure (Plomin & Deary, 2015; Turkheimer, 2000). The fundamental question no longer is whether simple additive effects influence behavioral traits, but rather how the interaction between neurobiology and sociocultural environment yields the wide diversity of traits manifested across human populations (Meany, 2010; Sasaki & Kim, 2016; Tabery, 2014)?

A parallel debate emerged among psychologists and centered on the preeminence of universalism versus the cultural relativism of behavioral traits (Adamopoulos & Lonner, 1994; Berry, Poortinga, Breugelmans, Chasiotis, & Sam, 2011; Norenzayan & Heine, 2005). Should psychological phenomena be construed to reflect common attributes shared by members of all cultures (universalism)? Or should behavior be understood solely through the lens of the specific social and cultural environment in which it is expressed (relativism)? Proponents of universalism consider psychological constructs to be invariant, biologically bound phenomena that are independent of the social and ecological environment. Culture exists outside of the individual and exerts its influence on measurable behavioral traits in an antecedent–consequent relationship. In contrast, proponents of relativism consider all psychological constructs to be culturally bound; culture lies inside the individual, and it is the intersection between culture and person that confers meaning to behavioral traits. To an extent, universalist and relativist perspectives correspond to the etic and emic distinction in anthropology. Coined by Pike (1954) and derived from the study of phonetics and phonemics in linguistics, an emic unit is a "physical or mental item or system treated by insiders as relevant to their system of behavior and as the same emic unit despite etic variability" (Pike, 1990, p. 28). In an emic approach, behavior is seen from the perspective of cultural insiders and recorded using qualitative methods that avoid the researcher's external point of view, whereas an etic approach regards behavior from the perspective of cultural outsiders who deem constructs as equivalent and universal.[1]

From these two philosophical debates (nature and nurture, universalism and relativism), it is possible to trace the conceptual and methodological developments that have shaped the contemporary study of culture, psychology, and neuroscience. In tracing those developments, this introductory chapter considers the theoretical assumptions implicit in the clinical and experimental analysis of culture and its bidirectional effect upon brain and behavior, thus laying the foundation for the subsequent contributions in this volume. First, a brief historical survey of the study of culture in psychology transports us to its roots in 19th-century European academic scholarship and anthropological field studies. The chapter then explores the construct of *culture*, considers whether an operational definition of the term is feasible or desirable, and examines the modes of vertical, horizontal, and oblique cultural transmission. We evaluate the quest for core universal attributes in cognition and behavior, with a consideration of the so-called WEIRD problem—the overabundance of scientific findings in psychology and neuroscience derived from Western, educated, industrialized, rich, and democratic subject samples. Lastly, the chapter reviews the ubiquitous East versus West (typically North American vs. East Asian) research paradigm present in cross-cultural psychology and cultural neuroscience, outlining its limitations as well as key replicable findings.

BRIEF HISTORICAL CONTEXT

Interest in the interaction between human behavior and its cultural and ecological context is not new. A pioneer in this endeavor was William H. R. Rivers, a psychiatrist and anthropologist at Cambridge University and founding member of the British Psychological Society (Adamopoulos & Lonner, 2001; Hart, 2017; Whittle, 2000). In 1898, Rivers took part in an expedition to the Torres Strait, a group of islands lying between Australia and New Guinea, where he conducted

---

1. The diffusion of the etic–emic concept beyond linguistics and anthropology and into psychology is surveyed in detail by Headland (1990).

psychophysical investigations among its inhabitants. Prevalent yet untested notions at the time suggested that islanders would demonstrate higher sensory and perceptual acuity than Europeans, presumably because they had to expend more energy on basic mental processes to survive in their natural environment. Rivers and his colleagues applied systematic and ingenious methods to collect data and, contrary to those prevalent ideas, found no discernible difference between Europeans and non-Europeans in any sensory modality. In subsequent years, Rivers traveled to southern India and performed detailed ethnographic studies of the Toda people but would return to Melanesia throughout his career while retaining his post as a lecturer in experimental psychology at Cambridge. For his ethnographic fieldwork, Rivers today is remembered as one of the fathers of British social anthropology. His work on sensory and perceptual differences later served as inspiration for the seminal cross-cultural studies by Segall, Campbell, and Herskovits (1966) on visual perception.

Wilhelm Wundt, a contemporary of Rivers, explored the role of language, myth, customs, and morals in communal life in his 10-volume treatise *Völkerpsychologie* (usually translated as *Folk Psychology*) and his single-volume *Elements of Folk Psychology* (1916). Wundt understood human behavior to be embedded within a sociocultural framework that extends beyond sensation and perception, and for which the experimental laboratory and methods of introspection that he developed earlier in his career would be insufficient (Diriwächter, 2012). Wundt (1916) proposed that "all phenomena with which mental sciences deal are, indeed, creations of the social community" (p. 2). Folk psychology, as opposed to individual psychology, stressed the analysis of complex mental functions in specific social and historical contexts. This emphasis on the social environment traced its ancestry to the ideas of Giambattista Vico in the 18th century and Moritz Lazarus and Heymann Steinthal in the mid-19th century, but it was Wundt's emphasis on the mutual interdependence between individual psychology and *Völkerpsychologie* that anticipated the interactionism present in the modern study of psychology and culture (Jahoda & Krewer, 1997). Although Wundt's conceptualization of culture was malleable throughout his initial writings—at times alluding to intellect, creativity, or art and at other times to broader notions of civilization—by the publication of his 10th volume in 1920 he conceived of *Völkerpsychologie* as analogous to cultural psychology. In that respect Wundt firmly broadened the scope of psychological inquiry beyond the laboratory and toward social anthropology and ethnology (Diriwächter, 2012; Jahoda, 1993, 2012).

The historical antecedents for the systematic and experimental study of culture and cognition traditionally are traced to the former Soviet Union, where beginning in the 1920s Lev Vygotsky, Aleksandr Luria, and Alexei Leontiev developed a cultural-historical psychology founded upon Marxist ideas (Puente & Agranovich, 2004; van der Veer, 2012). In 1924, Vygotsky joined the Moscow Institute of Psychology, where Luria and Leontiev were developing a conjugate motor method to study the association between emotional states and motor reactions (Kuzovleva, 1999; Moskovich, Bougakov, DeFina, & Goldberg, 2012). Vygotsky arrived with an encyclopedic knowledge of literature and social sciences and soon emerged as the intellectual leader of the group. Luria considered the young Vygotsky a genius and years later would credit Vygotsky with the historical development of his neuropsychological principles (Akhutina, 2003; Luria, 1979). Vygotsky accepted Marx's notion that ideas (i.e., higher mental functions) are the products of the material circumstances of life and, hence, could be understood only within an ontogenetic framework that explores the sociohistorical origins of those ideas (Nell, 1999). Of particular interest is the child's developmental acquisition of psychological tools, particularly language, that are forged and molded by the sociocultural environment in which the child lives. Higher cognitive processes thus share a social origin and become internalized during language acquisition. According to Vygotsky (1997), "every higher mental function was external because it was social before it became an internal, strictly mental function" (p. 105). Luria extended this notion into a triadic and interconnected formulation in which higher cognitive functions are sociocultural in origin, systemic in structure, and dynamic in development (Luria, 1965; Akhutina & Pylaeva, 2011).

Luria knew of the sensory-perceptual studies performed by Rivers at the turn of the century but observed that those studies failed to address the role of sociocultural factors on higher, not elementary, mental functions. Wundt had proposed a sociocultural origin for higher mental functions but believed that it lay outside the realm of experimentation (van der Veer & van Ijzendoorn, 1985). In contrast, Luria and Vygotsky embraced an experimental approach and "conceived the idea of carrying out the first far-reaching study of intellectual functions among adults from a non-technological non-literate, traditional society" (Luria, 1979, p. 60). During the 1920s and 1930s, the Soviet Republics were in the midst of broad and rapid transformation from agrarian to industrialized societies. Luria and Vygotsky considered the modernization of the Soviet Republics as a natural laboratory to investigate the impact of literacy on cognitive operations and thus meticulously planned a scientific expedition (Ardila, 2016; Luria, 1979; Nell, 1999).

As detailed by Tupper (Chapter 2 in this volume), in 1931 Luria and several colleagues departed for Uzbekistan and Kyrgyzstan (formerly Kirghizia) to conduct field studies among the local population. The subjects ranged from illiterate peasants living in remote villages to female students who had two to three years of formal study. The experimental sessions were conducted in relaxed settings, such as a teahouse or around a campfire on mountain pastures, after a period of casual conversation aimed to establish rapport. The cognitive tasks were developed in such a manner as to be meaningful for the subjects and open to several solutions, which would elucidate some aspect of their cognitive activity. In one of the experimental tasks, Luria presented the subjects with multiple series of drawings (e.g., hammer, saw, log, hatchet; glass, saucepan, spectacles, bottle). Illiterate subjects tended to categorize the drawings based upon concrete operations and rules used in practical life, whereas those with some education often relied on abstract categorization. In another series of experiments in which subjects were presented with logical syllogisms, illiterate subjects made excellent judgments about facts that would be of direct concern to them, demonstrating "worldly intelligence."

However, they struggled when presented with theoretical premises if those premises did not arise out of their personal experience, often treating the statements as a reflection of particular phenomena instead of universal propositions with a logical structure. Luria's observations during this expedition led him to conclude that the capacity for mental abstraction is directly linked to an individual's cultural and educational environment (Luria, 1979).

Culture as a subject worthy of academic pursuit fell into relative obscurity in the years immediately preceding and following World War II. In 1960, Luria delivered a series of lectures across North American universities that stimulated interest in the theories and research that he and Vygotsky had championed. Luria established a close working relationship with numerous psychologists in Europe and North America, notably Jerome Bruner, Karl Pribram, and Michael Cole. In 1962, Cole traveled to Moscow to study with Luria and would later edit Luria's biography *The Making of Mind* (Kuzovleva, 1999). That same year witnessed the first English translation of Vygotsky's *Thought and Language*, which remains his best-known work (van der Veer & Yasnitsky, 2011).

Outside of neuropsychology, the organizational roots of cross-cultural psychology arguably can be traced to the same period, when in 1966 the University of Nigeria in Ibadan hosted the International Conference on Social Psychological Research in Developing Countries. The conference attracted over 50 official delegates and numerous observers to discuss the generalizability of Western psychological principles to non-Western cultures (Adamopoulos & Lonner, 2001; Gibb, 1967). Thereafter, the Center for Cross-Cultural Research was established at Western Washington University and was instrumental in the development of the *Journal of Cross-Cultural Psychology*, which published its inaugural issue in 1970. The International Association for Cross-Cultural Psychology was founded in 1972, adopting the *Journal of Cross-Cultural Psychology* as its flagship publication and holding biennial international congresses to the present day.

The past two decades have witnessed unprecedented interest in the study of culture across all subdisciplines of psychology and, particularly, within the nascent field of cultural neuroscience. Perhaps spurred by immigration trends and the growing societal discourse on multiculturalism versus pluralism in Europe and North America, the study of culture and its bidirectional impact on cognition, emotion, and social behavior is flourishing. This burgeoning interest is evidenced by the number of chapters and textbooks, peer-reviewed manuscripts, academic programs, and scholarly conferences dedicated to the study of culture and psychological phenomena. For instance, an inspection of peer-reviewed studies indexed in the PubMed database from 1966 through 2016 using the truncated terms *cultur** and *psycholog** shows a modest increase in publications through 2000, followed by rapid expansion (Figure 1.1a). Similarly, a search using the truncated terms *cultur** and *neuropsycholog** yields scant publications prior to 1995, consistent with the observations offered at that time by Puente and Perez-Garcia (2000), followed by a notable increase in recent years (Figure 1.1b). Naturally, Figures 1.1a and 1.1b do not depict the exact aggregate amount of culture-related publications in each subdiscipline, but they serve as useful illustrations of the

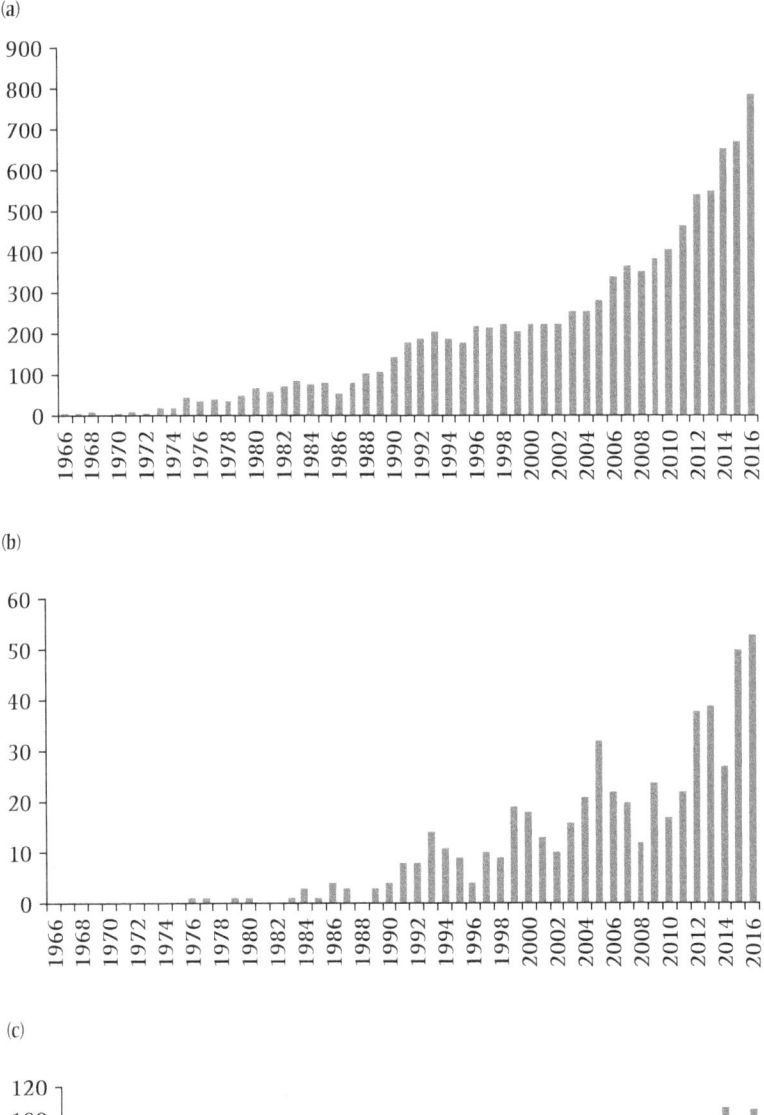

Figure 1.1. Peer-reviewed manuscripts indexed in PubMed using truncated terms (a) *cultur** and *psycholog**, (b) *cultur** and *neuropsycholog**, or (c) *cultur** and *neuroscien**.

growing scientific interest in the subject. (Comparable historical trends are evident using the PsycInfo database.)

Recently, advances in neuroimaging techniques, molecular genetics, and evolutionary theory have coalesced into an emerging experimental discipline known as cultural neuroscience. Cultural neuroscience aims to understand how cultural traits shape neurobiology and behavior and how neural and genetic mechanisms facilitate the emergence and transmission of cultural traits (Chiao, 2011; Seligman, Choudhury, & Kirmayer, 2016). Cultural neuroscience views the "human mind as biologically prepared and, yet, supplemented, transformed and fully completed through active participation and engagement in the eco-symbolic environment called culture" (Kitayama & Park, 2010, p. 112). Consonant with the surging interest in culture across related fields, the number of peer-reviewed publications indexed in PubMed using the terms *cultur** and *neuroscien** has quadrupled since the start of the current millennium (Figure 1.1c).

Together, these historical trends reflect the growing enthusiasm for the study of culture, brain, and behavior and underscore the fundamental necessity for cross-disciplinary communication and collaboration as new ideas are formulated and evidence-based findings become established.

THE CULTURAL MOSAIC

What is culture? Definitions of the construct have been offered and debated for nearly as long as its existence, with claims to operational legitimacy within sociology, anthropology, and psychology and yet no consensual agreement reached (Berry, Poortinga, Breugelmans, Chasiotis, & Sam, 2011; Heine, 2016; Kroeber & Kluckhohn, 1952; Matsumoto & Juang, 2013). The first modern use of the term in the English language is traced to Tylor (1871), who defined culture to encompass the "knowledge, belief, art, morals, laws, customs and any other capabilities and habits acquired by man as a member of society" (p. 1). Tylor's inclusive definition first shifted the concept away from the aesthetic connotations held by his contemporaries, who equated culture with intellect or artistic refinement (i.e., "high culture"). However, Tylor's anthropology was suffused with the evolutionary schemes of the time, in which societies were conceived to progress in hierarchical stages from "savagery" to "barbarism" and "civilization." To account for the observed variation across cultures, social evolutionists postulated that different societies were located at dissimilar evolutionary stages.

Contemporary definitions eschew the evolutionist, vertical array of societies and the implicit value judgments. Culture represents the historical accumulation of traditions, the embodiment of symbols in artifacts, and the normative emphasis on shared rules governing the activities of people. Culture also comprises the attitudes, values, and beliefs of particular groups transmitted from generation to generation through shared learning. In broad terms, two constituent dimensions characterize culture: the explicit material objects and observable behaviors of a group of people, and their implicit and shared ideas, values, and attitudes transmitted over time (Ferraro & Andreatta, 2012; Heine, 2016; Matsumoto & Juang,

2013). Steeped in the positivist stance permeating most of the 20th century, anthropologists initially focused on the first set of characteristics. Culture was deemed to be situated outside the person, framed by an objective, concrete reality that could be measured and studied using quantitative, structured instruments (Berry, Poortinga, Breugelmans, Chasiotis, & Sam, 2011). As noted by Shweder and Sullivan (1993), "real" anthropology consisted of ethnographic fieldwork that explored the rituals and kinships of a group of people. With the advent of postmodern criticism during the 1970s, the ideological pendulum swung toward subjectivity and deep skepticism of science (Spiro, 1996). Culture became situated inside the person, conceptualized as a web of ideas and symbols to be revealed through textual analysis or qualitative methods of data collection (Geertz, 1973). In its extreme formulation, which asserts the elusiveness of all meaning and knowledge, postmodernist perspectives questioned the existence of—and dismissed any attempts to identify—cultural groups. These theoretical shifts led to an epistemological crisis within cultural anthropology, a crisis that psychology and, by extension, neuroscience was able to withstand (Greenfield, 2000b).

Weathering through the postmodernist critique does not imply that psychology and neuroscience have not wrestled with comparable definitional struggles. For many years, psychologists adopted the initial anthropological perspective and emphasized the explicit and quantifiable characteristics of culture. The physical environment and social conditions in which people live were considered antecedent to psychological phenomena (Berry, Poortinga, Breugelmans, Chasiotis, & Sam, 2011; Lonner & Adamopoulos, 1997). This perspective, in which culture and human activity are seen as separable, became associated with *cross-cultural psychology* (Greenfield, 2000a). Methodologically, the modus operandi of cross-cultural psychology has been to select a particular construct from one culture, use instruments with known psychometric properties, and study the construct's comparability in another culture. The abiding question concerns generalizability—the degree to which psychological constructs are invariant across cultures or belong to a specific sociocultural context (Adamopoulos & Lonner, 2001; Lonner, 2015). The classic example of this paradigm is the cross-cultural study of intelligence, using tests developed in North America that are translated and administered to members of cultures throughout the rest of the world.

As culture began to be defined in terms of subjective meaning, psychologists shifted their focus of analysis from externalities to the internal and shared experiences among groups of people. Culture, mind, and behavior became indistinguishable, and the individual and sociocultural context mutually constitutive (Shweder & Sullivan, 1993; Shweder, 2007). In *cultural psychology*, the central theoretical processes are the social construction of shared meaning, whereby culture provides the lens or filter through which human beings interpret the world, and the socialization of individuals throughout their development into adult members of their cultural group (Greenfield, 2000a). Cultural psychologists favor an emic approach and refrain from using methodological procedures that are not derived from the culture being studied, particularly if those procedures do not incorporate that culture's values, beliefs, and modes of communication. Consequently, cross-cultural comparisons become secondary to a deeper understanding of the culture as experienced in the minds of its people. A related tradition, *indigenous psychology*, further asserts that psychological constructs and principles, not just data collection methods, should emanate from the values, myths, and traditions within each culture (Chakkarath, 2012).

In recent years, a measure of rapprochement between cross-cultural and cultural psychologists reflects the recognition that it is possible to conceptualize culture as both explicit and implicit, as "out there" and "in here" (Berry, Poortinga, Breugelmans, Chasiotis, & Sam, 2011; Ellis & Stam, 2015). It is feasible to conduct empirical studies that integrate qualitative methods from cultural psychology and quantitative instruments from cross-cultural psychology and thus secure a deeper understanding from psychometric results in real-world ecological contexts (Greenfield, 2000a). Still, the vast majority of studies continue to emphasize either the explicit or the implicit aspect of culture, and hence the type of data that are collected and analyzed (Berry et al., 2011).

Clinical neuropsychology and cultural neuroscience have not wrestled directly with these definitional issues. Instead, the explicit and implicit dimensions of culture are readily acknowledged, and where it becomes an object of study the focus may be on explicit and quantifiable factors such as language, education, or socioeconomic status or implicit and subjective factors such as values and beliefs (Ardila, 2007; Northoff, 2016). In neuropsychology, the overriding objective has been to determine whether cultural factors moderate test score differences that otherwise might be associated with clinical or medical conditions. In cultural neuroscience, the use of neuroimaging techniques and study of gene-culture interactions help elucidate group differences in behavior and neural function among nonclinical populations. While culture is often treated as an independent variable in each discipline, we are reminded by van de Vijver and Leung (1997) that individuals cannot be assigned randomly to cultural groups. Hence, any study involving culture represents a quasi-experimental research design and the interpretation of between-group differences in an outcome measure must be rendered with care.

Adding to this complexity, it is difficult to establish clear boundaries that demarcate groups of people along cultural characteristics. Culture is too often conflated with ethnicity or nationality, which precludes the opportunity of a richer characterization of its relevant components (Choudhury & Kirmayer, 2009; Losin, Dapretto, & Iacoboni, 2010; Wong, Strickland, Fletcher-Janzen, Ardila, & Reynolds, 2000). While it is certainly possible to compare Canadians with Argentinians on any number of measures, Canadians are the most culturally heterogeneous nation in the Western hemisphere while Argentinians are the most homogeneous (Gören, 2013). Moreover, variation among individuals within a cultural group can be nontrivial and meaningful (Freeman, 2013). Populations that are highly differentiated on sociological grounds (e.g., class, religion, socioeconomic status) will necessitate finer cultural mapping and, hence, consideration of subcultural groups (Avruch, 1998).

In sum, culture is constituted of explicit and implicit dimensions denoting shared learning, shared meaning, and shared context among groups of people.

Whenever possible, cross-cultural studies should define the construct clearly and operationally to capitalize on methodological approaches, maximize opportunities for replication, and, if relevant, help establish causation (Losin, Dapretto, & Iacoboni, 2010).

CULTURAL TRANSMISSION

Cultures are inherently dynamic. Cultural attributes such as values, beliefs, norms, and skills change over time, often at dissimilar rates across various subgroups, and are conveyed within or between populations through mechanisms of social learning (Durham, 1991). D'Andrade (1981) encapsulates this process as a

> good part of what any person knows is learned from other people. The teaching by others can be formal or informal, intended or unintended, and the learning can occur through observation or by being taught rules. However accomplished, the result is a body of learnings, called culture, transmitted from one generation to the next. (p. 179)

Gene-Culture Coevolution

Culture represents one of two systems for human beings to transmit information across time and space; the other system is genes (Durham, 1991). Through their language, teaching, imitation, and prosociality, humans alone in the animal kingdom possess the precise transmission means to develop complex cumulative culture (Dean, Kendal, Schapiro, Thierry, & Laland, 2012; Tomasello, 1999). Knowledge is transferred from generation to generation with sufficiently high fidelity for culture to function as an inheritance system (Lewis & Laland, 2012). Through a process known as the *ratchet effect*, cultural elements learned by individuals are modified over time, and those modified elements are transmitted to subsequent generations in a repetitive cycle that optimizes the survival of those ideas (Heine & Ruby, 2010; Tennie, Call, & Tomasello, 2009). Culture is therefore conceived as an inheritance system of descent with modification (Richerson, Boyd, & Henrich, 2010).

Although cultural traits need not conform to strict Mendelian forms of inheritance, strong parallels exist between cultural and genetic models of evolution (Creanza & Feldman, 2016; Mesoudi, 2017; Mesoudi, Whiten, & Laland, 2004, 2006). The mechanisms of biological evolutionary change (natural selection, mutation, drift) are discernible in analogue form to produce cultural evolutionary change. For instance, random errors when copying manufactured artifacts or innovation that arises when cultural traits are combined in novel ways are akin to mutation, insofar as the ideas transmitted from one person to another become modified in subsequent iterations (Creanza, Kolodny, & Feldman, 2017; Kempe, Lycett, & Mesoudi, 2012; Richerson, Boyd, & Henrich, 2010). The frequency distribution of baby names in the United States throughout the 20th century has obeyed a power law analogous to the infinite allele model of population genetics with random drift (Hahn & Bentley, 2003). Numerous additional examples suggest that it is reasonable to think of cultural evolution as a system of inherited variation, with selective retention of favorable culturally transmitted traits (Mesoudi, 2017).[2]

Gene-culture coevolution, also known as dual inheritance theory, provides the blueprint for this dynamic process and posits that genes and culture constitute two interacting forms of inheritance (Boyd & Richerson, 1985; Cavalli-Sforza & Feldman, 1981). The human genome constrains and promotes individual differences in brain development and function and influences psychological propensities on cultural learning. Concurrently, culturally transmitted information modifies selection pressures acting back on the genome (Kim & Sasaki, 2014; Meany, 2010; Laland, Odling-Smee, & Myles, 2010). An underlying assumption of the model is the existence of differential susceptibility to aspects of the environment; that is, particular genes may predispose carriers to be influenced and respond in particular ways to environmental input (Kim & Sasaki, 2014). In this formulation, culture is no longer just an emergent property of humans, but a fundamental cause of what makes humans who they are and will be (Laland, 2008).

Numerous empirical studies support the gene-culture coevolutionary model (Laland, Odling-Smee, & Myles, 2010; Mesoudi, 2017; Moya & Henrich, 2016), although agreement with the theoretical assumptions is not universal (Borsboom, 2006; Lewens, 2015; Read, 2006). The most extensively investigated example is the coevolution of dairy farming and adult lactose tolerance. Throughout their early history, human beings did not produce the enzyme necessary to digest lactose and the DNA record of Europeans prior to the Neolithic period does not show evidence of the lactose-tolerance allele. With the emergence of cattle domestication and the expansion of pastoralists from the Middle East to Europe and North Africa, populations who adopted a culture of dairy farming and milk consumption as their main source of nutrition induced selection pressure for lactose tolerance (Myles et al., 2005). A single nucleotide polymorphism located near the lactase gene (*LCT*) has been shown to be associated with lactose tolerance in Europeans, and a strong correlation between dairy farming and lactose tolerance exists across cultures (Laland et al., 2010). Another example that supports gene-culture coevolution is the pattern of variation in handedness (Laland, Kumm, Van Horn, & Feldman, 1995). Approximately 90% of humans are right-handed, suggesting a history of selection during recent human evolution. However, exclusively genetic models fail to account for the similarity in concordance rates among monozygotic and dizygotic twins and the known cultural influences on handedness (Corballis, 1997). In countries where left-handedness is considered a mental defect or the society stresses social conformity, the frequencies of left-handers is substantially less

---

2. This parallel between genetic and cultural evolution should not be conflated with the social evolutionist model from Tylor and his contemporaries discussed earlier, which held that societies advance or "evolve" in sequential stages from primitive to civilized.

than 10% (Ida & Bryden, 1996; Singh & Bryden, 1994; Teng, Lee, & Chang, 1976). Laland et al. (1995) estimate that, all other factors being equal, 78% of the human population should be right-handed. However, parental influence through imitation or direct instruction exerts a cultural bias to shape the child's handedness to resemble their own. The probability that a child will be right-handed increases to 92% if both parents are right-handed, and decreases to 64% if both parents are left-handed.

Enculturation, Socialization, and Acculturation

Whereas genetic transmission is primarily vertical, modes of cultural transmission may be vertical, oblique, or horizontal (Cavalli-Sforza & Feldman, 1981). Vertical transmission typically represents the transfer of cultural elements from parents to their children, from one generation to the next. Because biological parents and cultural parents are frequently the same, it is important in research studies to disentangle the confounding of genetic and vertical cultural transmission (Boyd & Richerson, 1985). Oblique transmission occurs when cultural elements are transferred from other members of the parents' generation (nonparental adults) and horizontal transmission takes place between members of the same generation (peer to peer). Under most conditions, vertical transmission is the predominant mechanism of cultural transfer for younger children, while social learning from teachers, friends, and other members of society becomes increasingly salient as children grow up (McElreath & Strimling, 2008; Nielsen, Cucchiaro, & Mohamedally, 2012). Occasionally, vertical transmission can be bidirectional, and information may be transferred upwards from child to parent, for instance, when children learn emerging technologies faster than their parents and teach them how to use new electronic devices. The relative importance of these three forms of cultural transmission also varies as a function of societal characteristics (Cavalli-Sforza & Feldman, 1981). Vertical transmission is more common in small and traditional societies, tends to be slower, and conserves much of the parental cultural traits. In contrast, horizontal transmission is increasingly relevant in modern, economically developed societies and allows for the rapid diffusion of new cultural traits and ideas (i.e., innovation; Mesoudi, 2010).

Cultural transmission (vertical, oblique, or horizontal) within the same culture occurs through one of two processes that transfer cultural traits from one person to another: *enculturation* and *socialization*. Enculturation takes place whenever an individual is enfolded by their culture (Berry & Georgas, 2009). Through passive modeling and social learning, typically within the structure of the child's own family, the developing individual gradually acquires the values, language, and other core elements of the immediate and primary culture. Socialization also transpires within the individual's primary culture but takes place through formal instruction or the deliberate shaping of the child's behaviors and beliefs. Socialization ensures that the members of the cultural group learn those attributes deemed by the broader society to be essential to their way of life. The net result of

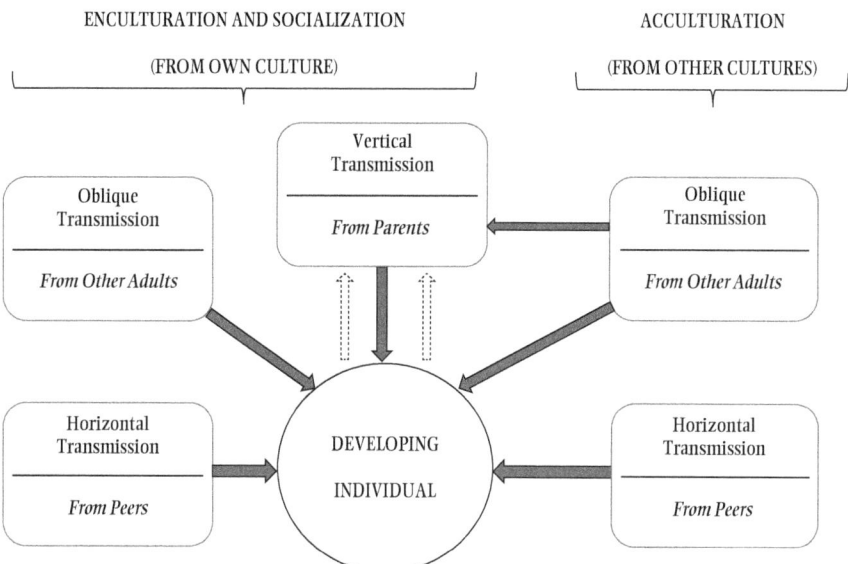

Figure 1.2. Cultural transmission framework. Adapted from J. W. Berry & J. Georgas, 2009, An ecocultural perspective on cultural transmission: The family across cultures, in U. Schönpflug (Ed.), *Cultural transmission: Psychological, developmental, social, and methodological aspects* (pp. 95–125). New York, NY: Cambridge University Press.

enculturation and socialization is the accumulation of similar behavioral characteristics within a specific culture (Berry & Georgas, 2009).

If the exchange of cultural elements occurs between members of different cultures, the appropriate term is *acculturation* (Berry, Poortinga, Breugelmans, Chasiotis, & Sam, 2011). Acculturation is the bidirectional process of change and accommodation that occurs when two or more cultural groups come into direct contact (Sam, 2006). It can lead to mutual adaptation and stems from large-scale and long-term shifts in sociocultural contact, often due to migration, military occupation, or colonization. The term was first formulated to encapsulate cultural adaptation at the broader group or institutional level, where acculturation involves changes to the social structure. However, psychologists promulgated a process of acculturation at the individual level that involves changes in behavioral repertoire (Berry, 2005; Lopez-Class, González Castro, & Ramirez, 2011). Examples of acculturation at the individual level include the partial or complete adoption of particular linguistic features, culinary preferences, and modes of fashion and recreation as one cultural group comes into contact with another cultural group. The dynamic framework that incorporates enculturation, socialization, and acculturation within and between cultures is illustrated in Figure 1.2.

Acculturation can be defined further along two dimensions: a relative preference for one's own cultural heritage and, conversely, integration with the dominant or receiving culture (see Figure 1.3). The resulting outcome yields a range of acculturation strategies: (a) *assimilation* (adoption of the receiving culture and

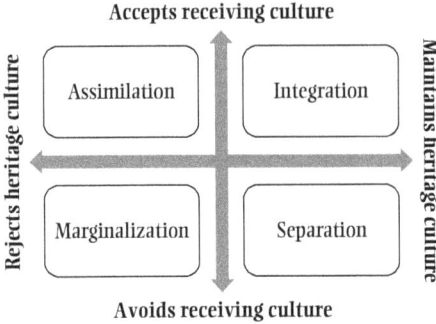

Figure 1.3. Acculturation strategies.

rejection of the heritage culture), (b) *separation* (rejection of the receiving culture and retention of the heritage culture), (c) *integration* or biculturalism (relative adoption of the receiving culture and relative retention of the heritage culture), and (d) *marginalization* (rejection of the receiving and heritage cultures). Marginalization, although conceptually useful, is practically nonexistent because most individuals conform to one of the first three acculturation strategies (Schwartz, Unger, Zamboanga, & Szapocznik, 2010). Among these strategies, cultural integration has been associated with better psychological adjustment and social outcomes.

Acculturation research typically focuses on immigrants and refugees, and in this context the degree of acculturation is modulated by a person's age at migration, migrational category (e.g., voluntary immigrant, political refugee, asylum seeker), race or ethnicity, language, and socioeconomic status (Alba & Nee, 2003; Sam & Berry, 2010; Segal, Mayadas, & Elliott, 2010; Schwartz, Montgomery, & Briones, 2006). For instance, individuals who migrate as older adults have vivid memories of their native country and often struggle to adopt the cultural practices of the receiving society. In most countries, ethnic minority immigrants are regarded more negatively than nonminority immigrants and disproportionately are more likely to experience prejudice, discrimination, and hostility (Saxton & Benson, 2003). Language can mitigate this adversity, as immigrants who share a language with the receiving culture experience less acculturative stress and have a higher tendency to integrate (Schwartz, Unger, Zamboanga, & Szapocznik, 2010). Furthermore, socioeconomic status is strongly associated with the degree of acculturation. Across most societies, immigrant groups from the lower socioeconomic strata experience more difficulty assimilating or integrating into the receiving society than immigrants with greater socioeconomic resources.

Collectively, these theoretical and empirical models underscore the dynamic nature of culture. Through mechanisms of social learning, cultural traits are transferred from parents to children, from nonparental adults to children, and from children to children within and between cultural bounds and, over time, become modified through various processes analogous to genetic change.

THE QUEST FOR UNIVERSALS

Psychological universals are core attributes conceptually shared by all or nearly all human beings across cultures (Norenzayan & Heine, 2005). The existence of universal cognitive, behavioral, emotional, and perceptual phenomena has long been an implicit assumption in psychology and related disciplines (Brown, 1991; Norenzayan & Heine, 2005) and the subject of substantial and substantive debate, particularly within linguistics (e.g., Dąbrowska, 2015; Chomsky, 2000; Dunn, Greenhill, Levinson, & Gray, 2011; Greenberg, 1963). Universalism seeks convergence in psychological phenomena across cultures and, in its absolute form, presupposes that cultural factors bear no influence on such phenomena. Alternatively, absolute relativism proposes that psychological phenomena are fully mediated by culture. Yet, it is untenable to presume that psychological attributes exist outside of the sphere of neurobiology, and the link between culture and neurobiology implies that questions regarding cultural differences are also questions of neurobiology (Meany, 2010; Sasaki & Kim, 2016). Hence, transactional models of culture and psychological attributes should be expected and considered more probable than absolute models.

Indeed, evidence for strict universalism is sparse, and it is no longer reasonable to expect that most psychological phenomena are immune from cultural factors. Human aggression may represent a universal insofar as the historical record provides undeniable evidence across time and space of aggressive behaviors in nearly all known populations, but aggression is manifested in vastly different forms across different societies (Adamopoulos & Lonner, 1994; Fry, 1998). Emotional expression, long considered a biologically constrained universal attribute, instead is determined by a complex interaction of psychobiological and sociocultural factors (Scherer, Clark-Polner, & Mortillaro, 2011). Consequently, the degree to which culture interacts with psychological attributes is best conceptualized as a dimensional continuum ranging from extreme universalism (minimal impact of culture) to extreme relativism (few objective commonalities in human behavior; Berry, Poortinga, Breugelmans, Chasiotis, & Sam, 2011). A position of moderate relativism emphasizes the interaction between person and cultural context and, in many respects, is the theoretical orientation most closely aligned with cultural psychology. In turn, moderate universalism accepts this interactionism but presupposes an antecedent-consequent (causal) relationship between cultural context and psychological attributes, an orientation that is usually associated with cross-cultural psychology.

Clinical neuropsychology and cultural neuroscience are similarly positioned within the spectrum of moderate relativism–universalism. Grounded in neurobiology, both disciplines acknowledge a shared neural architecture and physiology among human beings, while regarding the brain's functional connectivity and emergent properties (e.g., cognition, emotion, perception, behavior) to be malleable and contingent on cultural experience. Clinical neuropsychologists routinely consider the impact of cultural factors, particularly an examinee's language, ethnicity, and education, on measurable behaviors reflected in test score performance (Manly, 2008; Pedraza & Mungas, 2008). Cultural neuroscientists directly challenge the universalizing trends in neuroscience by exploring gene-brain-environment interactions and revealing culturally influenced neural pathways and mechanisms (Chiao, 2011; Kitayama & Park, 2010; Seligman, Choudhury, & Kirmayer, 2016).

The recognition that culture plays a pivotal role in the determination and expression of biologically constrained psychological attributes represents a relatively recent perspective. Psychology, particularly clinical psychology, emerged into the 20th century as a predominantly North American, monocultural enterprise, and not too long ago the role of culture in clinical neuropsychology was ignored, if not misunderstood (Lee & Sue, 2001; Puente & Agranovich, 2004). Psychiatry also failed to appreciate the impact of cultural factors on diagnostic and treatment considerations in mental health. Thus, during the development of the third and revised edition of the *Diagnostic and Statistical Manual of Mental Disorders* (DSM-III-R), commentary on cultural issues were compressed into two paragraphs in the introductory section (Sue, 1996). It was not until the fourth edition (DSM-IV) that culture was explicitly considered throughout the text of the manual (Mezzich et al., 1999). With the emergence of cultural and cross-cultural psychology as distinct subdisciplines and the increased societal spotlight on race relations in the United States, investigators began to explore the relationship between cultural factors and psychological phenomena as well as the role of culture in the provision of mental health services (Lee & Sue, 2001).

Methodologically, however, psychology and neuroscience continue to wrestle with a critical problem—the limited generalizability of research results when participant recruitment is drawn predominantly from Western, educated, industrialized, rich, and democratic (WEIRD) subject samples (Henrich, Heine, & Norenzayan, 2010). Academic research in the brain sciences typically relies on undergraduate college students, and the results from these samples often lead to broad (universal) inferences about human cognition and behavior. Not only are the samples nonrepresentative, inasmuch as they are individualistic, affluent, secular, and analytic with respect to the rest of the world, but they also constitute uneven geographic representation (Norenzayan & Heine, 2005). The majority of research citations in psychology come from the United States, and this percentage is higher than the corresponding figures from other scientific disciplines (May, 1997). The reliance on culturally homogenous, convenience samples from Western countries thus hinders empirical efforts to inform the investigation of true universal attributes across cultures.

The past two decades have witnessed increased awareness and concerted efforts to mitigate universalizing assumptions. Clinical neuropsychologists recognize that cultural factors are not tangential to the administration and interpretation of cognitive tests but are linked to the process of assessment on theoretical and methodological grounds. The validity of clinical inferences derived from cognitive tests may be rendered suboptimal and lead to elevated risk of diagnostic errors if normative reference data from relatively homogenous WEIRD, particularly Caucasian, samples are used in the evaluation of individuals who are dissimilar from those samples (Arango-Lasprilla, 2015; Manly et al., 1998; Norman, Evan, Miller, & Heaton, 2000; Smith, Ivnik, & Lucas, 2008). Fortunately, large normative studies throughout South America (Arango-Lasprilla, 2015), Spain (Peña-Casanova et al., 2009), Netherlands (De Vent et al., 2016), and China (Lee & Wang, 2010), among other countries, are contributing to an expanding database of reference values that more closely reflects the cultural and demographic characteristics of the local examinees. In cross-cultural psychology and neuroscience, experimental paradigms are explicitly challenging those universilizing assumptions by contrasting WEIRD samples to samples from small, nonindustrialized societies or non-Western societies (Henrich, Heine, & Norenzayan, 2010). Many of those studies have demonstrated remarkable between-group differences in a large number of psychological domains, including spatial cognition, social decision-making, self-concepts, and reasoning.

THE EAST–WEST PARADIGM

One of the most productive research paradigms in cross-cultural psychology and cultural neuroscience has been the study of psychological traits in individuals from Eastern and Western societies. For instance, socially normative behaviors in North American and Western European cultures include the unrestricted expression of one's opinions and the egalitarian treatment of social peers. Alternatively, East Asian societies emphasize cultural rules designed to achieve social consensus and harmony based on deference to group elders and hierarchically organized ways of communication (Kitayama & Park, 2010). The distinction in psychological characteristics between Eastern and Western cultures is not new and arguably can be traced back to the ancient Greek and Chinese civilizations of the eighth to third centuries BCE (Nisbett, Peng, Choi, & Norenzayan, 2001). The ancient Greeks imbued their daily lives with a sense of personal agency and speculated regularly about the objects and events around them, constructing *analytical* models that described and categorized such objects based on their properties and causal relationships. The ancient Chinese valued a sense of collective agency and in-group harmony and emphasized the *holistic* relationship between objects and events as situated within their respective fields. In thinking about their worlds, the ancient Greeks were inclined to focus on central objects and their attributes and created rules and categories to help them understand those objects whereas the ancient Chinese were inclined to focus on the relatedness of all things, placing a strong emphasis on context when making decisions about objects, events, and individuals (Nisbett, Peng, Choi, & Norenzayan, 2001; Nisbett & Masuda, 2003).

Contemporary research suggests that the cultural distinctions between the ancient Chinese and Greek civilizations persist in modern East Asian and Western societies (Kitayama & Park, 2010; Markus & Kitayama, 1991). People from Western cultures (generally inferred to include predominantly middle-class North American and Western European persons) engage in analytic thinking and place greater value on *individualism* and independence, are motivated predominantly by personal goals and desires, and respond emotionally in concert with those personal needs, and their self-construal as independent agents influences their social perception and attentional biases. Alternatively, people from East Asian cultures (usually persons from China, Korea, and Japan) engage in holistic thinking and place greater value on *collectivism* and interdependence, are motivated predominantly by social goals and concerns, and respond emotionally to such social concerns, and their interdependent self-construal influences their social perception and attentional biases. Furthermore, these differences in cultural views of the self are consistent with systematic variation in neural representations. Substantial research demonstrates medial prefrontal cortex activation when individuals from Western cultures think about the self (Craik et al., 1999; Denny, Kober, Wager, & Ochsner, 2012; Lieberman, Jarcho, & Satpute, 2004). In a study involving Western and Chinese college students, subjects were asked to make judgments about the self, their mother, or others when viewing a series of adjectives (Zhu, Zhang, Fan, & Han, 2007). Consistent with prior studies, Western subjects showed medial prefrontal cortex activation when making judgments about the self but not about their mothers, whereas Chinese subjects showed medial prefrontal cortex activation when their judgments were directed at both the self and their mother. Presumably, this discrepancy reflects the emphasis on relational unity among the Chinese, with tight and interconnected views of the self and family members manifested in overlapping neural representations (Kitayama & Park, 2010).

Cultural beliefs also influence visual attention and perception. Behavioral studies suggest that individuals from Western cultures perform better at visuoperceptual tasks during absolute conditions that demand focus on a focal object while simultaneously ignoring contextual information, while individuals from East Asian cultures perform better during relative conditions in which contextual information must be incorporated during the perception of a focal object (Chiao, 2011). Hedden, Ketay, Aron, Markus, and Gabrieli (2008) asked subjects to perform visual line size judgments and found that judgment decisions that were incongruent with the subject's cultural preferences led to more effortful attention and greater recruitment of frontal and parietal regions. Thus, European American subjects showed increased frontal–parietal activation when they attended to both object and context, while Asian subjects showed increased frontal–parietal activity when they ignored contextual information and attended solely to the object. Goh, Li, Tu, and Dallaire-Théroux (Chapter 6 in this volume) points out that cultural bias in visual processing extends beyond relative line length judgments and processing of foreground objects and backgrounds to processing of human facial features. Relative to a control condition consisting of house images, Western subjects viewing images of faces showed increased bilateral fusiform face area (FFA) activation, whereas East Asian subjects showed a lateralized pattern of greater right FFA activation (Goh et al., 2010). As noted by Goh, this finding is consistent with past neuroimaging studies suggesting that the left FFA is specialized for processing facial features while the right FFA performs more holistic face analysis. Consistent with these findings, behavioral studies using eye-movement technology have shown that Western subjects have a predilection to direct their gaze toward specific facial features, particularly the eyes, while East Asian subjects tend to fixate their gaze near the center of faces (Caldara, Zhou, & Miellet, 2010).

Culture-specific patterns of neural activation are evident also in the expression and regulation of emotional experiences. Individuals from Western cultures value high arousal emotions (e.g., enthusiasm, excitement) to a greater degree and are less likely to suppress their emotional expression than individuals from East Asian cultures (Butler, Lee, & Gross, 2007; Lim, 2016; Matsumoto et al., 2008; Tsai, 2007). Moreover, individuals from both Western and East Asian cultures demonstrate an in-group advantage in the communication of emotions, whereby such communication is more accurate when the persons expressing and receiving the information belong to the same cultural group (Elfenbein, & Ambady, 2002). Chiao et al. (2008) showed faces of Japanese or Caucasian American models with fearful, happy, angry, or neutral expressions to native Japanese and Caucasian American subjects and asked them to make an emotion categorization judgment. Results showed cultural specificity in amygdala response to fearful faces, with increased amygdala activation to fearful faces when those faces matched the subject's own cultural group but not the other cultural group. As noted by Chiao, Cheon, Pornpattanangkul, Mrazek, and Blizinsky (2013), this association is likely modulated by the degree of acculturation. In a study involving Asian adults living in Europe, exposure to the foreign culture and duration of stay were associated with the amygdalar response to facial expressions of emotions among Asian subjects compared with age-matched European counterparts (Derntl et al., 2009). Although research exploring the relationship between culture and emotional expression and regulation remains in its nascent state, convergent evidence suggests that culture wields meaningful influence on the amygdala in response to affective stimuli.

Taken together, a substantial body of research demonstrates remarkable and replicable differences in psychological traits as a function of cultural characteristics (Na et al., 2010). Western cultures lean toward a social orientation that values independence and encourages behaviors that affirm autonomy, while Eastern cultures emphasize an orientation that values interdependence and harmonious relationships. The cognitive style of Western cultures is analytic, with a tendency toward taxonomic object categorization and detachment of focal objects from their perceptual field. In contrast, Eastern cultures are holistic and accentuate the thematic associations among objects and their contextual relationship to the entire perceptual field. The two dimensions (independent/interdependent and analytic/holistic) are correlated at the group level and, in general, cultures that are relatively interdependent manifest a relatively holistic reasoning style. The East–West research paradigm therefore has yielded important information about the complex relationship between cultural factors, psychological traits, and neural architecture and opened additional avenues for investigation. For instance, do cultural experiences serve to modulate pre-existing neural patterns, or do they exert a constitutional influence that determines which particular brain regions are recruited during specific cognitive tasks? (Han & Northoff, 2008).

The pervasiveness of the East–West paradigm in cultural neuroscience research also serves to highlight some of the caveats and limitations in this body of investigation. First, overreliance on East–West distinctions has contributed to a relatively narrow research focus on self-construal and reasoning styles (individualism vs. collectivism, independence vs. interdependence, and analytic vs. holistic) at the expense of many other psychological traits and cultural constructs. The vast majority of cultural neuroscience research hinges on the study of East–West differences in self-construal (Seligman, Choudhury, & Kirmayer, 2016). Reliance on this construct in binary East–West comparisons marginalizes the potential study of analogous constructs in other cultures, including Latin American, Middle Eastern, and South Asian cultures. For example, Cohen (2009) highlighted multiple aspects of culture, including religion, socioeconomic status, and variation across geographic regions within countries, that remain understudied. Second, it bears remembering that any large cultural group, whether Eastern or Western, is inherently heterogeneous. Substantial variability exists within a given cultural group on any construct of study, and substantial overlap is to be expected between cultural groups. Third, research that compares non-Western groups to Western groups may invoke implicit Eurocentric world views and stereotypes and suggests in subtle ways that Western cultures are distinctively linked to values that reflect modernity or freedom (Martínez Mateo, Cabanis, Stenmanns, & Krach, 2013). Furthermore, many cultural neuroscience studies continue to appeal to biology or ancestry in their demarcation of ethnic or racial groups, and cross-cultural researchers and cultural neuroscientists must be vigilant about biological reductionism and reification of social disparities as racial categories based on neurobiological variables (Martínez Mateo, Cabanis, Cruz de Echeverría Loebell, & Krach, 2012; Stein, Chiao, & van Honk, 2016).

Culture, genes, and environment interact and shape the development and expression of cognitive, emotional, and behavioral traits across the lifespan. Understanding how they interact and engender the vast array of human psychological traits can yield enormous insight into the pathology, epidemiology, and evaluation and treatment of clinical disorders.

AIMS AND ORGANIZATION OF THE BOOK

The principal aim of this volume is to provide clinicians and researchers with an overview of contemporary topics relevant to the study of culture in psychology and neuroscience. While comprehensive volumes dedicated to cultural or cross-cultural psychology, cultural neuropsychology, and cultural neuroscience are readily available, the accumulated theoretical and empirical findings remain relatively sequestered within each of those academic subspecialties. In recent decades, cultural neuroscientists have steadily advanced our knowledge of the neuroanatomic correlates and genomic underpinnings of culture–neurobiology interactions, while recognizing that much about these interactions remains elusive. Cultural and cross-cultural psychologists have refined their assessment instruments, while clinical neuropsychologists have expanded the range of normative reference cohorts necessary for the valid evaluation of cognitive abilities within and across cultures. With the designation *clinical cultural neuroscience*, the current volume intends to provide the reader with a translational thematic framework that brings together topics of shared interest for clinicians and researchers alike. To this end, clinical cultural neuroscience represents a multidisciplinary approach to understand human behavior and culture along the spectrum of health and disease across multiple levels of analysis, from molecular genetics to sociocultural environment.

The volume is thematically organized into two sections that broadly reflect the cultural neuroscience and clinical neuropsychology approaches to the study of culture and behavior. To foster the translational intent, those sections are not presented separately and, instead, the chapters are listed in alternating thematic order. The goal is to encourage the reader to navigate seamlessly from clinically to experimentally focused topics, and vice versa. The current introductory chapter is devoted to examine core foundational characteristics and extract the common conceptual elements that span across the subsequent chapters in the volume. Hence, many of the subsequent chapters directly or indirectly incorporate assumptions about universalism versus relativism, consider the role of neurobiology versus social environment in the transmission and expression of psychological traits, and incorporate or evaluate Western-influenced research paradigms.

In Chapter 2, Tupper reviews the seminal contributions of Lev Vygotsky and Aleksandr Luria to the theoretical and experimental study of culture in cognitive psychology and neuropsychology. Tupper surveys the 1931–1932 expeditions to Central Asia that were intended to investigate higher-order reasoning and problem-solving skills among rural residents with varying degrees of literacy and education and describes the interpretations of the original results and subsequent controversies. In Chapter 3, Rabin, Brodale, Elbulok-Charcape, and Barr expand the findings from their survey of clinical neuropsychologists practicing in the United States and Canada, initially reported in Rabin, Paolillo, and Barr (2016). In the chapter in this volume, Rabin et al. highlight the ethnic, cultural, and linguistic diversity of the patient population receiving neuropsychological services and underscore the challenges to training and clinical practice in the current shifting environment of population trends and demographic characteristics. The impact of cultural factors on human memory is reviewed in detail by Yang, Wong, and Li in Chapter 4. Drawing predominantly from the vast research literature on East Asian and Western cultural differences, Yang et al. outline the mechanisms through which a person's cultural background can affect the quantity and quality of the information stored as memories and describe the differences between groups in recall accuracy for item-based versus context-based stimuli. In Chapter 5, Dotson, Levy, O'Shea, McLaren, and Szymkowicz review the epidemiology of mood disorders among ethnic minorities in the United States and the particular impact of cultural factors on symptomatic presentation among African Americans, Hispanic Americans, Asian Americans, and Native Americans. Dotson et al. also provide a useful and comprehensive review of the assessment instruments often used in the clinical evaluation of mood disorders and highlight the need for more structural and functional neuroimaging studies of depression in ethnic minority adults. In Chapter 6, Goh, Li, Tu, and Dallaire-Théroux describe the cultural biases inherent in the visual perception and attention to faces, scenes, and other visual stimuli and review how cultural influences might modulate the neurobiological mechanisms underpinning visual cognition. In keeping with the translational aims of this volume, Goh et al. then consider potential clinical aspects of the influence of culture on visual perception and attention, followed by a brief discussion on modern technological advances in computer and mobile technology and the rapid access to culture-specific visual information unbounded by physical or temporal distance.

In Chapter 7, Gold reviews the concept of cognitive reserve and the available evidence linking bilingualism to cognitive reserve as a mechanism that can delay the symptomatic onset of Alzheimer's disease. Gold summarizes those findings and hypothesizes that bilingualism may delay Alzheimer's disease symptom onset through its impact on executive control pathways, while acknowledging that the precise neural mechanisms for this association are yet to be established. In Chapter 8, Santos, Fujii, and Pedraza discuss the substantial linguistic diversity in the United States, the clinical and neural aspects of bilingualism, and important considerations for neuropsychologists when using translated instruments or relying on interpreters. The authors then discuss several factors that can influence the process and content dynamics of a clinical evaluation with a non-English speaker, with a particular focus on the two largest immigrant groups (Latin American and Asian-Pacific Islanders), followed by a review of several assessment instruments tailored toward the assessment of non-English speakers. In Chapter 9, Chou and Booth describe the neural organization and representation of semantic knowledge in children with or without autism spectrum disorder. The authors then contrast the neurocognitive development of semantic knowledge and processing in children proficient in English versus Chinese. In Chapter 10, Bordes Edgar and Romero further explore the impact of culture and linguistic diversity in children's development, with a focus on the pediatric neuropsychological examination. The authors highlight the challenges and opportunities inherent in the clinical evaluation of bilingual children, with an additional discussion of the important role of advocacy by clinicians and healthcare providers at the local and national level. In Chapter 11, Zuelsdorff, Barnes, and Okonkwo review the epidemiology of Alzheimer's disease, with an emphasis on the incidence and prevalence of the disease among African American adults. The authors summarize our current understanding of the pathophysiology of Alzheimer's disease in African American compared with non-Hispanic White individuals and consider the genetic, vascular, educational, and other psychosocial factors that contribute to health disparities and can affect the symptomatic manifestation of Alzheimer's disease. This volume concludes with a chapter by Pedraza and van de Vijver that addresses the impact of bias and measurement inequivalence on the valid evaluation of individuals across cultures, with a review of specific strategies aimed to identify and minimize the risk of bias in newly developed tests and assessment instruments or mitigate the cost of bias for tests already in existence.

REFERENCES

Adamopoulos, J., & Lonner, W. J. (1994). Absolutism, relativism, and universalism in the study of human behavior. In W. J. Lonner & R. Moypass (Eds.), *Psychology and culture* (pp. 129–134). Boston: Allyn and Bacon.

Adamopoulos, J., & Lonner, W. J. (2001). Culture and psychology at a crossroad: Historical perspective and theoretical analysis. In D. Matsumoto (Ed.), *The handbook of culture and psychology* (pp. 11–34). New York, NY: Oxford University Press.

Akhutina, T. V. (2003). L. S. Vygotsky and A. R. Luria: Foundations of neuropsychology. *Journal of Russian and East European Psychology, 41,* 159–190. https://doi.org/10.2753/RPO1061-0405410304159

Akhutina, T. V., & Pylaeva, N. M. (2011). L. Vygotsky, A. Luria, and developmental neuropsychology. *Psychology in Russia: State of the Art, 4,* 155–175. https://doi.org/10.11621/pir.2011.0009

Alba, R., & Nee, V. (2003). *Remaking the American mainstream: Assimilation and contemporary immigration.* Cambridge, MA: Harvard University Press.

Arango-Lasprilla, J. C. (2015). Commonly used neuropsychological tests for Spanish speakers: Normative data from Latin America. *Neurorehabilitation, 37,* 489–491. https://doi.org/10.3233/NRE-151276

Ardila, A. (2007). The impact of culture on neuropsychological test performance. In B. P. Uzzell, M. O. Pontón, & A. Ardila (Eds.), *International handbook of cross-cultural neuropsychology* (pp. 23–44). Mahwah, NJ: Erlbaum.

Ardila, A. (2016). L. S. Vygotsky in the 21st century. *Psychology in Russia: State of the Art, 9*(4), 4–15. https://doi.org/10.11621/pir.2016.0401

Avruch, K. (1998). *Culture and conflict resolution.* Washington, DC: U.S. Institute of Peace.

Berry, J. W. (2005). Acculturation: Living successfully in two cultures. *International Journal of Intercultural Relations, 29,* 697–712. https://doi.org/10.1016/j.ijintrel.2005.07.013

Berry, J. W., & Georgas, J. (2009). An ecocultural perspective on cultural transmission: The family across cultures. In U. Schönpflug (Ed.), *Cultural transmission: Psychological, developmental, social, and methodological aspects* (pp. 95–125). New York, NY: Cambridge University Press.

Berry, J. W., Poortinga, Y. H., Breugelmans, S. M., Chasiotis, A., & Sam, D. L. (2011). *Cross-cultural psychology: Research and applications* (3rd ed.). New York, NY: Cambridge University Press.

Borsboom, D. (2006). Evolutionary theory and the riddle of the universe. *Behavioral and Brain Sciences, 29,* 351. https://doi.org/10.1017/S0140525X06269085

Boyd, R., & Richerson, P. J. (1985). *Culture and the evolutionary process.* Chicago, IL: University of Chicago Press.

Brown, D. (1991). *Human universals.* San Francisco, CA: McGraw-Hill.

Butler, E. A., Lee, T. L., & Gross, J. J. (2007). Emotion regulation and culture: Are the social consequences of emotion suppression culture-specific? *Emotion, 7,* 30–48. https://doi.org/10.1037/1528-3542.7.1.30

Caldara, R., Zhou, X., & Miellet, S. (2010). Putting culture under the "spotlight" reveals universal information use for face recognition. *PLoS One, 5*(3), e9708. https://doi.org/10.1371/journal.pone.0009708

Cavalli-Sforza, L. L., & Feldman, M. (1981). *Cultural transmission and evolution: A Quantitative approach.* Princeton, NJ: Princeton University Press.

Chakkarath, P. (2012). The role of indigenous psychologies in the building of basic cultural psychology. In J. Valsiner (Ed.), *The Oxford handbook of culture and psychology* (pp. 71–95). New York, NY: Oxford University Press.

Chiao, J. Y. (2011). Cultural neuroscience: Visualizing culture-gene influences on brain function. In J. Decety & J. T. Cacioppo (Eds.), *The Oxford handbook of social neuroscience* (pp. 742–761). New York, NY: Oxford University Press.

Chiao, J. Y., Cheon, B. K., Pornpattananngkul, N., Mrazek, A. J., & Blizinsky, K. D. (2013). Cultural neuroscience: Progress and promise. *Psychological Inquiry, 24,* 1–19. https://doi.org/10.1080/1047840X.2013.752715

Chiao, J. Y., Iidaka, T., Gordon, H. L., Nogawa, J., Bar, M., Aminoff, E., . . . Ambady, N. (2008). Cultural specificity in amygdala response to fear faces. *Journal of Cognitive Neuroscience, 20,* 2167–2174. https://doi.org/10.1162/jocn.2008.20151.

Chomsky, N. (2000). *New horizons in the study of language and mind.* Cambridge, England: Cambridge University Press.

Choudhury, S., & Kirmayer, L. J. (2009). Cultural neuroscience and psychopathology: Prospects for cultural psychiatry. *Progress in Brain Research, 178,* 263–283. https://doi.org/ 10.1016/S0079-6123(09)17820-2

Cohen, A. B. (2009). Many forms of culture. *American Psychologist, 64,* 194–204. https://doi.org/10.1037/a0015308

Corballis, M. C. (1997). The genetics and evolution of handedness. *Psychological Review, 104,* 714–727. https://doi.org/10.1037/0033-295X.104.4.714

Craik, F. I. M., Moroz, T. M., Moscovitch, M., Stuss, D. T., Winocur, G., Tulving, E., & Kapur, S. (1999). In search of the self: A positron emission tomography study. *Psychological Science, 10,* 26–34. https://doi.org/10.1111/1467-9280.00102

Creanza, N., & Feldman, M. W. (2016). Worldwide genetic and cultural change in human evolution. *Current Opinion in Genetics & Development, 41,* 85–92. https://doi.org/10.1016/j.gde.2016.08.006

Creanza, N., Kolodny, O., & Feldman, M. W. (2017). Cultural evolutionary theory: How culture evolves and why it matters. *Proceedings of the National Academy of Sciences, 114,* 7782–7789. https://doi.org/10.1073/pnas.1620732114

D'Andrade, R. G. (1981). The cultural part of cognition. *Cognitive Science, 5,* 179–195. https://doi.org/10.1207/s15516709cog0503_1

Dąbrowska, E. (2015). What exactly is universal grammar, and has anyone seen it? *Frontiers in Psychology, 23*(6), 852. https://doi.org/10.3389/fpsyg.2015.00852

De Vent, N. R., Agelink van Rentergem, J. A., Schmand, B. A., Murre, J. M. J., ANDI Consortium, & Huizenga, H. M. (2016). Advanced Neuropsychological Diagnostics Infrastructure (ANDI): A normative database created from control datasets. *Frontiers in Psychology, 7,* 1601. https://doi.org/10.3389/fpsyg.2016.01601

Dean, L. G., Kendal, R. L., Schapiro, S. J., Thierry, B., & Laland, K. N. (2012). Identification of the social and cognitive processes underlying human cumulative culture. *Science, 335,* 1114–1117. https://doi.org/10.1126/science.1213969

Denny, B. T., Kober, H., Wager, T. D., & Ochsner, K. N. (2012). A meta-analysis of functional neuroimaging studies of self and other judgments reveals a spatial gradient for mentalizing in medial prefrontal cortex. *Journal of Cognitive Neuroscience, 24,* 1742–1752. https://doi.org/10.1162/jocn_a_00233

Derntl, B., Habel, U., Robinson, S., Windischberger, C., Kryspin-Exner, I., Gur, R. C., & Moser, E. (2009). Amygdala activation during recognition of emotions in a foreign ethnic group is associated with duration of stay. *Social Neuroscience, 4,* 294–307. https://doi.org/10.1080/17470910802571633

Diriwächter, R. (2012). Völkerpsychologie. In J. Valsiner (Ed.), *The Oxford handbook of culture and psychology* (pp. 43–57). New York, NY: Oxford University Press.

Dunn, M., Greenhill, S. J., Levinson, S. C., & Gray, R. D. (2011). Evolved structure of language shows lineage-specific trends in word-order universals. *Nature 473*, 79–82. https://doi.org/10.1038/nature09923

Durham, W. H. (1991). *Coevolution: Genes, culture, and human diversity*. Stanford, CA: Stanford University Press.

Elfenbein, H. A., & Ambady, N. (2002). Is there an in-group advantage in emotion recognition? *Psychological Bulletin, 128*, 243–249. https://doi.org/10.1037/0033-2909.128.2.243

Ellis, B. D., & Stam, H. J. (2015). Crisis? What crisis? Cross-cultural psychology's appropriation of cultural psychology. *Culture & Psychology, 21*, 293–317. https://doi.org/10.1177/1354067X15601198

Ferraro, G., & Andreatta, S. (2012). *Cultural anthropology: An applied perspective* (9th ed.). Belmont, CA: Wadsworth.

Freeman, J. B. (2013). Within-cultural variation and the scope of cultural neuroscience. *Psychological Inquiry, 24*, 26–30. https://doi.org/ 10.1080/1047840X.2013.767069

Fry, D. P. (1998). Anthropological perspectives on aggression: Sex differences and cultural variation. *Aggressive Behavior, 24*, 81–95. https://doi.org/10.1002/(SICI)1098-2337(1998)24:2<81::AID-AB1>3.0.CO;2-V

Galton, F. (1874). *English men of science: Their nature and nurture*. London, England: Macmillan.

Geertz, C. (1973). *The interpretation of cultures*. New York, NY: Basic Books.

Gibb, C. (1967). Summary report: International Conference on Social Psychological Research in Developing Countries. *Australian Psychologist, 2*, 40–45.

Goh, J. O. S., Leshikar, E. D., Sutton, B. P., Tan, J. C., Sim, S. K. Y., Hebrank, A. C., & Park, D. C. (2010). Culture differences in neural processing of faces and houses in the ventral visual cortex. *Social Cognitive and Affective Neuroscience, 5*, 227–235. https://doi.org/10.1093/scan/nsq060

Gören, E. (2013). *Economic effects of domestic and neighbouring countries' cultural diversity*. Zentra Working Papers in Transnational Studies No. 16/2013. https://doi.org/10.2139/ssrn.2255492

Greenberg, J. H. (1963). Some universals of grammar with particular reference to the order of meaningful elements. In J. H. Greenberg (Ed.), *Universals of language* (pp. 73–113). Cambridge, MA: MIT Press.

Greenfield, P. M. (2000a). Three approaches to the psychology of culture: Where do they come from? Where can they go? *Asian Journal of Social Psychology, 3*, 223–240. https://doi.org/10.1111/1467-839X.00066

Greenfield, P. M. (2000b). What psychology can do for anthropology, or why anthropology took postmodernism on the chin. *American Anthropologist, 102*, 564–576. https://doi.org/10.1525/aa.2000.102.3.564

Hahn, M. W., & Bentley, R. A. (2003). Drift as a mechanism for cultural change: An example from baby names. *Proceedings of the Royal Society of London B, 270*(Suppl), S120–S123. https://doi.org/10.1098/rsbl.2003.0045

Han, S., & Northoff, G. (2008). Culture-sensitive neural substrates of human cognition: A transcultural neuroimaging approach. *Nature Reviews Neuroscience, 9*, 646–654. https://doi.org/10.1038/nrn2456

Hart, K. (2017). *W. H. R. Rivers is our forgotten founding father*. Open Anthropology Cooperative Press Working Paper Series No. 24. Retrieved from www.openanthcoop.net/press

Headland, T. N. (1990). Introduction: A dialogue between Kenneth Pike and Marvin Harris on emics and etics. In T. N. Headland, K. L. Pike, & M. Harris (Eds.), *Etics and emics: The insider/outsider debate* (pp. 13–27). Newbury Park, CA: SAGE.

Hedden, T., Ketay, S., Aron, A., Markus, H. R., & Gabrieli, J. D. (2008). Cultural influences on neural substrates of attentional control. *Psychological Science, 19,* 12–17. https://doi.org/10.1111/j.1467-9280.2008.02038.x

Heine, S. J. (2016). *Cultural psychology* (3rd ed.). New York, NY: Norton.

Heine, S. J., & Ruby, M. B. (2010). Cultural psychology. *Wiley Interdisciplinary Reviews: Cognitive Science, 1,* 254–266. https://doi.org/10.1002/wcs.7

Henrich, J., Heine, S. J., & Norenzayan, A. (2010). The weirdest people in the world? *Behavioral and Brain Sciences, 33,* 61–83. https://doi.org/10.1017/S0140525X0999152X

Ida, Y., & Bryden, M. P. (1996). A comparison of hand preference in Japan and Canada. *Canadian Journal of Experimental Psychology, 50,* 234–239. https://doi.org/10.1037/1196-1961.50.2.234

Jahoda, G. (1993). *Crossroads between culture and mind: Continuities and change in theories of human nature.* Cambridge, MA: Harvard University Press.

Jahoda, G. (2012). Culture and psychology: Words and ideas in history. In J. Valsiner (Ed.), *The Oxford handbook of culture and psychology* (pp. 25–42). New York, NY: Oxford University Press.

Jahoda, G., & Krewer, B. (1997). History of cross-cultural and cultural psychology. In J. W. Berry, Y. H. Poortinga, & J. Pandey (Eds.), *Handbook of cross-cultural psychology, Vol. 1: Theory and Method* (2nd ed., pp. 1–42). Needham Heights, MA: Allyn and Bacon.

Kempe, M., Lycett, S., & Mesoudi, A. (2012). An experimental test of the accumulated copying error model of cultural mutation for Acheulean handaxe size. *PLoS One, 7*(11), e48333. https://doi.org/10.1371/journal.pone.0048333

Kim, H. S., & Sasaki, J. Y. (2014). Cultural neuroscience: Biology of the mind in cultural contexts. *Annual Review of Psychology, 65,* 487–514. https://doi.org/10.1146/annurev-psych-010213-115040

Kitayama, S., & Park, J. (2010). Cultural neuroscience of the self: Understanding the social grounding of the brain. *Social, Cognitive, and Affective Neuroscience, 5,* 111–129. https://doi.org/10.1093/scan/nsq052.

Kroeber, A. L., & Kluckhohn, C. (1952). *Culture: A critical review of concepts and definitions.* Cambridge, MA: Harvard University.

Kuzovleva, E. (1999), Some facts from the biography of A. R. Luria. *Neuropsychology Review, 9,* 53–56. https://doi.org/10.1023/A:1025695021621

Laland, K. N. (2008). Exploring gene-culture interactions: Insights from handedness, sexual selection and niche-construction case studies. *Philosophical Transactions of the Royal Society B, 363,* 3577–3589. https://doi.org/10.1098/rstb.2008.0132

Laland, K. N., Kumm, J. Van Horn, J. D., & Feldman, M. W. (1995). A gene-culture model of human handedness. *Behavior Genetics, 25,* 433–445. https://doi.org/10.1007/BF02253372

Laland, K. N., Odling-Smee, J., & Myles, S. (2010). How culture shaped the human genome: Bringing genetics and the human sciences together. *Nature Reviews Genetics, 11,* 137–148. https://doi.org/10.1038/nrg2734

Lee, J., & Sue, S. (2001). Clinical psychology and culture. In D. Matsumoto (Ed.), *The handbook of culture and psychology* (pp. 287–305). New York, NY: Oxford University Press.

Lee, T. M. C., & Wang, K. (2010). *Neuropsychological measures: Normative data for Chinese* (2nd ed.). Hong Kong: Laboratory of Neuropsychology, University of Hong Kong.

Lewens, T. (2015). *Cultural evolution: Conceptual challenges.* New York, NY: Oxford University Press.

Lewis, H. M., & Laland, K. N. (2012). Transmission fidelity is the key to the build-up of cumulative culture. *Philosophical Transactions of the Royal Society B, 367,* 2171–2180. https://doi.org/10.1098/rstb.2012.0119

Lieberman, M. D., Jarcho, J. M., & Satpute, A. B. (2004). Evidence-based and intuition-based self-knowledge: An fMRI study. *Journal of Personality and Social Psychology, 87,* 421–435. https://doi.org/10.1037/0022-3514.87.4.421

Lim, N. (2016). Cultural differences in emotion: Differences in emotional arousal level between the East and the West. *Integrative Medicine Research, 5,* 105–109. https://doi.org/10.1016/j.imr.2016.03.004

Lonner, W. J. (2015). Half a century of cross-cultural psychology: A grateful coda. *American Psychologist, 70,* 804–814. https://doi.org/10.1037/a0039454

Lonner, W. J., & Adamopoulos, J. (1997). Culture as antecedent to behavior. In J. W. Berry, Y. H. Poortinga, & J. Pandey (Eds.), *Handbook of cross-cultural psychology, Vol. 1: Theory and Method* (2nd ed., pp. 43–84). Needham Heights, MA: Allyn and Bacon.

Lopez-Class, M., González Castro, F., & Ramirez, A. G. (2011). Conceptions of acculturation: A review and statement of critical issues. *Social Science & Medicine, 72,* 1555–1562. https://doi.org/ 10.1016/j.socscimed.2011.03.011

Losin, E. A., Dapretto, M., & Iacoboni, M. (2010). Culture and neuroscience: Additive or synergistic? *Social Cognitive and Affective Neuroscience, 5,* 148–158. https://doi.org/10.1093/scan/nsp058

Luria, A. R. (1965). L. S. Vygotsky and the problem of functional localization. *Neuropsicologia 3,* 387–392. https://doi.org/10.2753/RPO1061-0405400117

Luria, A. R. (1979). *The making of mind.* Cambridge, MA: Harvard University Press.

Manly, J. J. (2008). Critical issues in cultural neuropsychology: Profit from diversity. *Neuropsychology Review, 18,* 179–183. https://doi.org/10.1007/s11065-008-9068-8

Manly, J. J., Jacobs, D. M., Sano, M., Bell, K., Merchant, C. A., Small, S. A., & Stern, Y. (1998). Cognitive test performances among nondemented elderly African Americans and Whites. *Neurology, 50,* 1238–1245. https://doi.org/10.1212/wnl.50.5.1238

Markus, H. R., & Kitayama, S. (1991). Culture and the self: Implications for cognition, emotion, and motivation. *Psychological Review, 98,* 224–253. https://doi.org/10.1037/0033-295X.98.2.224

Martínez Mateo, M., Cabanis, M., Cruz de Echeverría Loebell, N., & Krach, S. (2012). Concerns about cultural neurosciences: A critical analysis. *Neuroscience and Biobehavioral Reviews, 36,* 152–161. https://doi.org/10.1016/j.neubiorev.2011.05.006

Martínez Mateo, M., Cabanis, M., Stenmanns, J., & Krach, S. (2013). Essentializing the binary self: Individualism and collectivism in cultural neuroscience. *Frontiers in Human Neuroscience, 7,* 289. https://doi.org/10.3389/fnhum.2013.00289

Matsumoto, D., & Juang, L. (2013). *Culture and psychology* (5th ed.). Belmont, CA: Wadsworth.

Matsumoto, D., Yoo, S. H., Nakagawa, S., Alexandre, J., Altarriba, J., Anguas-Wong, A. M., . . . Zengeya, A. (2008). Culture, emotion regulation, and adjustment. *Journal of Personality and Social Psychology, 94,* 925–937. https://doi.org/10.1037/0022-3514.94.6.925

May, R. M. (1997). The scientific wealth of nations. *Science, 275,* 793–796. https://doi.org/ 10.1126/science.275.5301.793

McElreath, R., & Strimling, P. (2008). When natural selection favors imitation of parents. *Current Anthropology, 49,* 307–316. https://doi.org/10.1086/524764

Meany, M. J. (2010). Epigenetics and the biological definition of gene x environment interactions. *Child Development, 81,* 41–79. https://doi.org/ 10.1111/ j.1467-8624.2009.01381.x.

Mesoudi, A. (2010). Studying cultural innovation in the psychology lab. In M. J. O'Brien & S. J. Shennan (Eds.), *Cultural innovation: Contributions from evolutionary anthropology* (pp.175–191). Cambridge, MA: MIT Press.

Mesoudi, A. (2017). Pursuing Darwin's curious parallel: Prospects for a science of cultural evolution. *Proceedings of the National Academy of Sciences, 114,* 7853–7860. https://doi.org/10.1073/pnas.1620741114

Mesoudi, A., Whiten, A., & Laland, K. N. (2004). Perspective: Is human cultural evolution Darwinian? Evidence reviewed from the perspective of *The Origin of Species*. *Evolution, 58,* 1–11. https://doi.org/10.1111/j.0014-3820.2004.tb01568.x

Mesoudi, A., Whiten, A., & Laland, K. N. (2006). Towards a unified science of cultural evolution. *Behavioral and Brain Sciences, 29,* 329–383. https://doi.org/10.1017/S0140525X06009083

Mezzich, J. E., Kirmayer, L. J., Kleinman, A., Fabrega, H., Jr., Parron, D. L., Good, B. J., ... Manson, S. M. (1999). The place of culture in DSM-IV. *Journal of Nervous and Mental Disease, 187,* 457–464. https://doi.org/10.1097/00005053-199908000-00001

Moskovich, L., Bougakov, D., DeFina, P., & Goldberg. E. (2012). A. R. Luria: Pursuing neuropsychology in a swiftly changing society. In A. Y. Stringer, E. L. Cooley, & A. L. Christensen (Eds.), *Pathways to prominence in neuropsychology: Reflections of twentieth-century pioneers* (pp. 49–62). New York, NY: Psychology Press.

Moya, C., & Henrich, J. (2016). Culture–gene coevolutionary psychology: Cultural learning, language, and ethnic psychology. *Current Opinion in Psychology, 8,* 112–118. https://doi.org/ 10.1016/j.copsyc.2015.10.001

Myles, S., Bouzekri, N., Haverfield, E., Cherkaoui, M., Dugoujon, J. M., & Ward, R. (2005). Genetic evidence in support of a shared Eurasian-North African dairying origin. *Human Genetics, 117,* 34–42. https://doi.org/10.1007/s00439-005-1266-3

Na, J., Grossmann, I., Varnum, M. E., Kitayama, S., Gonzalez, R., & Nisbett, R. E. (2010). Cultural differences are not always reducible to individual differences. *Proceedings of the National Academy of Sciences, 107,* 6192–6197. https://doi.org/10.1073/pnas.1001911107

Nell, V. (1999). Luria in Uzbekistan: The vicissitudes of a cross-cultural neuropsychology. *Neuropsychology Review, 9,* 45–52. https://doi.org/10.1023/A:1025643004782

Nielsen, M., Cucchiaro, J., & Mohamedally, J. (2012). When the transmission of culture is child's play. *PLoS One, 7*(3), e34066. https://doi.org/ 10.1371/journal.pone.0034066

Nisbett, R. E., & Masuda, T. (2003). Culture and point of view. *Proceedings of the National Academy of Sciences, 100,* 11163–11175. https://doi.org/10.1073/pnas.1934527100

Nisbett, R. E., Peng, K., Choi, I., & Norenzayan, A. (2001). Culture and systems of thought: Holistic versus analytic cognition. *Psychological Review, 108,* 291–310. https://doi.org/10.1037/0033-295X.108.2.291

Norenzayan, A., & Heine, S. J. (2005). Psychological universals: What are they and how can we know? *Psychological Bulletin, 131,* 763–784. https://doi.org/10.1037/0033-2909.131.5.763

Norman, M. A., Evan, J. D., Miller, S. W., & Heaton, R. K. (2000). Demographically corrected norms for the California Verbal Learning Test. *Journal of Clinical and Experimental Neuropsychology, 22,* 80–94. https://doi.org/10.1076/1380-3395(200002)22:1;1-8;FT080

Northoff, G. (2016). Cultural neuroscience and neurophilosophy: Does the neural code allow for the brain's enculturation? In J. Y. Chiao, S. Li, R. Seligman, & R. Turner (Eds.), *The Oxford handbook of cultural neuroscience* (pp. 21–39). New York, NY: Oxford University Press.

Pedraza, O., & Mungas, D. (2008). Measurement in cross-cultural neuropsychology. *Neuropsychology Review, 18,* 184–193. https://doi.org/ 10.1007/s11065-008-9067-9

Peña-Casanova, J., Blesa, R., Aguilard, M., Gramunt-Fombuena, N., Gómez-Ansón, B., Oliva, R., . . . Sol, J. M. (2009). Spanish multicenter normative studies (NEURONORMA Project): Methods and sample characteristics. *Archives of Clinical Neuropsychology, 24,* 307–319. https://doi.org/10.1093/arclin/acp027

Pike, K. L. (1954). *Language in relation to a unified theory of the structure of human behavior, Part I* (Preliminary ed.). Glendale, CA: Summer Institute of Linguistics.

Pike, K. L. (1990). On the emics and etics of Pike and Harris. In T. N. Headland, K. L. Pike, & M. Harris (Eds.), *Etics and emics: The insider/outsider debate* (pp. 28–47). Newbury Park, CA: SAGE.

Plomin, R., & Deary, I. J. (2015). Genetics and intelligence differences: Five special findings. *Molecular Psychiatry, 20,* 98–108. https://doi.org/10.1038/mp.2014.105

Plomin, R., Shakeshaft, N. G., McMillan, A., & Trzaskowski, M. (2014). Nature, nurture, and expertise. *Intelligence, 45,* 46–59. https://doi.org/10.1016/j.intell.2013.06.008

Puente, A. E., & Agranovich, A. V. (2004). The cultural in cross-cultural neuropsychology. In G. Goldstein, S. R., Beers, & M. Hersen (Eds.), *Comprehensive handbook of psychological assessment, Vol. 1: Intellectual and neuropsychological assessment* (pp. 321–332). Hoboken, NJ: Wiley.

Puente, A. E., & Perez-Garcia, M. (2000). Psychological assessment of ethnic minorities. In G. Goldstein & M. Hersen (Eds.), *Handbook of psychological assessment* (3rd ed., pp. 527–552). New York, NY: Pergamon.

Rabin, L. A., Paolillo, E., & Barr, W. B. (2016). Stability in test-usage practices of clinical neuropsychologists in the United States and Canada over a 10-year period: A follow-up survey of INS and NAN members. *Archives of Clinical Neuropsychology, 31,* 206–230. https://doi.org/10.1093/arclin/acw007

Read, D. W. (2006). Cultural evolution is not equivalent to Darwinian evolution. *Behavioral and Brain Sciences, 29,* 361.

Richerson, P. J., Boyd, R., & Henrich, J. (2010). Gene-culture coevolution in the age of genomics. *Proceedings of the National Academy of Sciences, 107*(Suppl 2), 8985–8992. https://doi.org/10.1073/pnas.0914631107

Sam, D. L. (2006). Acculturation: Conceptual background and core components. In D. L. Sam & J. W. Berry (Eds.), *The Cambridge handbook of acculturation psychology* (pp. 11–26). Cambridge, England: Cambridge University Press.

Sam, D. L., & Berry, J. W. (2010). Acculturation: When individuals and groups of different cultural backgrounds meet. *Perspectives on Psychological Science, 5,* 472–281. https://doi.org/10.1177/1745691610373075

Sasaki, J. Y., & Kim, H. S. (2016). Nature, nurture, and their interplay: A review of cultural neuroscience. *Journal of Cross-Cultural Psychology, 48,* 1–19. https://doi.org/ 10.1177/0022022116680481

Saxton, G. D., & Benson, M. A. (2003). The origins of socially and politically hostile attitudes toward immigrant outgroups: Economics, ideology, or national context? *Journal of Political Science, 31,* 101–137.

Scherer, K. R., Clark-Polner, E., & Mortillaro, M. (2011). In the eye of the beholder? Universality and cultural specificity in the expression and perception of emotion. *International Journal of Psychology, 46*, 401–435. https://doi.org/10.1080/00207594.2011.626049

Schwartz, S. J., Montgomery, M. J., & Briones, E. (2006). The role of identity in acculturation among immigrant people: Theoretical propositions, empirical questions, and applied recommendations. *Human Development, 49*, 1–30. https://doi.org/10.1159/000090300

Schwartz, S. J., Unger, J. B., Zamboanga, B. L., & Szapocznik, J. (2010). Rethinking the concept of acculturation: Implications for theory and research. *American Psychologist, 65*, 237–251. https://doi.org/10.1037/a0019330

Segall, M., Campbell, D., & Herskovits, M. J. (1966). *The influence of culture on visual perception*. Indianapolis, IN: Bobbs-Merrill.

Segal, U. A., Mayadas, N. S., & Elliott, D. (2010). The immigration process. In U. A. Segal, D. Elliott, & N. S. Mayadas (Eds.), *Immigration worldwide: Policies, practices, and trends* (pp. 3–16). New York, NY: Oxford University Press.

Seligman, R., Choudhury, S., & Kirmayer, L. J. (2016). Locating culture in the brain and in the world: From social categories to the ecology of mind. In J. Y. Chiao, S. Li, R. Seligman, & R. Turner (Eds.), *The Oxford handbook of cultural neuroscience* (pp. 3–20). New York, NY: Oxford University Press.

Shweder, R. A. (2007). An anthropological perspective: The revival of cultural psychology—some premonitions and reflections. In S. Kitayama & D. Cohen (Eds.), *Handbook of cultural psychology* (pp. 821–836). New York, NY: Guilford.

Shweder, R. A., & Sullivan, M. A. (1993). Cultural psychology: Who needs it? *Annual Review of Psychology, 44*, 497–523. https://doi.org/10.1146/annurev.ps.44.020193.002433

Singh, M., & Bryden, M. P. (1994). The factor structure of handedness in India. *International Journal of Neuroscience, 74*, 33–43. https://doi.org/10.3109/00207459408987227

Smith, G. E., Ivnik, R. J., & Lucas, J. A. (2008). Assessment techniques: Tests, test batteries, norms, and methodological approaches. In J. Morgan & J. Ricker (Eds.), *Textbook of clinical neuropsychology* (pp. 38–57). New York, NY: Taylor & Francis.

Spiro, M. E. (1996). Postmodernist anthropology, subjectivity, and science: A modernist critique. *Comparative Studies in Society and History, 38*, 759–780. https://doi.org/10.1017/S0010417500020521

Stein, D. J., Chiao, J. Y., & van Honk, J. (2016). Cultural neuroscience in South Africa: Promises and pitfalls. In J. Y. Chiao, S. Li, R. Seligman, & R. Turner (Eds.), *The Oxford handbook of cultural neuroscience* (pp. 143–151). New York, NY: Oxford University Press.

Sue, S. (1996). Measurement, testing, and ethnic bias: Can solutions be found? In G. R. Sodowsky & J. C. Impara (Eds.), *Multicultural Assessment in Counseling and Clinical Psychology* (pp. 7–36). Lincoln, NE: Buros Institute of Mental Measurements.

Tabery, J. (2014). *Beyond versus: The struggle to understand the interaction of nature and nurture*. Cambridge, MA: MIT Press.

Teng, E. L., Lee, P., & Chang, P. C. (1976). Handedness in a Chinese population: Biological, social and pathological factors. *Science, 193*, 1148–1150. https://doi.org/10.1126/science.986686

Tennie, C., Call, J., & Tomasello, M. (2009). Ratcheting up the ratchet: On the evolution of cumulative culture. *Philosophical Transactions of the Royal Society B, 364*, 2405–2415. https://doi.org/10.1098/rstb.2009.0052

Tomasello, M. (1999). *The cultural origins of human cognition*. Cambridge, MA: Harvard University Press.
Tsai, J. L. (2007). Ideal affect: Cultural causes and behavioral consequences. *Perspectives on Psychological Science, 2,* 242–259. https://doi.org/10.1111/j.1745-6916.2007.00043.x
Turkheimer, E. (2000). Three laws of behavior genetics and what they mean. *Current Directions in Psychological Science, 9,* 160–164. https://doi.org/10.1111/1467-8721.00084
Tylor, E. B. (1871). *Primitive culture* (2 vols.). London, England: Murray.
van de Vijver, F. J. R., & Leung, K. (1997). *Methods and data analysis for cross-cultural research*. Thousand Oaks, CA: SAGE.
van der Veer, R. (2012). Cultural-historical psychology: Contributions of Lev Vygotsky. In J. Valsiner (Ed.), *The Oxford handbook of culture and psychology* (pp. 58–68). New York, NY: Oxford University Press.
van der Veer, R., & van Ijzendoorn, M. H. (1985). Vygotsky's theory of the higher psychological processes: Some criticisms. *Human Development, 28,* 1–9. https://doi.org/10.1159/000272931
van der Veer, R., & Yasnitsky, A. (2011). Vygotsky in English: What still needs to be done. *Integrative Psychological & Behavioral Science, 45,* 475–493. https://doi.org/10.1007/s12124-011-9172-9
Vygotsky, L. S. (1997). Genesis of higher mental functions. In R. W. Rieber (Ed.), *The collected works of L. S. Vygotsky. Vol. 4: The history of the development of higher mental functions* (pp. 97–119). New York, NY: Plenum.
Whittle, P. (2000). W. H. R. Rivers and the early history of psychology at Cambridge. In A. Saito (Ed.), *Bartlett, culture and cognition* (pp. 21–35). London, England: Psychology Press.
Wong, T. M., Strickland, T. L., Fletcher-Janzen, E., Ardila, A., & Reynolds, C. R. (2000). Theoretical and practical issues in the neuropsychological assessment and treatment of culturally dissimilar patients. In E. Fletcher-Janzen, T. L. Strickland, & C. R. Reynolds (Eds.), *Critical issues in neuropsychology: Handbook of cross-cultural neuropsychology* (pp. 3–18). New York, NY: Kluwer Academic/Plenum.
Wundt, W. M. (1916). *Elements of folk psychology: Outlines of a psychological history of the development of mankind* (E. L. Schaub, Trans.). London, England: George Allen & Unwin.
Zhu, Y., Zhang, L., Fan, J., & Han, S. (2007). Neural basis of cultural influence on self-representation. *Neuroimage, 34,* 1310–1316. https://doi.org/ 10.1016/j.neuroimage.2006.08.047

2

Vygotsky, Luria, and the Cultural-Historical Approach to Cognitive Psychology and Neuropsychology

Experimental and Theoretical Beginnings

DAVID E. TUPPER

INTRODUCTION

It is almost a truism, but culture is now known to influence the development and structure of cognition, neuropsychological test findings, and the development and functional organization of cerebral regions. This book is, in fact, a recognition and testament to these topics in contemporary cognitive neuroscience and psychology.

 Consideration of culture as a topic in cognitive psychology and neuroscience is not a recent phenomenon, despite the proliferation of contemporary writings on the topic (Ardila, 2018; Bennett & Hacker, 2013; Cagigas & Bilder, 2009; Kozulin, 2011; Nell, 2000; Yamada & Lamberty, 2016). In human history—for instance, in antiquity through the Middle Ages to the Renaissance—people were aware of differences among humans and emphasized the existence of groups of people that differ from them by their behavior and other characteristics. A more scientific approach concerning research on human similarities and dissimilarities started to develop as late as the end of 19th and the beginning of 20th centuries (see Jahoda, 2012). For instance, Emile Durkheim (social scientist) and Pierre Janet (pioneering psychologist) proposed early theories, attempting to explain dissimilarities in behavior of people with origins from different cultures. These purely speculative theories were later followed by empirical although mostly observational studies (see Lonner, 2013); for example, Sir Francis Galton (scientist and early psychometrist) was one of the first investigators to conduct comparative studies of intelligence. Findings obtained during the early psychological, ethnological, and anthropological studies of the 19th and early 20th centuries led many scientists of that time to the idea that a key issue for understanding human behavior is the interaction between an individual, their personal biology, and the cultural environment in which they find themselves (Irvine & Berry, 1988; Nisbett & Norenzayan, 2002).

It is beyond the scope of this chapter to review the distant and diverse history of culture as a factor in sociology, anthropology, psychology, and neuroscience. For instance, the idea of a cultural psychology is older than the discipline of psychology itself (Cole, 1996). Nevertheless, almost 100 years ago a theory of cultural-historical development was proposed, which remains relevant to contemporary cognitive psychology, neuropsychology, and neuroscience (see Toomela, 2014b) and which will be the focus of this chapter. This popular theory serves as an early model of how cultural and historical (individual/developmental) factors can influence not only cognition and cerebral development, but also the conduct and presentation of experimental research. In fact, this theory as proposed by several prominent psychologists prompted them to conduct an early cross-cultural investigation of cognition and culture that has led to further research on this topic, as well as recent controversy. This chapter will briefly review the theoretical basis of these ideas, the early interpretations of data gathered from this early 20th-century research on these ideas and the continued historical understanding of this early cultural research.

CONTEXT AND DEVELOPMENT OF A CULTURAL THEORY

There are well-known historical (anthropological, social scientific, and even popular) examples of writings about foreign peoples and their customs and beliefs—what we may broadly call culture. One may conjecture that perceived differences in encounters with "foreign" peoples are triggers for curiosity about culture and human diversity (Jahoda, 2012). A growing interest in culture in a specialty appears to occur when increasing intercultural encounters, and subsequent recognition of cultural differences, are combined with circumstances in which tangible benefits or understanding can result from exchanges of resources with people from these other backgrounds (Irvine & Berry, 1988) and can lead to subsequent cross-cultural investigations.

Cultural psychology is the study of how cultures reflect and shape psychological processes. The main tenet of cultural psychology is that mind and culture are inseparable and mutually constitutive, meaning that people are shaped by their culture and their culture is also shaped by them (Cole, 1996; Kozulin, 2011). Cultural psychology is distinct from cross-cultural psychology in that cross-cultural research generally uses culture as a means of testing the universality of psychological processes rather than determining how local cultural practices shape psychological processes. As differences in customs entailed differences in beliefs and behaviors, these features were often noted and even emphasized as early as Herodotus more than two millennia ago (Jahoda, 2012). Some writers on culture such as Michael Cole (Cole, 1996; Cole & Engeström, 1993) have said that culture and mind are really different facets of the same phenomenon. In a similar manner, cultural neuroscience explores the interplay between the social transmission of knowledge and the functional systems of the brain (Seligman, Choudhury, & Kirmayer, 2016).

Theories often reflect the social context and historical period in which they were written, and the cultural-historical theory discussed here is no exception. Clearly, the context for development of a cultural theory was very different 100 years ago than today. Cultural-historical psychology is connected with the name of Lev Vygotsky and originated in the 1920s in what was then the Soviet Union. Lev S. Vygotsky (1896–1934) was a Russian psychologist-teacher who laid the groundwork for the sociocultural or cultural-historical theory of cognitive development; this theory has risen to the forefront of the fields of child development and education around the world in the last three decades and has had considerable influence in cultural and cognitive psychology (Berk & Harris, 2003; Daniels, 2008; Gielen & Jeshmaridian, 1999; Leontiev & Luria, 1968; Van der Veer, 1996, 1997; Van der Veer & Valsiner, 1991; Zavershneva & Van der Veer, 2018). Sociocultural theory emphasizes the connection between the individual's social and psychological worlds, by regarding communication between children and more expert, adult members of their culture as the source of consciousness, higher cognitive processes, and of the ability to regulate thought, action, and behavior. Vygotsky's primary goals in the development of this theory were to create a unifying perspective in psychology that would resolve some of the theoretical contradictions of his time (e.g., reflexes vs. psyche) and to address practically the serious problems of Soviet society at that time, which included a high rate of illiteracy and limited care for children with disabilities. Large-scale social changes following the Russian revolution set the stage for and energized Vygotsky's new perspective on child and cognitive development in Russia, and the theory was accepted initially as generally consistent with Marxist principles (although later Vygotsky's work was suppressed by Soviet authorities). Vygotsky considered child development to occur via social interaction and participation in culturally meaningful activities, much as the social movements and authoritarian organization of labor and production were occurring at that time.

Vygotsky advanced a set of ideas that was quite unlike that of contemporaries such as Karl and Charlotte Bühler, Kurt Koffka, Kurt Lewin, Jean Piaget, Wilhelm Stern, and Heinz Werner (Cole & Wertsch, 1996; Nisbett & Norenzayan, 2002) or Ivan Pavlov for that matter. He posited a unique theory about the merging of natural (biological-maturational) and cultural dimensions in child development (Berk & Harris, 2003; Van der Veer, 2007; Yasnitsky & Ferrari, 2008). The cultural dimension was based on the acquisition of cultural means that transformed the child's functioning. Applying the socially acquired cultural means to the self, the child becomes conscious of his or her mental functioning. The mastering of cultural means takes years and is only finished in adolescence. At first, higher mental functioning needs the support of concrete material actions and objects, but gradually the child becomes able to operate on the purely mental or representational level. According to Vygotsky, the most important cultural means (instrument) is speech, with its variants of social, egocentric, and inner speech (although recent authors have questioned Vygotsky's emphasis on speech to the exclusion of the more concrete means and actions; see Van der Veer, 2012). The inclusion of speech into mental functioning (first from parents or caregivers, and only later internalized) implies that the interrelationships between different mental functions become changed and form a dynamic mental system. On the cortical level, this probably means that different cortical centers become connected, disconnected, or reconnected depending on life experiences. A primary role in intellectual development is played by education, which teaches children a logically connected world view. Children can profit from education when it falls in their zone of proximal development, a key concept in Vygotsky's educational framework. To the extent that cultures offer different cultural means inside or outside education, individuals from different cultures will display different modes of thinking. Children (or adults) from different cultures may have the same intellectual potential, but they may end up thinking in fundamentally different ways.

Alexander R. Luria (1902–1977) was a well-known Russian psychologist and neurologist who is recognized as one of the pre-eminent neuropsychologists of the 20th century (Akhutina & Shereshevsky, 2014; Homskaya, 2001; Tupper, 2003). Luria's life was changed in January 1924 when he met Vygotsky at the Second All-Russian Congress on Psychoneurology in Leningrad. Luria was so impressed with Vygotsky's approach to psychology that he helped arrange an invitation for Vygotsky to work in Moscow with him and other associates. From the time that they met to the time of Vygotsky's premature death (from tuberculosis) in 1934, Luria and Vygotsky, along with Alexei Leontiev, a like-minded colleague, worked together to create a practical and Marxist-oriented Soviet psychology using the cultural-historical theory as its basis (Gielen, & Jeshmaridian, 1999; Luria, 1935; Radzikhovskii & Khomskaya, 1982; Yasnitsky, 2012a). The main principle that united Luria and Vygotsky in the new theory was the desire to show that psychology could view higher psychological functions only in the context of their development in historical (individual) and cultural processes and demonstrated via objective principles of brain function (the systemic organization of cerebral processes).

In the years prior to Vygotsky's death, these Russian researchers believed that higher mental processes are formed initially between people and in social interaction and only later become internalized in individuals as inner speech and other cognitive regulatory structures. All mental functions are therefore initially culturally determined and mediated. Vygotsky worked vociferously to expand and develop his theory while he was ill, and Luria found the cultural-historical approach to represent exactly what he was looking for as a young clinician and researcher. Although Vygotsky's complete elaboration of this theory was cut short by his death, Luria remained true to a dynamic, cultural-historical theory for the rest of his life, even in his neuropsychological practice (Akhutina & Shereshevsky, 2015; Cole, 2005; Homskaya, 2001).

A basic tenet of the cultural-historical approach to cognition is that mental functioning in the present emerges from the interplay of different developmental factors including the individual's cultural history, ontogeny and maturation, and microgenesis or the temporal unfolding of cerebral processes (Cole, 2005; Laboratory of Comparative Human Cognition, 1979; Toomela, 2014a). This type of psychological development supports individual self-regulation and higher-level psychological processing.

Vygotsky and Luria in the 1930s proposed historical support for this early theory as well as limited experimental research on the role of culture in structuring an individual's cognitive functioning. The following section will review their main experimental research endeavors that they felt supported the cultural-historical approach, namely their expeditions to Central Asia. As will be seen, the early interpretation of this research is now controversial and, because of the more complete current understanding of the context and findings of the expeditions, serves as an important reminder about the difficult nature of cross-cultural research into cognitive functioning and psychological processes.

Ultimately, although certainly not unchallenged, Vygotsky's theory has served as the foundation for an expanding literature on cultural variation in cognitive development. Sociocultural or cultural-historical theory is important in granting social experience a fundamental role in cognitive development, has inspired research that has deepened our understanding of everyday cognitive development in children, and has been incorporated into neuropsychological and neuroscientific theorizing (Christensen, 2009; Goldberg, Akhutina, Melikyan, Mikadze, Mervis, & Bisoglio, 2015; Kotik-Friedgut & Ardila, 2014; Puente & Agranovich, 2003). Cultural-historical theory complements more traditional approaches to cognitive development, which have focused largely on solitary experimentation and laboratory research on isolated factors using a more static and biological organism. In fact, in a later writing Luria (1971) reiterated that he feels psychology is primarily a cultural and historical science.

TESTING OF THE THEORY: EXPEDITIONS TO CENTRAL ASIA

To consider the context of the Central Asia expeditions, it is necessary to understand a little about the forces operating in the Soviet Union in the early part of the 20th century (see also Sirotkina & Smith, 2012). Following the Russian revolution, General Secretary Joseph Stalin implemented a series of five-year plans meant to improve the Soviet domestic situation through rapid industrialization, nation-wide efforts for literacy development, and the collectivization of agriculture. The first five-year plan began in 1928, and from 1929 to 1932 the Soviet authorities conducted two operations concurrently that included the collectivization of agriculture and the elimination of *kulaks* or independent farmers. The idea was that individual farming would be replaced by farming in collective farms (*kolkhozes*) and that the property of kulaks—who were supposedly exploiting their less prosperous fellow farmers—would be impounded by the state, thus moving toward a more socialist society. In addition, rapid industrialization and support of literacy development were emphasized; for instance, Russian was imposed as the country-wide language. Ultimately, the plan was ineffective due to the short amount of time needed to achieve the goals as well as the major social changes required, and it led to widespread loss of life through famine, major conflicts among populations (due to the cultural changes required for rapid industrialization), and passive resistance in parts of the union. This rapid transformation of the country nevertheless created situations in which less developed communities would need to demonstrate rapid gains in productivity, literacy, and other cultural advances. Thus, Vygotsky and Luria saw these changes as creating a potentially important and unique cultural milieu that could be studied experimentally for its effects on individual cognition and psychological function, particularly with regard to the cultural-historical theory they were developing.

The central idea prominent at that time, upon which the rationale for experimentation was based, was that lower psychological processes would be the same for all living people while higher psychological processes would differ between persons belonging to different cultures depending on their level of advancement (Zinchenko, 1997). This idea was based on informal evidence gathered by Thurnwald, Werner, Lévy-Bruhl, Durkheim, and others (considered comparatively by Vygotsky & Luria, 1930/1993; see also Luria, 1979) as well as the theory developed initially by Vygotsky. It was only natural—in view of the limited nature of the previous cross-cultural evidence—that Vygotsky and Luria felt the need to witness these cognitive similarities and differences themselves in a carefully designed psychological study (Van der Veer & Valsiner, 1991).

Along with the previously described nationalist changes creating a milieu for potential research, Vygotsky and Luria had expanded the scope of their professional connections. In 1925, they made separate trips to various laboratories in Europe during which they became personally acquainted with Gestalt scholars, exposing them to other types of research and creating close professional interrelationships among the Russians and international scholars (Yasnitsky, 2012a, 2012b). A former student and colleague of Vygotsky and Luria, Alexander Zaporozhets, made a trip to the far east of Russia in the spring and summer 1929, focused on Soviet educational policies and experiments on the development of children (Proctor, 2013). Vygotsky also took a trip to Tashkent (Uzbekistan) in April and May 1929, during which time he gave lengthy seminars, visited remote regions, and conducted pilot experimental research on their theory (Vygodskaya & Lifanova, 1996). This trip likely set up a direct link and potential sites for the Moscow researchers in their subsequent Central Asia expeditions.

Experiments in the early 20th century were different than present day (Danziger & Ballantyne, 1997) and were primarily performed through naturalistic observation. During the early 1930s, Luria took the lead role and organized and conducted two psychological expeditions to distant, presumably less developed (what were called primitive) regions in Central Asia in an attempt to further support the cultural-historical theory; he initially considered a third expedition. The research efforts were planned with a number of prominent investigators, including Vygotsky, who was too ill to actually join the trips. The purposes of the cross-cultural investigations were to study the influence of the dramatic changes in modernization on the development of psychological processes due to the rapid sociocultural change taking place in the Soviet republics of central Asia at that time, and to use newer experimental methods to study those changes. Luria felt that both the choice of the location (small villages that remained "primitive") and the time period (the period of collectivization and industrialization under the first five-year plan) were particularly suitable for conducting such cross-cultural research. As previously noted, the expeditions were partly successful in supporting the theory, although the findings and all the controversy surrounding the expeditions were not clear initially. The following will present a brief description of the expeditions and summary of findings as originally described, and the more recent understanding and controversies about the process and outcomes of the expeditions will be discussed.

The first expedition to Central Asia was conducted in the summer of 1931 during the early timeframe of the Russian cultural revolution. Luria organized a team of mostly Soviet researchers from Moscow to conduct the initial investigations (see Figure 2.1), which were essentially pilot studies for the subsequent more detailed research planned later. The 1931 expedition, according to limited contemporary publications by Luria (1931a, 1931b, 1932), was organized by the Uzbek Research Institute of Samarkand and the Moscow Institute of Experimental Psychology to investigate the variations in thought and other psychological processes of people living in a very primitive economic and social environment and to record the changes that develop from the introduction of higher and more complex forms of economic life, theoretically raising the "general cultural level" (Luria, 1932,

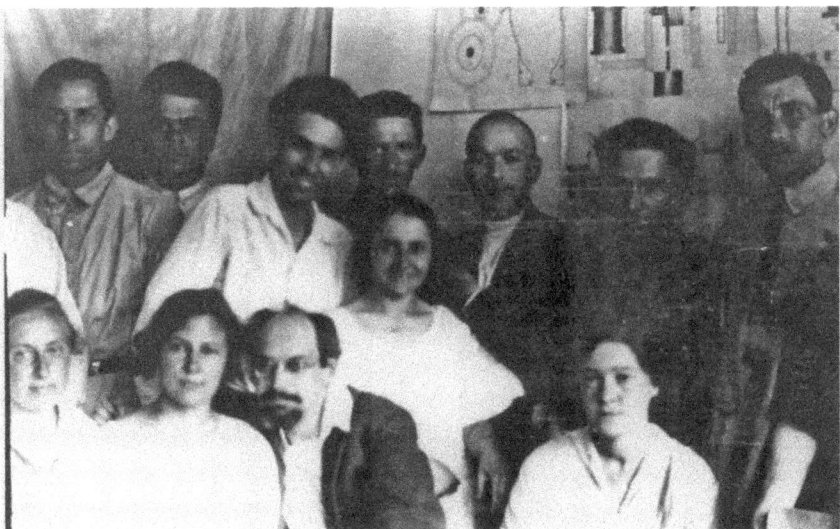

Figure 2.1. Luria (back row, third from left) and Vygotsky (back row, third from right) with colleagues at the time of the Central Asia expeditions. Photo courtesy of Gita Vygodskaya.

p. 241). The expedition, in July 1931, followed a two-month seminar provided by Luria in Samarkand in May and June 1931. This first expedition included various groups of investigators studying the structure of perception (color, form, and optical illusions), visual thinking, elementary intellectual processes, verbal-logical thinking, concept formation, causal thinking, and personality. Using adult subjects from the native population, the expedition began in Uzbekistan; the "primitive" group was gathered from people living in nomadic conditions in the Alai Mountain region and in the districts (kishlaks) of Shamimardan, Yordan, and adjacent uplands. A control group from the Naryn River region (Kyrgyzstan) with a very active cotton-raising industry and highly developed collective farming, but still described as "backward culturally," was also studied. Limited findings were described in Luria's 1931 and 1932 papers, although it was emphasized that a second expedition with a more international character was to be planned. The knowledge collected during the first expedition was nevertheless significant, including an estimated 600 individual case reports, and the material concerning the organization (visual and practical) of intellectual processes was unique for that time period.

Apparently, the results of the first expedition seemed to Luria and Vygotsky so supportive, interesting, and promising for their theory that they immediately began planning a second expedition. Luria was inspired and thus wanted to make the second expedition to Uzbekistan an international undertaking. He therefore wrote to their German Gestalt psychologist colleagues Wolfgang Köhler, Kurt Lewin, and Kurt Koffka to ask them to participate. Luria listed in his letters his desire to study a variety of mental processes such as deductive thinking, metaphor and symbol comprehension, logical operations, perception of shapes, colors and optical-geometrical illusions, drawing, and specifics of remembering and counting (Yasnitsky, 2013a, 2013c). Unfortunately, Köhler was not able to participate because of illness. Luria's objective was fairly clear: The expedition was originally conceived as a large-scale international study with the participation of major Western scholars, so that the 1931 expedition publications had the purpose of raising publicity and luring foreign experts to participate in the second expedition. In parallel with the publication of the first expedition's reports, Luria continued to correspond with these Western scientists, hoping to entice them in the new project and to obtain their agreement to participate. According to Luria (see Yasnitsky, 2013c), the second expedition was to become, on the one hand, a less important expedition, meant to check the data obtained during the first expedition and, on the other hand, an expedition that would prepare the next, most ambitious third expedition in the summer of 1933 (which ultimately never occurred).

Although Köhler and Lewin could not participate in the second expedition to Central Asia, Kurt Koffka (1886–1941), a German-American Gestalt scholar living at the time in Northampton, Massachusetts, did accept Luria's invitation and joined the research team of the second expedition. According to Luria's subsequent summary publications (Luria, 1933, 1934), the 1932 expedition was organized by Moscow State University, the Ukrainian Psychoneurological Academy

(Kharkov), and the Uzbek Pedagogical Academy, with support from the government of Uzbekistan and the People's Commissariat of Education of the Uzbek Socialist Soviet Republic.

As in the previous expedition, the researchers were interested in confirming changes in perception, problem-solving, memory, and other aspects of cognitive functioning associated with the historic changes in economic activity and schooling; a number of naturalistic experiments with both literate and illiterate people were implemented during the expedition, and a broad range of test measures were utilized. Because one of the hypotheses that the higher mental processes would change under the influence of the social reforms was due to the influence of schooling, the second expedition also included children and youth among the subjects. The investigators again used observational and clinical assessment methods to test their hypothesis that the structure of human cognitive processes differs according to the ways in which the various social groups live out their lives. The researchers expected to find that people whose lives are dominated by concrete, practical activities have a different method of thinking from people whose lives require abstract, verbal, and theoretical approaches to reality. Luria was especially interested in studying "the very dynamics of the transition from the more elementary psychological laws to the more complex processes" (Luria, 1934, p. 255). The extended team consisted of Koffka, three researchers from Moscow—Luria, Shemyakin, and Mordkovich—and an Uzbek group of researchers from the local Pedagogical Academy directed by P. Leventueff.

The preparations for this second expedition to Central Asia were more thorough, and it was presumably better funded and recognized. The groups formed more specific research teams studying various cognitive processes, including Luria's group focused on situational thinking, Koffka's group investigated perception, Leventueff's group studied causative thinking, Shemyakin and his group looked at the understanding of symbols, Mordkovich's group focused on symbolic posters in situational thinking, as well as a related group studying counting in complex thinking (Luria, 1934). It was felt that two main groups of the population would represent the research subjects: illiterate women (*ichkary*), on the one hand, and representatives of the *kolkhozes*, who had undergone some educational training, on the other hand. Both groups were assessed with similar testing procedures to evaluate these processes.

On his way to Uzbekistan, Koffka traveled to Europe from Northampton, and having waited in Berlin for the necessary papers, he arrived in Moscow at the end of May 1932 (see Harrower, 1983; Yasnitsky, 2012b, 2013b, 2013c). Koffka stayed in Moscow for a short time and gave a presentation at the local Institute of Psychology on May 29, 1932. It was during his visit to Moscow that Koffka first met Vygotsky, who served as an interpreter (reportedly quite capably) for the lecture because Koffka's talk was presented in German.

Similar to the first expedition, the second expedition took place in the *kishlaks* and *dzhailaus* (mountain pastures) of Uzbekistan and Kirghizia (Kyrgyzstan). The expedition was carried out in June and July 1932 (see Figure 2.2). It began

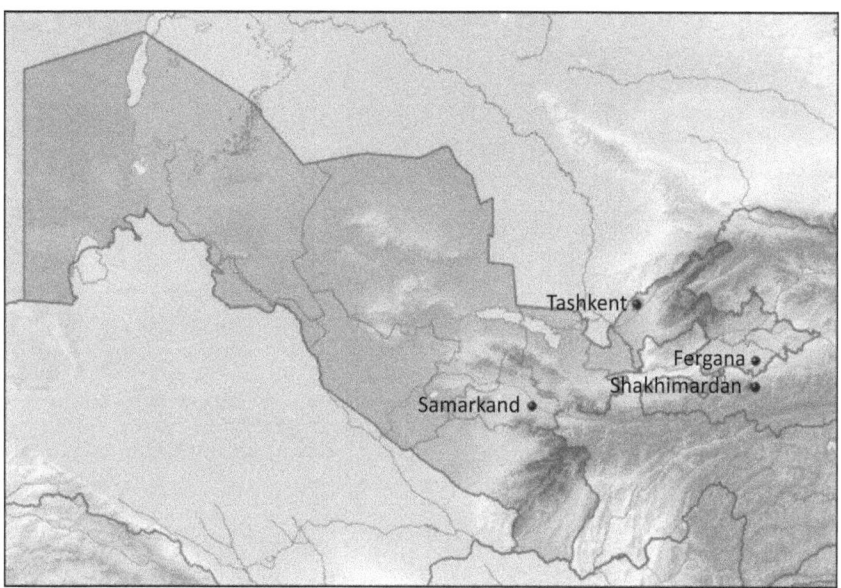

Figure 2.2. Map of the 1932 Central Asia expedition sites in Uzbekistan. Adapted from PsyAnima, Society for Psychology & Allied Sciences website. http://psyanimajournal.livejournal.com/9395.html

with a departure from Moscow on June 1 to Tashkent, and after a short delay in Tashkent (June 5–9), the investigators stayed in Samarkand for a one-week preparatory seminar (June 9–16). After the seminar in Samarkand, the extended research team departed for the remote areas of eastern Uzbekistan, first to the relatively "civilized" and mostly collectivized Fergana Valley (June 16–July 1) and then to the mountain areas of Shakhimardan, formerly part of Kyrgyzstan (July 1 to the end of July). Shakhimardan had a predominantly traditional population, which had not been collectivized and actively resisted the economic and social innovations of the Bolsheviks. The data collected in these two areas of Eastern Uzbekistan formed the material for the comparative analysis of the changes in mental functioning (a score sheet from the expedition can be seen in Figure 2.3). By late July or early August 1932, the members of the expedition had completed the fieldwork and began to return to their homes.

Only a brief summary of an illustrative portion of the findings, mostly from the second expedition, will be provided here, with more details available primarily in Luria's writings (Luria, 1932, 1933, 1934, 1976). Luria's own monograph (1974, 1976) of the results presents only a fraction of the data and appears to intermix findings from both expeditions. In that book, he does not refer at all, for example, to the study of religious thinking that was carried out (Yasnitsky, 2013a). The primary topics discussed here will be the study of visual or optic illusions (carried out mostly under the direction of Koffka; see the following discussion for the associated controversy) and the studies of more abstract thinking.

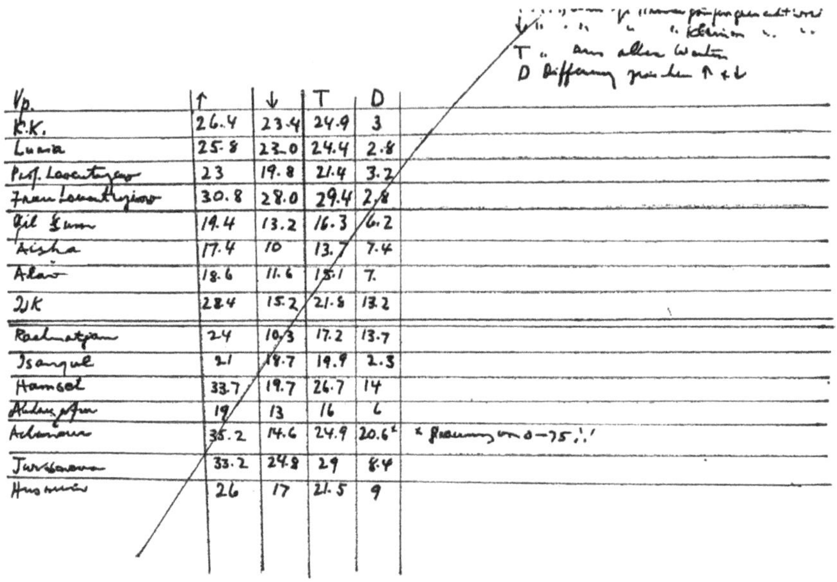

Vp.	↑	↓	T	D
K.K.	26.4	23.4	24.9	3
Luria	25.8	22.0	24.4	2.8
Perf. Lacent...	23	19.8	21.4	3.2
Frau Lacent...	30.8	28.0	29.4	2.8
Gil Sum	14.4	13.2	16.3	6.2
Aisha	17.4	10	13.7	7.4
Alam	18.6	11.6	15.1	7.
JK	28.4	15.2	21.8	13.2
Rael...jam	24	10/3	17.2	13.7
Isanyul	21	18.7	19.9	2.3
Hamsel	33.7	19.7	26.7	14
Hiday p...	19	13	16	6
Achmoura	35.2	14.6	24.9	20.6ᵉ * Praum... on 0-75,!'
Jurs...	33.2	24.8	29	8.4
Hus...	26	17	21.5	9

Figure 2.3. Scoring sheet from the 1932 Central Asia expedition. Courtesy of L. Mecacci. Adapted from L. Mecacci, 2005, Luria: A unitary view of human brain and mind. *Cortex, 41*, 819. Note the listings of the initials K.K., which represent Kurt Koffka and Luria, who served as control subjects.

It is no surprise given Koffka's involvement that the study of visual illusions (such as the Poggendorf) was directly influenced by Gestalt psychology. Gestalt psychologists had claimed that the Gestalt perceptual principles were the result of enduring characteristics of the brain and not bound up with culturally transmitted meanings of objects. However, Vygotsky and Luria assumed that at least some of these visual processes partially rested upon semantic, interpretative processes and would, therefore, differ between subjects belonging to different cultures. The general understanding of the results since the time of the study is that Luria found that illiterate/primitive people were not able to recognize the same visual illusions as the control subjects and were not able to distinguish depth in either photographs or drawings (Luria, 1976; despite comments made in his 1934 publication indicating that the recognition of the illusions varied based on the "attitude of the testees towards the experimenter" [Luria, 1934, p. 257]). His later summary monograph essentially repeats the same conclusions he reached 40 years before (omitting the doubts regarding the data that were noted originally), stating that it may be considered proven that "even relatively simple processes involved in perception of colors and geometrical shapes depend to a considerable extent on the subjects' practical experience and their cultural milieu" (Luria, 1976, p. 45). There is even an interesting and notorious anecdote passed along by Luria's subsequent colleagues concerning the visual illusion findings (e.g., Homskaya, 2001, p. 26), in which Luria reportedly sent Vygotsky a telegram saying "Uzbeks don't have any illusions" and which reportedly was intercepted by governmental officials and interpreted in the political sense, ultimately leading to a moratorium on the continuation of the research (see also Allik, 2013; Goncharov, 2013; Yasnitsky, 2013b).

The 1932 expedition studies also included investigations of higher-level reasoning, problem-solving and classification/generalization capabilities (Luria, 1974, 1976). To investigate methods of classifying objects, Luria essentially followed the same procedure as in the study of perception where subjects were divided into groups of different "educational level" or "degree of primitivism," and these groups were presented with various classification or generalization tasks. The following subject groups were identified by Luria (1976): (a) *ichkary* women living in remote villages who were illiterate and not involved in modern social activities; (b) peasants in remote villages, who continued to maintain an individual economy, were illiterate, and were very primitive; (c) women who attended short-term courses and who had no formal education and almost no literacy training; (d) *kolkhoz* workers; and (d) female students admitted to a teachers' training college. The subjects were presented with geometrical figures and other stimuli, and Luria found that only the "most culturally advanced" group of subjects named geometrical figures by categorical names such as circle, triangle, etc. (Luria, 1976, p. 32). The most "primitive" subjects designated all the figures with the names of objects: a circle, for example, would be called a plate, watch, moon, etc. The more "primitive" subjects refused to combine figures belonging to the same geometrical category into one group, seeing them as very concrete objects that had nothing to do with each other. He reported that the culturally more "advanced" subjects had no problem with this task. Roughly similar findings were noted with regard to cultural variations in methods of grouping using either graphic or categorical classifications (see Table 2.1).

A famous qualitative example of the research on deduction and inference using syllogisms by the primitive or unsophisticated subjects was also provided by Luria (1976, pp. 108–109). The syllogism presented to the subject was "In the Far North,

Table 2.1. RESULTS OF THE GROUPINGS AND CLASSIFICATION EXPERIMENT IN THE CENTRAL ASIA EXPEDITION

Group	Number of subjects	Graphic method of grouping	Graphic and categorical methods of grouping	Categorical classification
Illiterate peasants from remote villages	26	21 (80%)	4 (16%)	1 (4%)
Collective-farm activists (barely literate)	10	0	3 (30%)	7 (70%)
Young people with one or two years' schooling	12	0	0	12 (100%)

From A. R. Luria, 1976, *Cognitive development: Its cultural and social foundations.* Cambridge, MA: Harvard University Press. p. 78.

where there is snow, all bears are white. Novaya Zemlya is in the Far North and there is always snow there. What color are the bears there?"

The subject replied, "There are different sorts of bears. I don't know; I've seen a black bear, I've never seen any others. . . . Each locality has its own animals; if it's white, they will be white; if it's yellow, they will be yellow."

Despite the potential importance of the cross-cultural results, only brief initial findings from both expeditions were published by Luria at the time (1930s), and a fuller description of the results was not available for many years. Luria and his fellow researchers concluded their research in the summer of 1932, and Luria reportedly completed a draft monograph about the results of the expedition by the end of 1933 or the beginning of 1934 (see Luria, 1974, p. 4, where he mentions that Vygotsky died soon after its completion), but the monograph remained unpublished or uncompleted for 40 years. It has been interpreted that further publications in the Soviet Union were probably restricted due to the political climate where such research, particularly research that could implicate issues with the Soviet government or their plans, was not allowed. It is also apparent that a large-scale expedition in 1933 did not take place and that the second expedition to Central Asia was the final one.

Because of Vygotsky's death and the social pressures on such cross-cultural research, Luria was not able to publish more than his very early brief descriptions of the 1931 and 1932 studies at that time. It was not until the 1970s when similar pressures were not present that Michael Cole could encourage and persuade Luria to put into more complete form the findings and interpretations of the distant research. The results found by Luria and his colleagues have nevertheless been generally supported by modern research, although they have been interpreted by various researchers in various ways (e.g., Cole & Scribner, 1974). Even Cole, for instance, does not believe that the Uzbekis in Luria's investigations really acquired new modes of thought, but rather he interprets the data as the result of "changes in the application of previously available modes to the particular problems and contexts of discourse represented by the experimental setting" (Cole, cited in Luria, 1976, p. xv). Clearly, the interpretation of cross-cultural cognitive research findings is complex based on many factors.

In more recent years, the Central Asia expeditions and resulting publications and interpretations have become a matter of controversy, mostly as the result of further analysis of the context and surrounding written materials that have become available. Anton Yasnitsky and colleagues (see Lamdan & Yasnitsky, 2016; Yasnitsky, 2012a, 2013a, 2013b, 2013c) have been at the forefront of reinterpretation of these issues. As examples of the controversy, two interrelated major issues will be discussed briefly. The first issue has been called the Luria–Koffka controversy and centers particularly around issues surrounding the optical illusion findings and interpretations, as well as Kurt Koffka's role in the second expedition. The second controversy relates to the possible reasons that the main research findings were delayed by 40 years.

It is worth noting that Koffka's participation in the second expedition was rather short-lived. He is reported to have developed some type of illness during the expedition and had to leave to return home prematurely, before the full completion of the expedition in August 1932; the complete explanation for the illness is unknown, although there is some suspicion that discomfort with the overall cultural differences was a factor (Harrower, 1983; Lamdan & Yasnitsky, 2016). More significant, it appears that Koffka had a very different interpretation of the optical illusions results than did Luria, as evidenced by the early (1934) comments about preliminary results of no difference between Uzbekis and Western control subjects in recognizing illusions once situational factors were accounted for (it turns out that the findings in Luria's 1934 summary article were provided by Koffka and that later writings by Luria did not recognize that interpretation; see Lamdan, 2013; Lamdan & Yasnitsky, 2016). Luria, in fact, tended to downplay or disregard any contradictory results or interpretations; Koffka is not mentioned in the text or listed in the Index of Luria (1976). Koffka reached a different conclusion than Luria; he thought that the Uzbekis did succumb to optical illusions like the Poggendorf illusion and were able to see perspective in drawings just as the control subjects. In addition, Koffka's remark about the role of the subjects' attitude toward the experimenter is quite relevant. As noted by Lamdan and Yasnitsky (2016), the subjects, at times distrustful and hostile, from remote mountainous areas of Shakhimardan were observed by both Luria and Koffka and constituted the very same group of primitive subjects who allegedly succumbed to a relatively smaller number of optical illusions in the first expedition in 1931. Luria observed the fear and distrust of his subjects in these regions in Uzbekistan and Kyrgyzstan but apparently never believed these factors might be of any relevance in the context of the research. In contrast, Koffka not only observed the fear and anxiety but also attributed the difference in performance of the two groups of subjects primarily to the difference in their emotional state during the experiment (see Harrower, 1983).

In addition to these differences, there is another factor that likely explains the different interpretations. As mentioned previously, the expedition was well organized and supported. The team of researchers was actually accompanied by presidents of the two Central Asian Soviet republics and by the local head of the secret police. Thus, strangers associated with state authorities and accompanied by a convoy of security forces visited the remote villages to perform some type of scientific experiments. This situation alone would be enough to cause feelings of suspicion and anxiety in the local population; some people even fled back into the mountains to get away (see Lamdan & Yasnitsky, 2016). Yasnitsky (2013b) feels that Koffka, who was more of an outsider to the expedition, was less biased by Marxist thinking and the focus on sociocultural theories and thus more sensitive to the actual conditions of the experimental subjects. Unlike Luria, Koffka did not dismiss the social situation of the experiment but considered it as an utterly essential factor, which allowed him to realize that the emotional state of the subjects of the study notably affected their psychological performance. It is unfortunate, but it is not clear that Koffka and Luria ever resolved these differences. Yasnitsky and colleagues feel that it is rather ironic that the Luria–Koffka controversy provides a clear-cut example of how Luria and Vygotsky, who are considered to be the founders of cultural-historical psychology missed these cultural and personal components of the psychological research they conducted in Central Asia (see also Yasnitsky, 2012b, where a more complete cultural-historical Gestalt psychology is presented).

Aside from this controversy, the original findings of both Central Asia expeditions were incompletely presented in the 1930s when they occurred and, in fact, were not published at all in the Soviet Union. As noted, the traditional explanation for this delay was that such research was shut down for political reasons at the time. It is clear that the Soviet Union was undergoing significant policy changes and that any research that demonstrated limited benefits of the governmental "affirmative action" industrialization and collectivization plans would be considered anti-Soviet. As Lamdan and Yasnitsky (2016) point out, the Central Asia findings that significant psychological differences still existed in the Soviet Union of 1932 would place the researchers of such an ethno-psychological investigation in the crosshairs as a target of political criticism. Thus, Luria and Vygotsky's research as well as the cultural-historical approach could be considered as antinational and likely made Luria and the investigators reluctant to publish the findings. The differences between the major conclusions drawn by Luria and Koffka (a well-known international scholar already) of some of the results may have been an additional factor in Luria's reluctance to more completely present the results at that time or to consider alternative explanations.

RELEVANCE AND CONTEMPORARY IMPLICATIONS

Despite the vicissitudes of the Central Asia expeditions and the surrounding controversies, the early Luria–Vygotsky theoretical and empirical findings have been mostly accepted as supporting cultural influences on cognitive development and have stimulated subsequent research (Ghassemzadeh, Posner, & Rothbart, 2013; Hyman, 2012; Khan, 2014; Nell, 1999). Contemporary cognitive psychology and neuroscience have increasingly incorporated the understanding of cultural differences into models of cognition and cerebral organization, whether or not the cultural-historical theory has been its basis (Agranovich & Puente, 2007; Toomela, 2014b).

As mentioned previously, there has been a recent revisionist revolution in thinking about Vygotsky's cultural-historical theory with regard to cultural psychology, child development, cognitive psychology, and education in the past 20 or more years (Moll, 1990; Yasnitsky & Van der Veer, 2016). This revolution has been especially prominent with regard to education and child development, where Vygotsky's thinking has been influential and where there has been an explosion of writings and applications. Much of the impact of the cultural-historical theory in education has been with regard to two aspects of the theory: first, the fundamental role that growing conceptual thinking (via mediated speech and other mechanisms) plays in the education and development of children, and, second, the leading role of school and instruction more generally in stimulating children's mental development (Van der Veer, 2012). As mentioned previously with regard to this last role, Vygotsky conceived of teaching as providing a new zone of proximal development for children that helps expand the nature of their cognitive processing beyond just a focus on biological-maturational changes. It is this zone that emphasizes scaffolding of instruction as a useful tool in the instructional realm that has led to its popularity in education (see also Daniels, 2008; Oliveira & Rego, 2010).

Given the importance of speech and internalized mental processes in the cultural-historical theory, language development, and especially literacy have been particular topics of research in cross-cultural cognitive psychological research and neuropsychology. Huettig and Mishra (2014) have emphasized that at present more than one fifth of humanity remains unable to read and write and that adult literacy remains a tangible problem despite generalized socio-economic development across the world. These authors note that both literacy and schooling serve as needs for the further development of cognitive functioning around the world that go beyond just the processing of written words and sentences (see also Schubert, 1983). More specifically, Ardila (2018) has emphasized the importance of literacy as a major factor in individual neuropsychological assessment, and he has been a prolific researcher who has provided normative data for use with many cultural populations (see also other cultural data available in Ferraro, 2016).

Michael Cole has been a major figure in 20th-century cross-cultural investigations in psychology. Having trained with Luria, Cole has investigated the development and implementation of cognitive activities in many varied cultural contexts. For instance, he has studied everyday problem-solving and schooling using syllogisms in the Yucatan (see Cole, 1996) and the cognitive consequences of formal and informal education (Scribner & Cole, 1973). Cole and colleagues performed a classic cross-cultural study using anthropological, linguistic, and experimental-psychological techniques in the study of cognitive processes in the Kpelle, a Liberian tribal group in Western Africa (Cole, Gay, Glick, & Sharp, 1971). This field research was initiated because the investigators noted the Liberian tribal children experienced a great deal of difficulty with Western-style mathematics, and they wondered what kind of mathematics knowledge they bring to school. The investigators sought to find intragroup variation in these types of everyday activities that could be related to cognitive activity in the Kpelle as well as contrasts between the Kpelle and other cultural groups. In this "New Mathematics Project" in Liberia, they found, for example, that the tribal people were exceptionally good at estimating various amounts of rice but were both inconsistent and inaccurate at measuring lengths, which, on the surface, appeared closely related skills. The investigators discovered that rice farming is central to the Kpelle culture while length measurement is a very specific activity dependent on the thing being measured by the Kpelle (e.g., the metric for measuring cloth is not the same as for rice). To summarize their ultimate findings, they reported that the Kpelle were significantly inferior to Americans in some cognitive tasks and clearly superior to Americans in others. Cole et al. concluded that ultimately a person will be good at doing the things that are important to them and that they have occasion to do often. Cole's conceptualization of cultural-historical theory sees emergent psychological processes as being culturally constituted, with mediation of the future and the past in the present by the individual and, therefore, requiring a process-based view of human activities for research investigations (Cole, 1990, 1995).

Neuropsychology and cognitive psychology have benefited from the influence of Luria's continuing work after Vygotsky's death, including his longer history of writings as a neuropsychologist (Luria, 1979; Moskovich, Boougakov, DeFina, & Goldberg, 2002). Luria's use of the cultural-historical approach has informed his brain-behavior theorizing in many ways, particularly with regard to the notion that the brain is developed and organized as a dynamic system dependent on cultural exposure and individual experience, as well as his research on self-regulatory and frontal-executive functioning (Luria, 1971; Meccaci, 2005). Das (2003) has been a prominent researcher and writer with regard to the differences and common factors in Luria's neuropsychological systemic (and cultural) model of cerebral organization. He has provided a more contemporary theory of how cognitive processing is organized cerebrally and how the brain utilizes a culturally developed knowledge base within the context of information processing and executive functions for a more representative model of intelligence. With regard to cultural neuropsychological and neuroscientific thinking, Kostyanaya and Rossouw (2013) also point out that Luria's cultural-historical approach to neuropsychology has led to many investigators developing both bottom–up and top–down models of cerebral organization via dynamic, systemic models (see also Kotik-Friedgut & Ardila, 2014). Like Vygotsky and Luria, who were both interested in problems of mental disability, the cultural-historical approach has much to offer for a more holistic understanding of normal as well as abnormal cognitive functioning (Lamdan & Yasnitsky, 2013).

Especially relevant to the current chapter, Janna Glozman of Moscow University (and of Luria's actual laboratory and clinic) recently replicated and extended Luria and Vygotsky's original 1931 and 1932 studies of the influence of cultural change on psychological processes using two groups of adults from a remote area of the Russian Far East (northern Kamchatka peninsula). Glozman collaborated with other researchers from Moscow, Belgorod, and Petropavlovsk universities in this cultural replication study (Glozman, 2018). All of the participants had completed elementary school, but some worked as nomadic herdsmen in the tundra, while others lived and worked in an almost inaccessible village. The assessment used many of Luria's original classification and generalization measures, as well as more formal neuropsychological and projective assessments. The results revealed that despite the fact that all of the participants had a modest level of education, the villagers performed better than the herdsmen on a variety of cognitive and neuropsychological tasks, similar to more Western subjects, indicating the varied influence of social experience on cognitive functioning that Luria and Vygotsky had reported in a similar fashion more than eight decades earlier. Although there are limitations to the study, Glozman recognized specifically that the life attitudes and values of the Kamchatka inhabitants, who live in difficult living conditions, differ from people from central regions of Russia. In particular, the inhabitants show the unique character of their own culture, national values and the common interests of their own ethnic group with others living in the same region. Michael Cole interprets the new findings as confirming the effectiveness of Luria's cultural-historical approach in advancing both theoretical and practical understanding of the development of higher psychological functions (Cole, 2018).

Unfortunately, Luria and Vygotsky, despite their valiant efforts in coordinating and conducting a cross-cultural study at an important time in history, were swayed by the cultural context in which they themselves worked and ultimately missed one of the important lessons in such research—that experimenters (or clinicians) need to consider carefully their own impact on participants and the cultural differences they bring to their efforts. It is hoped that 21st-century cross-cultural cognitive and neuropsychological research can further benefit from a look back at these early beginnings.

REFERENCES

Agranovich, A. V., & Puente, A. E. (2007). Do Russian and American normal adults perform similarly on neuropsychological tests? Preliminary findings on the relationship between culture and test performance. *Archives of Clinical Neuropsychology, 22*, 273–282. http://dx.doi.org/10.1016/j.acn.2007.01.003

Akhutina, T. V., & Shereshevsky, G. (2014). Cultural-historical neuropsychological perspective on learning disability. In A. Yasnitsky, R. van der Veer, & M. Ferrari (Eds.), *The Cambridge handbook of cultural-historical psychology* (pp. 350–377). Cambridge, England: Cambridge University Press.

Akhutina, T. V., & Shereshevsky, G. (2015). Alexander Luria: A brief biography. In G. J. Rich & U. P. Gielen (Eds.), *Pathfinders in international psychology* (pp. 105–117). Charlotte, NC: Information Age.

Allik, J. (2013). Do primitive people have illusions? *PsyAnima, Dubna Psychological Journal, 6*(3), 40–42.

Ardila, A. (2018). *Historical development of human cognition: A cultural-historical neuropsychological perspective*. New York, NY: Springer.

Bennett, M. R., & Hacker, P. M. S. (2013). *History of cognitive neuroscience*. Malden, MA: Wiley-Blackwell.

Berk, L. E., & Harris, S. (2003). Vygotsky, Lev. In L. R. Nadel (Ed.), *Encyclopedia of cognitive science* (Vol. 4, pp. 532–536). London, England: Macmillan.

Cagigas, X., & Bilder, R. M. (2009). Where culture meets neuroimaging. In A.-L. Christensen, E. Goldberg, & D. Bougakov (Eds.), *Luria's legacy in the 21st century* (pp. 23–29). New York, NY: Oxford University Press.

Christensen, A.-L. (2009). Luria's legacy in the 21st century. In A.-L. Christensen, E. Goldberg, & D. Bougakov (Eds.), *Luria's legacy in the 21st century* (pp. 3–16). New York, NY: Oxford University Press.

Cole, M. (1990). Cognitive development and formal schooling: The evidence from cross-cultural research. In L. C. Moll (Ed.), *Vygotsky and education: Instructional implications and applications of sociocultural psychology* (pp. 89–110). New York, NY: Cambridge University Press.

Cole, M. (1995). Culture and cognitive development: From cross-cultural research to creating systems of cultural mediation. *Culture & Psychology, 1*, 25–54. https://doi.org/10.1177/1354067X951103

Cole, M. (1996). *Cultural psychology: A once and future discipline.* Cambridge, MA: Belknap Press of Harvard University Press.

Cole, M. (2005). A. R. Luria and the cultural-historical approach to psychology. In T. Akhutina, J. Glozman, L. Moskovich, & D. Robbins (Eds.), *A. R. Luria and contemporary psychology* (pp. 35–41). New York, NY: Nova Science.

Cole, M. (2018). Luria's legacy in cultural-historical psychology. *Psychology in Russia: State of the Art, 11*(2), 2–6. https://doi:10.11621/pir.2018.0200

Cole, M., & Engeström, Y. (1993). A cultural-historical approach to distributed cognition. In G. Salomon (Ed.), *Distributed cognitions: Psychological and educational considerations* (pp. 1–46). Cambridge, England: Cambridge University Press.

Cole, M., Gay, J., Glick, J. A., & Sharp, D. W. (1971). *The cultural context of learning and thinking: An exploration in experimental anthropology.* New York, NY: Basic Books.

Cole, M. & Scribner, S. (1974). *Culture and thought: A psychological introduction.* New York, NY: Wiley.

Cole, M. & Wertsch, J. V. (1996). *Contemporary implications of Vygotsky and Luria.* Worcester, MA: Clark University Press.

Daniels, H. (2008). *Vygotsky and research.* London, England: Routledge.

Danziger, K. & Ballantyne, P. (1997). Psychological experiments. In W. G. Bringmann, H. E. Lück, R. Miller, & C. E. Early (Eds.), *A pictorial history of psychology* (pp. 233–239). Chicago, IL: Quintessence.

Das, J. P. (2003). A look at intelligence as cognitive neuropsychological processes: Is Luria still relevant? *Japanese Journal of Special Education, 40,* 631–647. https://doi.org/10.6033/tokkyou.40.631

Ferraro, F. R. (Ed.). (2016). *Minority and cross-cultural aspects of neuropsychological assessment* (2nd ed.). New York, NY: Taylor & Francis.

Ghassemzadeh, H., Posner, M. I., & Rothbart, M. K. (2013). Contributions of Hebb and Vygotsky to an integrated science of mind. *Journal of the History of Neuroscience, 22,* 292–306. https://doi:10.1080/0964704X.2012.761071

Gielen, U. P., & Jeshmaridian, S. S. (1999). Lev S. Vygotsky: The man and the era. *International Journal of Group Tensions, 28,* 273–301. https://doi:10.1023/A:1021837200385

Glozman, J.M. (2018). A reproduction of Luria's expedition to Central Asia. *Psychology in Russia: State of the Art, 11*(2), 7–16. https://doi:10.11621/pir.2018.0201

Goldberg, E., Akhutina, T. V., Melikyan, Z. A., Mikadze, Y. V., Mervis, J. E., & Bisoglio, J. (2015). History of neuropsychology in Russia. In W. Barr & L. A. Bielauskas (Eds.), *The Oxford handbook of history of clinical neuropsychology.* New York, NY: Oxford University Press.

Goncharov, O. A. (2013). Commentary on A. Yasnitsky's article "Kurt Koffka: 'Uzbeks DO HAVE illusions!' The Luria-Koffka controversy." *PsyAnima, Dubna Psychological Journal, 6*(3), 34–36.

Harrower, M. (1983). The Russian-Uzbekistan expedition. In M. Harrower, *Kurt Koffka: An unwitting self-portrait* (pp. 143–164). Gainesville, FL: University Presses of Florida.

Homskaya, E. D. (2001). *Alexander Romanovich Luria: A scientific biography.* New York, NY: Kluwer Academic/Plenum.

Huettig, F., & Mishra, R. K. (2014). How literacy acquisition affects the illiterate mind—A critical examination of theories and evidence. *Language and Linguistics Compass, 8/10,* 401–417. https://doi:10.1111/lnc3.12092

Hyman, L. (2012). The Soviet psychologists and the path to international psychology. In J. Renn (Ed.), *The globalization of knowledge in history: Based on the 97th Dahlem Workshop* (pp. 631–668). Berlin, Germany: Edition Open Access.

Irvine, S. H., & Berry, J. W. (1988). The abilities of mankind: A revaluation. In S. H. Irvine & J. W. Berry (Eds.), *Human abilities in cultural context* (pp. 3–59). Cambridge, England: Cambridge University Press.

Jahoda, G. (2012). Culture and psychology: Words and ideas in history. In J. Valsiner (Ed.), *The Oxford handbook of culture and psychology* (pp. 25–42). New York, NY: Oxford University Press.

Khan, V. S. (2014). On one expedition for the study of cognitive processes among the peoples of Central Asia in the early 1930s. *International Journal of Central Asian Studies, 18,* 1–21.

Kostyanaya, M. I., & Rossouw, P. (2013). Alexander Luria: Life, research & contributions to neuroscience. *International Journal of Neuropsychotherapy, 1,* 47–55. https://doi:10.12744/ijnpt.2013.0047-0055.

Kotik-Friedgut, B., & Ardila, A. (2014). Cultural-historical theory and cultural neuropsychology today. In A. Yasnitsky, R. van der Veer, & M. Ferrari (Eds.), *The Cambridge handbook of cultural-historical psychology* (pp. 378–399). Cambridge, England: Cambridge University Press.

Kozulin, A. (2011). Cognitive aspects of the transition from a traditional to a modern technological society. In P. R. Portes & S. Sala (Eds.), *Vygotsky in 21st century society: Advances in cultural historical theory and praxis with non-dominant communities* (pp. 66–86). New York, NY: Peter Lang.

Laboratory of Comparative Human Cognition. (1979). What's cultural about cross-cultural cognitive psychology? *Annual Review of Psychology, 30,* 145–170. https://doi.org/10.1146/annrev.ps.30.020179.001045

Lamdan, E. (2013). Who had illusions? Alexander R. Luria's Central Asian experiments on optical illusions. *PsyAnima, Dubna Psychological Journal, 6*(3), 66–76.

Lamdan, E., & Yasnitsky, A. (2013). "Back to the future": Toward Luria's holistic cultural science of human brain and mind in a historical study of mental retardation. *Frontiers in Human Neuroscience, 7,* 1–3. https://doi:10.3389/fnhum.2013.00509.

Lamdan, E., & Yasnitsky, A. (2016). Did Uzbeks have illusions? The Luria–Koffka controversy of 1932. In A. Yasnitsky & R. Van der Veer (Eds.), *Revisionist revolution in Vygotsky studies* (pp. 175–200). London, England: Routledge.

Leontiev, A. N., & Luria, A. R. (1968). The psychological ideas of L. S. Vygotskii. In B. B. Wolman (Ed.), *Historical roots of contemporary psychology* (pp. 338–367). New York, NY: Harper & Row.

Lonner, W. (2013). Chronological benchmarks in cross-cultural psychology. Foreword to the *Encyclopedia of Cross-Cultural Psychology. Online Readings in Psychology and Culture, 1*(2). https://doi.org/10.9707/2307-0919.1124

Luria, A. R. (1931a). Psychological expedition to Central Asia. *Science, 74,* 383–384. http://dx.doi.org/10.1126/science.74.1920.383

Luria, A. R. (1931b). Psychologische expedition nach Mittalasien. *Zeitschrift für Angewandte Psychologie, 40,* 551–552.

Luria, A. R. (1932). Psychological expedition to Central Asia. *Pedagogical Seminary and Journal of Genetic Psychology, 40,* 241–242. https://doi.org/10.1080/08856559.1932.10534223

Luria, A. R. (1933). The second psychological expedition to Central Asia. *Science, 78,* 191–192. http://dx.doi.org/10.1126/science.78.191-a

Luria, A. R. (1934). The second psychological expedition to Central Asia. *Pedagogical Seminary and Journal of Genetic Psychology, 44,* 255–259. https://doi.org/10.1080/08856559.1934.10532497

Luria, A. R. (1935). L.S. Vygotsky. *Character and Personality, 3,* 238–240.

Luria, A. R. (1971). Towards the problem of the historical nature of psychological processes. *International Journal of Psychology, 6,* 259–272. http://dx.doi.org/10.1080/00207597108246692
Luria, A. R. (1974). *Ob istoricheskom razvitii poznavatel'nykh protsessov: Eksperimental'no psikhologichekoye issledovaniye* [On the historical development of psychological processes: Experimental psychological investigations]. Moskva, Russia: Nauka.
Luria, A. R. (1976). *Cognitive development: Its cultural and social foundations.* Cambridge, MA: Harvard University Press.
Luria, A. R. (1979). *The making of mind: A personal account of soviet psychology.* Cambridge, MA: Harvard University Press.
Mecacci, L. (2005). Luria: A unitary view of human brain and mind. *Cortex, 41,* 816–822. https://doi.org/10.1016/S0010-9452(08)70300-9
Moll, L. C. (Ed.). (1990). *Vygotsky and education: Instructional implications and applications of sociocultural psychology.* New York, NY: Cambridge University Press.
Moskovich, L., Bougakov, D., DeFina, P., & Goldberg, E. (2002). A. R. Luria: Pursuing neuropsychology in a swiftly changing society. In A. Y. Stringer, E.L. Cooley, & A.-L. Christensen (Eds.), *Pathways to prominence in neuropsychology: Reflections of twentieth-century pioneers* (pp. 49–62). New York, NY: Psychology Press.
Nell, V. (1999). Luria in Uzbekistan: The vicissitudes of cross-cultural neuropsychology. *Neuropsychology Review, 9,* 45–52. https://doi.org/10.1023/A:1025643004782
Nell, V. (2000). *Cross-cultural neuropsychological assessment: Theory and practice.* Hillsdale, NJ: Erlbaum.
Nisbett, R. E., & Norenzayan, A. (2002). Culture and cognition. In H. Pashler & D. Medin (Eds.), *Stevens' handbook of experimental psychology: Vol. 2, Memory and cognitive processes* (pp. 561–597). New York, NY: Wiley.
Oliveira, M. K., & Rego, T. C. (2010). Contributions to contemporary research of Luria's cultural-historical approach. *Educação e Pesquisa, 36,* 107–121. http://doi:10.1590/S1517-9702200000400009
Proctor, H. (2013). Kurt Koffka and the expedition to Central Asia. *PsyAnima, Dubna Psychological Journal, 6*(3), 43–52.
Puente, A. E., & Agranovich, A. V. (2003). The cultural in cross-cultural neuropsychology. In G. Goldstein & S. R. Beers (Eds.), *Comprehensive handbook of psychological assessment: Vol. 1, Intellectual and Neuropsychological Assessment* (pp. 321–332). New York, NY: Wiley.
Radzikhovskii, L. A., & Khomskaya, E. D. (1982). A. R. Luria and L. S. Vygotsky: Early years of their collaboration. *Soviet Psychology, 20*(1), 3–21. https://doi.org/10.2753/RPO1061-040520013
Schubert, J. (1983). The implications of Luria's theories for cross-cultural research on language and intelligence. In B. Bain (Ed.), *Sociogenesis of language and human conduct* (pp. 59–77). New York, NY: Plenum.
Scribner, S., & Cole, M. (1973). Cognitive consequences of formal and informal education. *Science, 182,* 553–559. http://doi:10.1126/science.182.4112.553
Seligman, R., Choudhury, S., & Kirmayer, L. J. (2016). Locating culture in the brain and in the world: From social categories to the ecology of mind. In J. Y. Chiao, S.-C. Li, R. Seligman, & R. Turner (Eds.), *The Oxford handbook of cultural neuroscience* (pp. 3–20). New York, NY: Oxford University Press.
Sirotkina, I., & Smith, R. (2012). Russian Federation. In D. B. Baker (Ed.), *The Oxford handbook of the history of psychology: Global perspectives* (pp. 412–441). New York, NY: Oxford University Press.

Toomela, A. (2014a). Methodology of cultural-historical psychology. In A. Yasnitsky, R. van der Veer, & M. Ferrari (Eds.), *The Cambridge handbook of cultural-historical psychology* (pp. 315–349). Cambridge, England: Cambridge University Press.

Toomela, A. (2014b). There can be no cultural-historical psychology without neuropsychology. And vice versa. In A. Yasnitsky, R. van der Veer, & M. Ferrari (Eds.), *The Cambridge handbook of cultural-historical psychology* (pp. 101–125). Cambridge, England: Cambridge University Press.

Tupper, D. E. (2003). Luria, Alexander R. In L. R. Nadel (Ed.), *Encyclopedia of cognitive science* (Vol. 2, pp. 965–969). London, England: Macmillan/Nature.

Van der Veer, R. (1996). The concept of culture in Vygotsky's thinking. *Culture & Psychology, 2,* 247–263. https://doi.org/10.1177/1354067X9600200302

Van der Veer, R. (1997). Lev Semyonovich Vygotsky. In W. G. Bringmann, H. E. Lück, R. Miller, & C. E. Early (Eds.), *A pictorial history of psychology* (pp. 352–355). Chicago, IL: Quintessence.

Van der Veer, R. (2007). Vygotsky in context: 1900–1935. In H. Daniels, M. Cole, & J. V. Wertsch (Eds.), *The Cambridge companion to Vygotsky* (pp. 21–49). Cambridge, University: Cambridge University Press.

Van der Veer, R. (2012). Cultural-historical psychology: Contributions of Lev Vygotsky. In J. Valsiner (Ed.), *The Oxford handbook of culture and psychology* (pp. 58–68). New York, NY: Oxford University Press.

Van der Veer, R., & Valsiner, J. (1991). *Understanding Vygotsky: A quest for synthesis.* Oxford, England: Blackwell.

Vygodskaya, G. L., & Lifanova, T. M. (1996). *Lev Semenovich Vygotskii: Zhizn', deyatel'nost', shtrikhi k portretu* [Lev Semenovich Vygotsky: Brushstrokes of a portrait]. Moskva: Smysl.

Vygotsky, L. S., & Luria, A. R. (1930/1993). *Studies on the history of behavior: Ape, primitive, and child.* Hillsdale, NJ: Erlbaum. [Original published in Russian as *Etudi po istorii povedeniya: Obeziana, primitiv, rebenok.* Moscow-Leningrad: GIZ, 1930]

Yamada, T., & Lamberty, G. J. (2016). Trends in the neuropsychological assessment of culturally diverse populations. In F. R. Ferraro (Ed.), *Minority and cross-cultural aspects of neuropsychological assessment* (2nd ed., pp. 1–10). New York, NY: Taylor & Francis.

Yasnitsky, A. (2012a). A history of cultural-historical Gestalt psychology: Vygotsky, Luria, Koffka, Lewin, and others. *PsyAnima, Dubna Psychological Journal, 1,* 98–101.

Yasnitsky, A. (2012b). Revisionist revolution in Vygotskian science: Toward cultural-historical Gestalt psychology. *Journal of Russian and East European Psychology, 50*(4), 3–15. https://doi.org/10.2753/RPO1061-0405500400

Yasnitsky, A. (2013a). Bibliografiia osnovnykh sovetskikh rabot po kross-kul'turnoi psikho-nevrologii I psikhologii natsional'nukh men'shinstv perioda kollejktivatazatsii, industrializatsii I kul'turnoi revoliutsii (1928–1932) [Bibiography of main Soviet publications on cross-cultural psychoneurology and psychology of national minorities during the period of collectivization, industrialization, and Cultural Revolution (1928–1932)]. *PsyAnima, Dubna Psychological Journal, 6*(3), 97–113.

Yasnitsky, A. (2013b). Kurt Koffka: "U uzbekov EST illiuzii!" Zaochnaia polemika mezdhu Luriei I Koffkoi [Kurt Koffka: "Uzbeks do have illusions!" The Luria–Koffka controversy. *PsyAnima, Dubna Psychological Journal, 6*(3), 1–25.

Yasnitsky, A. (2013c). Psychological expeditions of 1931–1932 to Central Asia: Chronicle of events in letters and documents. *PsyAnima, Dubna Psychological Journal, 6*(3), 114–166.

Yasnitsky, A., & Ferrari, M. (2008). From Vygotsky to Vygotskian psychology: Introduction to the history of the Kharkov school. *Journal of the History of the Behavioral Sciences, 44*(2), 119–145. https://doi.org/10.1002/jhbs.20303

Yasnitsky, A., & Van der Veer, R. (Eds.). (2016). *Revisionist revolution in Vygotsky studies* (pp. 175–200). London, England: Routledge.

Zavershneva, E., & Van der Veer, R. (Eds.). (2018). *Vygotsky's notebooks: A selection.* New York, NY: Springer.

Zinchenko, V. P. (1997). Russian psychology. In W. G. Bringmann, H. E. Lück, R. Miller, & C. E. Early (Eds.), *A pictorial history of psychology* (pp. 572–576). Chicago, IL: Quintessence.

3

Challenges in the Neuropsychological Assessment of Ethnic Minorities

LAURA A. RABIN, DONALD L. BRODALE,
MILUSHKA M. ELBULOK-CHARCAPE, AND WILLIAM B. BARR

INTRODUCTION

According to the most recent U.S. Census, population growth is fastest among ethnic minorities (U.S. Census Bureau, 2011). By 2044, the United States is expected to become a plurality nation, with no single racial or ethnic group representing more than 50% of the total population (Colby & Ortman, 2015). Shifts in population trends and immigration patterns, both locally and globally, create challenges and opportunities for clinical neuropsychologists who increasingly must consider issues of race, ethnicity, socioeconomic status, culture, language, and other demographic dimensions in the conduct of culturally sensitive and competent assessments.

Over the past decade, diversity issues have received increased attention in neuropsychology. A 2008 Diversity Summit, for example, addressed the challenges of assessing ethnic minorities and called for the development of a plan for the future of cross-cultural neuropsychology (Byrd, Razani, Suarez, LaFosse, Manly, & Attix, 2010; Romero et al., 2009). Publications emerging from this summit offered recommendations for the use of demographic corrections in neuropsychological assessment (Romero et al., 2009), a critique of methods used to evaluate non-English speaking patients, and solutions for the recruitment and retention challenges of ethnic minority neuropsychology students (Byrd et al., 2010). Additionally, summit participants articulated a need for improved public awareness of and access to neuropsychological services by ethnic minorities, improved dialogue between test publishers and practitioners on issues of demographic corrections, and increased recruitment of ethnic minorities into normative studies (Romero et al., 2009). Shortly thereafter, Rivera Mindt and colleagues presented a "call to action" for the field through specific provisions to improve multicultural knowledge, education, training, and research (Rivera Mindt, Byrd, Saez, & Manly, 2010). The authors acknowledged that many of their stated goals for improved neuropsychological practice with ethnic minorities would take time to realize but encouraged neuropsychologists to utilize the optimal available methods to promote individual and organizational change (Rivera Mindt et al., 2010).

These summit and call to action papers, alongside other position papers (e.g., Judd et al., 2009) and consensus recommendations (e.g., American Psychological Association, 2003, 2010; International Test Commission, 2010) were based on expert opinion about diversity issues in neuropsychology and the field of psychology more broadly. The current chapter presents a bottom–up perspective from practicing clinical neuropsychologists about the challenges faced when conducting neuropsychological assessments of ethnic and racial minorities. Data were derived from the 2011 neuropsychological assessment survey described in the following text (Rabin, Paolillo, & Barr, 2016). While survey findings related to minority and cross-cultural issues were reported previously (Elbulok-Charcape, Rabin, Spadaccini, & Barr, 2014), the current chapter reconsiders those data in the context of recent work in the field of minority and cross-cultural neuropsychology. Despite distinctions between terms such as *multicultural, cross-cultural, diversity,* and *minority (neuro)psychology* (see Geisinger, 2003), these terms tend to be used interchangeably in the neuropsychology literature and, therefore, will not be distinguished from one another in the current chapter. Moreover, while various diversity factors impact the neuropsychological assessment process (e.g., gender, age, sexual orientation, disability, class status, religious orientation), the current chapter focuses on racial and ethnic diversity and immediately attendant issues such as language. Also, consistent with previous researchers (e.g., Byrd et al., 2010; Rivera Mindt et al., 2010), we will refer to individuals with culturally, racially, or ethnically diverse backgrounds (e.g., African Americans, Hispanics/Latinos, Asians/Asian Americans, American Indians, and Alaska Natives) as *ethnic minorities.* The reader is referred to Ferraro (2016) for additional discussion of culture, race, and ethnicity, along with information about the historical development of these terms and their current application within the field of neuropsychology.

OVERVIEW OF SURVEY METHODS

Data were collected as part of a broader survey of neuropsychological assessment practices conducted in 2011 (http://rabin.us/nas-2011.pdf), which represented a 10-year follow-up study of neuropsychologists' test usage practices in the United States and Canada (Rabin, Barr, & Burton, 2005). A complete description of the survey method is provided in Rabin et al. (2016). In brief, potential participants were randomly selected doctorate-level members from the International Neuropsychological Society (INS) or the National Academy of Neuropsychology (NAN) who resided in the United States or Canada. The authors first combined the membership lists and eliminated overlap. The resulting list was then carefully screened to ensure that it did not contain members without a doctorate degree, incorrect or invalid addresses, those not practicing clinical neuropsychology, or the deceased.

Packets consisting of an explanatory letter, 73-item questionnaire, and a stamped self-addressed envelope were sent to 2,178 individuals—approximately one third of the combined INS and NAN membership. Potential participants received a reminder postcard 4 weeks later, and nonrespondents received a replacement survey 10 weeks after the initial mailing. Respondents were informed that the survey was voluntary and asked to return the physical questionnaire by mail or complete an online equivalent. To ensure anonymity, every questionnaire was assigned a randomly generated identification code, and responses were not directly linked to participants' names. The authors also removed any identifying information that accompanied returned questionnaires.

The questionnaire inquired about various issues such as test usage within specific cognitive domains, forensic practice issues, views about the use of psychometrists, perceived challenges associated with the selection and utilization of neuropsychological tests and test data, and use of computerized tests. Eleven survey items (two free response and nine selected response item formats) related specifically to cultural competence and neuropsychologists' assessment of ethnic minorities. These items were developed based on a review of the literature related to minority and cross-cultural neuropsychology. In addition, we solicited and received feedback from experts in multicultural neuropsychology and refined questions based on their suggestions. The resulting 11 items addressed participants' race/ethnicity (1 item); percentage of professional time spent with ethnic minority individuals (1 item); nature and extent of any training received in neuropsychological assessment of ethnic minorities or cross-cultural neuropsychology (3 items); assessments conducted in languages other than English (2 items); approaches to testing individuals not fluent in English (1 item); approaches to interpreting test scores of ethnic minorities (2 items); and perceived challenges associated with assessing ethnic minorities (1 item). The appendix presents the actual items and response options where relevant.

KEY FINDINGS AND DISCUSSION

Response Rate and Participant Characteristics

Of the 2,178 questionnaires initially mailed, 512 were suitable for data analysis. Sixty-eight questionnaires were returned undeliverable to the given addresses, and 87 were unusable because they were blank or came from individuals lacking a doctoral degree or who were inactive in neuropsychological practice. Thus, after accounting for these 155 unusable and undeliverable questionnaires, the 512 doctorate-level INS and NAN respondents represent a 25.7% usable response rate. Although a higher response rate would have been preferable, efforts were made to boost participation through multiple mailings and the provision of incentives. Surveyed participants were comparable, in terms of basic demographic and practice-related characteristics, to participants in another large survey of neuropsychologists conducted at roughly the same time, the TCN/AACN 2010 "Salary Survey" (Sweet, Meyer, Nelson, & Moberg, 2011).

The average survey respondent was close to 50 years of age, had practiced neuropsychology for 15 years, and performed approximately three neuropsychological assessments per week. The overwhelming majority of respondents held PhDs (81%), most commonly in clinical psychology (65%). The majority of respondents were female (54%). Only 21% of respondents were board certified in neuropsychology. Respondents reported spending the majority of their professional time (60%) conducting neuropsychological assessments, with considerably less time devoted to psychotherapy (11%) and research (7%). In terms of the setting in which respondents performed their neuropsychological work, the majority endorsed private or group practice (60%), with medical hospitals also commonplace (32%). The most common referral sources were neurologists, psychiatrists, general medical practitioners, and psychologists/neuropsychologists (each endorsed by >50% of respondents). The most frequently endorsed assessment referral question was determination of diagnosis (71% of respondents). On average, respondents reported spending the greatest percentage of professional time with adults (27%), followed by older adults (25%), young adults (21%), children (14%), and adolescents (13%). Also, respondents reported working with a variety of patient/diagnostic groups, most frequently those with head injury (55% of respondents), dementia (49% of respondents), and attention deficit/hyperactivity disorder (38% of respondents).

Ethnic Designations of Respondents and Patients

Although non-Whites comprised approximately 36% of the U.S. population at the time the survey was conducted (U.S. Census Bureau, 2017), less than 10% of neuropsychologists identified as non-White (Figure 3.1). These percentages are similar to those reported in the 2005 and 2012 TCN/AACN Salary Surveys

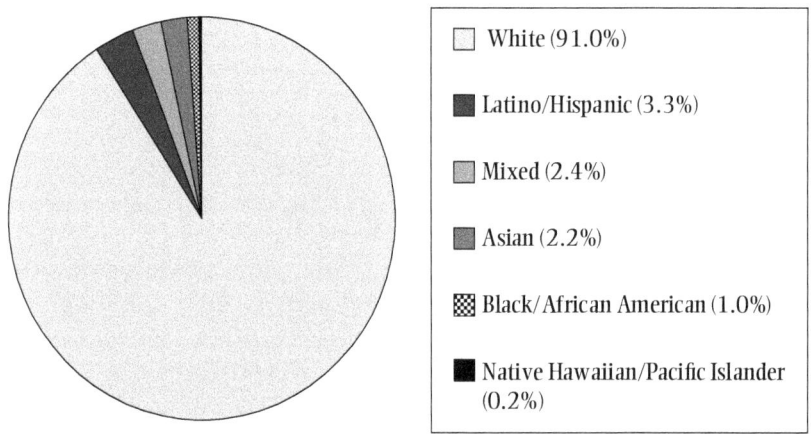

Figure 3.1. Neuropsychologist's race/ethnicity ($n = 510$)

(Sweet, Nelson, & Moberg, 2006; Sweet, Meyer, Nelson, & Moberg, 2011). By contrast, survey respondents reportedly work with diverse populations, with roughly one third of their professional time, on average, spent with non-White individuals, most commonly Black or African American individuals and Latino or Hispanic individuals (Figure 3.2).

Although the integration of neuropsychologists who identify as ethnic minorities into the field does not guarantee culturally competent assessments, there are important reasons for fostering an ethnically, culturally, and linguistically diverse body of neuropsychologists. Diverse students may be committed to serving those within their underserved communities (Proctor, Simpson, Levin, & Hackimer, 2014). Also, diverse colleagues can serve as mentors and role models to students and young professionals and bring new perspectives to theory development and clinical application (Byrd et al., 2010) including challenging the mono-cultural perspective and dominant cultural bias inherent to the field (Wong, Strickland, Fletcher-Janzen, Ardila, & Reynolds, 2000). As highlighted by members of the 2008 Multicultural Problem Solving Summit, various obstacles, which reflect larger systemic issues, deter minority individuals from pursuing or achieving a career in neuropsychology (Byrd et al.,2010). These issues include lack of an educational pipeline (e.g., limited exposure to the field of neuropsychology at the undergraduate level), rigid Graduate Records Exam (GRE) cut-off scores for graduate school admittance, socioeconomic issues (e.g., the need to enter the workforce early to pay back undergraduate loans or ease household financial needs), and a history of mistrust of psychology as a profession (Byrd et al., 2010; Rivera Mindt et al., 2010; Romero et al., 2009; Salinas, Bordes-Edgar, & Puente, 2016). In terms of graduate student retention, a convergence of research suggests the importance of adequate financial support, academic environments supportive of diversity, solid mentoring by both faculty and peers, and opportunities for students to work clinically with ethnic minority populations of interest (Proctor & Simpson, 2016).

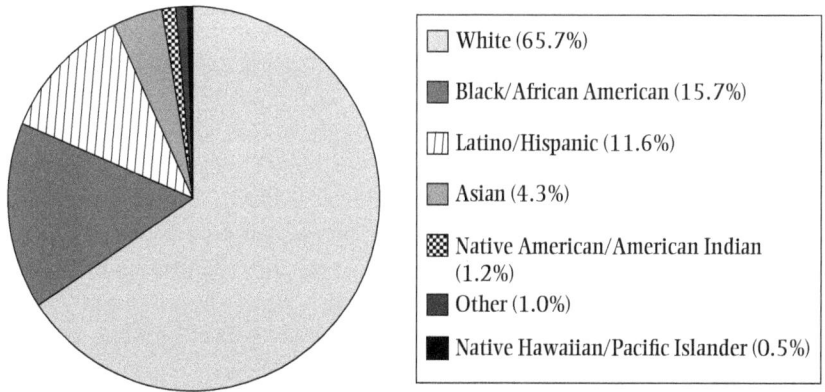

Figure 3.2. Mean neuropsychological practice time by patient race/ethnicity ($n = 497$)

Table 3.1. SETTINGS FOR PROFESSIONAL TRAINING IN NEUROPSYCHOLOGICAL ASSESSMENT OF ETHNIC MINORITIES (N = 351)

Setting	Responses n	%
Graduate or doctoral studies	224	63.8
Postdoctoral studies, fellowships, or residencies	199	56.7
Continuing education courses, certifications, or workshops	69	19.7
Internships	45	12.8
Conferences	10	2.8

Note: Sum of respondent percentages exceeds 100% because the survey did not constrain responses.

Training

Although roughly one third of respondents' professional time, on average, is spent with non-White individuals (Figure 3.2), more than one fourth (27.7%, $n = 139$) reported receiving no training in neuropsychological assessment of ethnic minority populations or cross-cultural neuropsychology. For those who did receive such training ($n = 351$, see Table 3.1), for the vast majority (77%) this training occurred through formal neuropsychology education only—that is, during graduate school, internship, or postdoctoral fellowship/residency.

Guidelines for practicum training in neuropsychology include essential competencies that doctoral students must achieve before transitioning to internship and fellowship (Nelson et al., 2015), including the integration of knowledge of diversity issues in assessment, research, treatment, and consultation. Notably, these are the same basic competencies required for specialty practice in neuropsychology at the professional level (Rey-Casserly, Roper, & Bauer, 2012). How best to incorporate multicultural training into existing psychology doctoral program curricula, however, is not readily known. Proctor and Simpson (2016) recommend moving beyond instruction of culture-specific knowledge alone to facilitation of students' critical thinking skills about how cultural values, beliefs, and behaviors interact with other factors, such as socioeconomic status, language, and religion, to impact cognition. They describe an "integration-separate" course model, where separate courses develop particular areas of competence, while content covering diversity issues is simultaneously spread throughout core coursework (Proctor & Simpson, 2016). As an example, a program with faculty expertise in aging could offer a course on cross-cultural issues in the neuropsychological assessment of dementia. Simultaneously, multicultural content would be infused into most other courses and diversity issues addressed in practica evaluations, comprehensive examinations, and subsequent licensure and board certification examinations (Allison, Echemendia, Crawford, & Robinson, 1996; Rivera Mindt et al., 2010). Students and faculty would be encouraged to join professional organizations that target multicultural issues (e.g., Hispanic Neuropsychological Society; American

Psychological Association [APA] Division 45: Society for the Psychological Study of Culture, Ethnicity and Race; Society for Indian Psychologists). Additional opportunities might include financial support for student and faculty research on multicultural issues, externship opportunities with ethnic minority clinical populations, and guest speakers or brown-bag talks on diversity topics. Through these efforts, doctoral students would be prepared to seek and obtain APA-accredited internships and postdoctoral fellowships that emphasize working with culturally diverse individuals.

Respondents (n = 81) also received multicultural neuropsychology training at the professional practice level through continuing education opportunities and/or self-education (see Table 3.1). Of these respondents, roughly half also had received formal training (during graduate school, internship, or fellowship) while the other half received multicultural training at the professional level only. Thus, for some individuals, multicultural training within formal neuropsychological education may be nonexistent or insufficient. With respect to the format of professional training, respondents reported various formal and informal methods including didactics, conference presentations, assigned readings, case conferences, consultation with colleagues, and group discussion (see Table 3.2). Those who attend neuropsychology conferences offered through major professional organizations (e.g., INS, NAN, APA Division 40: Society for Clinical Neuropsychology) will find continuing education programs, research talks, symposia, keynote speeches, and discussion groups focused on diversity issues that provide opportunities for advanced training and discussion. Neuropsychologists can also participate in webinars developed specifically to provide multicultural and diversity training (Sunderaraman, Love, & Madore, 2016). As with the design of educational curricula for doctoral and postdoctoral students, however, it is not entirely clear how to establish and monitor multicultural neuropsychological competence for professionals (Díaz-Santos & Hough, 2016).

Table 3.2. FORMAT OF PROFESSIONAL TRAINING IN NEUROPSYCHOLOGICAL ASSESSMENT OF ETHNIC MINORITIES (N = 366)

Format	Responses	
	n	%
Formal didactics	192	52.5
Conference presentations	139	38.0
Assigned readings	115	31.4
Case conferences	79	21.6
Consultation with colleagues	78	21.3
Group discussions	75	20.5
Other	13	3.6

Note: Sum of respondent percentages exceeds 100% because respondents could endorse 2 responses.

Language

Despite the growing need for evaluations in languages other than English, only 15% ($n = 78$) of respondents reported administering tests in languages other than English, with 11 of those 78 conducting assessments in two non-English languages. Reflecting the official bilingualism of their country, nearly every Canadian practitioner who employed a language other than English ($n = 11$) reported use of French (91%, $n = 10$). The only other languages reported in use (German and Farsi) were endorsed by one respondent each, further underscoring reported linguistic uniformity among Canadians. Although respondents in the United States collectively reported offering their services in numerous languages, Spanish was by far the most commonplace (see Figure 3.3). This finding is not surprising given the number of Hispanic and Latino individuals currently residing and seeking services in North America.

Respondents who conducted non-English assessments in the United States and Canada combined to report 89 instances of such language use, with varied degrees of sophistication: 39% of non-English languages were practiced with

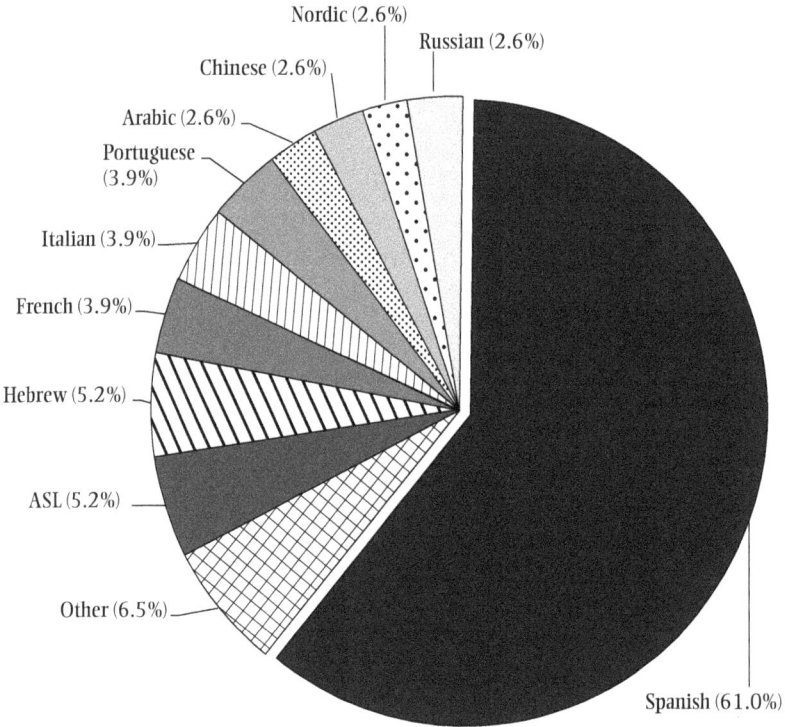

Figure 3.3. Non-English Languages Employed in Neuropsychological Assessments in the United States ($n = 77$). "Chinese" is a combination of Chinese and Cantonese responses (1 each); "Nordic" is a combination of Danish and Swedish responses (1 each); and "Other" is a combination of Finnish, German, Japanese, Korean, and Tagalog responses (1 each).

native or bilingual fluency, 24% with full professional proficiency, and 34% with only limited working proficiency. The remaining 3% of the languages reported did not specify any particular proficiency level. Review of the literature reveals differing opinions about what constitutes linguistic competence—including proficiency at the level of a native speaker who also has lived in the patient's country; completion of advanced education in the target language; or linguistic fluency combined with an understanding of the patient's culture and knowledge of how diversity variables impact neuropsychologicaltest performance (Hernández-Cardenache, Curiel, Raffo, Kitaigorodsky, & Burguera, 2016; Rivera Mindt et al., 2008). Admittedly, these standards represent ideals that are not always attainable. Consultation with expert colleagues could help determine whether the requisite levels of linguistic, educational, and professional competence have been achieved prior to conducting an evaluation (Brickman, Cabo, & Manly, 2006). Overall, our results suggest that some neuropsychologists who offer services in a language other than English lack recommended proficiency in the target language, given that over one third of respondents who conduct assessments in a foreign language reported limited working proficiency. Results also call attention to the need to standardize criteria for linguistic competence in the field.

Respondents were also asked about their typical approaches to assessing patients not fluent in English (see Table 3.3). The most common approach was referral to an appropriately trained neuropsychologist. In an ideal world, the neuropsychologist to which the patient is referred would perform an assessment in the patient's preferred language, using adequately validated, translated, and appropriately normed tests. Such referrals, however, are likely uncommon given the unfortunate shortage of multilingual professionals and validly translated neuropsychological measures (Salinas et al., 2016; Strutt, Burton, Resendiz, & Peery, 2016). Moreover, such referrals can present financial, logistical, and/or health-related hardships for some patients (Casas et al., 2012).

The next most commonly reported approach was use of an interpreter. Interpreters are intuitively appealing because they can communicate directly with patients and may educate neuropsychologists about clients' cultural issues,

Table 3.3. TYPICAL HANDLING OF NEUROPSYCHOLOGICAL ASSESSMENT FOR PATIENTS NONFLUENT IN ENGLISH (N = 491)

Response	Responses	
	n	%
Refer to neuropsychologists fluent in patient's language	337	68.6
Employ interpreters	199	40.5
Administer culturally unbiased tests	129	26.3
Employ bilingual technicians/psychometrists	47	9.6
Other	31	6.3
Administer translated test versions	24	4.9

Note: Sum of respondent percentages exceeds 100% because respondents could endorse two responses.

thereby facilitating the interpersonal exchange between neuropsychologist and client (Romero et al., 2009; Wong & Fujii, 2004). Ideally, interpreters are proficient in English and the target language—both in its spoken and written form—knowledgeable about patients' specific cultural groups, and trained in key principles of neuropsychological assessment such as standardization procedures and accurate reporting of responses. Given ethical issues related to dual relationships and confidentiality, interpreters should be trained professionals and not family members (Dugbartey, 2014). There are also cultural considerations related to social and gender matching of interpreters (Fujii, 2017). Unfortunately, it is often not possible to employ interpreters who meet ideal criteria, compromising the validity of the assessment (Hernández-Cardenache et al., 2016; Riccio, Yoon, & McCormick, 2014). Moreover, interpreter use can impact neuropsychological tests scores on verbal tests that tend to increase demands on interpreter abilities and attention (Casas et al., 2012) and on other tests due to social facilitation effects (Howe & McCaffrey, 2010).

Roughly one fourth of respondents reported that when assessing patients not fluent in English, they would administer tests designed to be culturally unbiased. However, research has consistently cautioned against the illusion of culture-free tests, an acknowledgment of the impossibility of creating cognitive ability tests not influenced by experience (Cole, 2008; Díaz-Santos & Hough, 2016). Even nonverbal tasks, traditionally perceived as being less biased than verbal tasks, are subject to influences such as educational opportunities and cultural experience (Shuttleworth-Edwards & Van der Merwe, 2016). Fortunately, in recent years there has been an increase in empirical research, book chapters, and compendia that provide descriptions of tests with validated translations or evidence for cross-cultural utility (e.g., Casals-Coll et al., 2013; Chan, Shum, & Cheung, 2003; Curiel et al., 2016; Fujii, 2017; Guàrdia-Olmos, Peró-Cebollero, Rivera, & Arango-Lasprilla, 2015; Lim, Collinson, Feng, & Ng, 2010; Riccio, Yoon, & McCormick, 2014; Strutt, Burton, Resendiz, & Peery, 2016; Thames, Karimian, & Steiner, 2016). While work in this area is encouraging, more is needed, as many languages still lack sufficiently validated neuropsychological tests and norms. Thus, while wholly culture-free measures may be unattainable, with appropriate research and effort, neuropsychologists should be able to construct batteries appropriate for most individuals who present for care (Fujii, 2017).

Other responses to the question of how to assess patients not fluent in English, each endorsed by less than 10% of respondents, included use of bilingual technicians, various other options, and administration of translated versions of tests. The use of bilingual technicians is difficult because current demand outweighs supply, and financial and logistical hardships interfere with access (Casas et al., 2012). Test translation is also challenging—neuropsychologists are advised to search for translated measures that have been validated for specific patient groups before attempting self-translation (Abedi, 2002; Fujii, 2017; Horton, Carrington, & Lewis-Jack, 2001; Van Gorp, Myers, & Drake, 2000). Simply stated, translated instruments are generally not equivalent to English counterparts. Translations are, at times, literal and lack the sophistication needed to convey regional idiosyncrasies in the target language (Bender, García, & Barr, 2010; Manly, 2006; Vilar-López & Puente, 2010). In many instances, terms, phrases, or grammatical constructs cannot be directly translated without changing the meaning of test items, which may subsequently confuse examinees, introduce scoring biases, and lower the construct validity of the measure (Perez-Arce & Puente, 1996; Strutt, Burton, Resendiz, & Peery, 2016). Tests created in the United States naturally reflect a U.S.-dominant cultural bias and therefore may be inappropriate for any person who does not fit the standardization sample and intended audience of the original measures.

In cases where translation is necessary, guidelines provided in the Standards for Education and Psychological Testing (American Educational Research Association, APA, & National Council on Measurement in Education, 2014) and International Test Commission (2010) should be followed. There are additional considerations for neuropsychological measures such as including words with roughly the same frequency or syllabic count as the original measure, for example when translating a list learning test (Strutt, Burton, Resendiz, & Peery, 2016). While these considerations may be straightforward for certain types of tests (e.g., memory, arithmetic), other types of tests (e.g., intelligence scales) pose great challenges in cross-cultural adaptation (Greenfield, 1997). Fujii (2017) recommends focusing on tests that require translation only of instructions, with responses consisting of performing a task or selecting from among multiple choices of nonverbal material (e.g., Block Design, Matrix Reasoning). Strutt et al. (2016) further suggest making slight cultural and linguistic modifications to existing measures that account for language idiosyncrasies while preserving the original factor structure of a test. With new approaches to test adaptation and with the assistance of advanced statistical approaches (e.g., item analysis, invariance analysis), it is increasingly possible to derive measures with similar construct validity across languages (Geisinger, 2003; Strutt et al., 2016), although this can be a costly endeavor (Romero et al., 2009). Strutt et al. conclude that with the emergence of these advanced methods and databases of cross-culturally validated neuropsychological tests, it is increasingly difficult to ethically justify the use of in-house or on-site translations that fail to meet rigorous standards.

Test Interpretation and Challenges

Roughly two thirds of respondents (67.5%) endorsed using special approaches to interpret the cognitive scores of ethnic minorities. The most common approach indicated was the use of separate norms based on a group that most closely matches the patient's ethnicity or cultural experience (see Table 3.4). Although ethnicity and race do not cause variability in cognitive test performance, they are markers for several nonbrain-related factors that do significantly impact performance such as acculturation, literacy, quality of education, stereotype threat, and test wiseness (Fujii, 2017; Manly, Jacobs, Touradji, Small, & Stern, 2002; Sayegh, 2016). Additionally, the use of norms derived from assessments of non-Hispanic Whites (who tend to be well-educated, native English-speakers, and middle to

Table 3.4. INTERPRETIVE APPROACHES TO COGNITIVE TEST SCORES OF ETHNIC MINORITIES (N = 346)

Approach	Responses	
	n	%
Apply norms derived from most similar cultural group(s)	285	82.4
Adjust scores for differences in years of education	157	45.4
Adjust scores for differences in educational experiences	79	22.8
Other	26	7.5
Employ regression-based normative techniques	14	4.0

Note: Sum of respondent percentages exceeds 100% because respondents could endorse two responses.

upper class) when interpreting data from ethnic minorities, can lead to artificially lower cognitive test scores and false-positive findings (Manly, 2006; Puente & Perez-Garcia, 2000; Wong & Fujii, 2004). For these reasons, the use of norms for separate ethnic, racial, or cultural groups is fairly common and intuitively appealing (Brickman et al., 2006).

While ethnicity- or race-specific norms are useful in augmenting the diagnostic utility of neuropsychological tests, they can also pose problems (Brickman et al., 2006; Pedraza & Mungas, 2008). For example, they do not resolve issues of test bias or cultural equivalence in neuropsychological tests (Brickman et al., 2006). They also do not explain the origin of performance differences between ethnic minorities and other populations, leading to harmful interpretations regarding supposed genetic limitations of ethnic minorities (Gasquoine, 2009; Manly, 2006; Nell, 2004; Romero et al., 2009). In addition, there is tremendous cultural, linguistic, and educational heterogeneity within ethnic minority subgroups (e.g., Cubans, Dominicans, and Puerto Ricans among Latinos/Hispanics), leading to complications when determining whom to include in a normative data set and for which individuals a given data set is deemed appropriate (Castaño, Biever, Gonzalez, & Anderson, 2007). Given the vast number of ethnic minority groups and subgroups, it is impossible to create data sets appropriate for each. In addition, most normative data sets were not created through population-based sampling, limiting their representativeness (Sayegh, 2016). These and other issues are discussed by Brickman et al. (2006), who suggest using race-specific normative data only when the following specific criteria are met: (a) Research has established a significant relation between race/ethnicity and performance on the target test; (b) the norms have adequate cell sizes; (c) the norms are appropriately stratified and capture demographic factors that contribute to test performance such as age, education, and sex; and (d) the patient appropriately matches the normative sample in terms of demographics and educational and cultural experiences.

Slightly under one fourth of respondents reported adjusting cognitive test scores for differences in years of education. Education is widely known to exert strong effects on cognitive ability, including performance on verbal and nonverbal neuropsychological tests (Ardila et al., 2010; Heaton, Ryan, & Grant, 2009), and accounting for education can lead to more precise detection of clinically significant impairment (Howieson, Loring, & Hannay, 2004). A common correction for educational attainment (Duff et al., 2003; Malec et al., 1992) involves placing raw or timed scores into a frequency distribution by education levels selected by the researcher (e.g., ≤11 years, 12 years, 13–15 years, and ≥16 years) and then assigning percentile ranks based on the scores' placement within that distribution. Next, percentile ranks are converted to scaled scores. However, corrections for quantity of education (i.e., stated years of education or grade attainment) can introduce bias because the level of education attained may not correspond to the quality of educational experience (Manly et al., 2002). Thus, the next most common approach, endorsed by roughly one fourth of respondents, was to adjust cognitive test scores for differences in educational experience by using estimates such as reading recognition. Notably, some research suggests that specific educational correctives (e.g., years or education or educational quality) might be optimally suited for specific ethnic groups to maximize accurate detection of cognitive impairment (Rohit et al., 2007). Thus, neuropsychologists should consult the literature before applying educational corrections to patients' test scores. It is also important to caution that these various corrective approaches address ethnic diversity by reducing its complexity to a singular concern with educational achievement.

Less than 10% of respondents spontaneously reported using other approaches to interpreting the test scores of ethnic minorities. These included the use of clinical judgment or subjective clinical interpretation of data, qualitative approach, process approach, or use of regular norms with cautionary statements in test reports. Although discussion of each of these approaches and their strengths and limitations is beyond the scope of this chapter, the reader is referred to Fujii (2017) and Lau (2014), who address these and other assessment and interpretive approaches. For example, recent research has highlighted the potential value of using individual comparison standards instead of separate norming tables (Gasquoine, 2009; Fujii, 2017). In this approach, the neuropsychologist uses an estimate of the patient's premorbid intellectual abilities (incorporating information about acculturation, socioeconomic status, quality of education, etc.) as a benchmark to compare current performances on neuropsychological tests (using conventional normative tables). Finally, less than 5% of respondents endorsed the use of regression-based norms, which generate individualized predicted scores based on multiple variables (e.g., age, sex, education) simultaneously (Heaton, Avitable, Grant, & Matthews, 1999). Regression-based norms can be useful in interpreting the cognitive test scores of ethnic minorities, yet limitations arise if assumptions of the multiple regression analyses are violated or if individuals' demographic characteristics fall outside the normative samples' range (Testa, Winicki, Pearlson, Gordon, & Schretlen, 2009). Also, it can be challenging to identify the most relevant demographic variables for a given individual (Van Breukelen & Vlaeyen, 2005), and clinicians might be unfamiliar with such norms or find them difficult to apply (Van der Elst, Van Boxtel, Van Breukelen, & Jolles, 2006).

Table 3.5. GREATEST PERCEIVED CHALLENGES TO NEUROPSYCHOLOGICAL ASSESSMENT OF ETHNIC MINORITIES (N = 499)

Challenge	Responses n	%
Lack of appropriate norms	337	67.5
Lack of appropriate tests	214	42.9
Difficulty locating consulting/referral colleague	177	35.5
Lack of trained neuropsychologists/psychometrists	98	19.6
Lack of training opportunities	43	8.6
Other	16	3.2

Note: Sum of respondent percentages exceeds 100% because respondents could endorse two responses.

A final questionnaire item asked respondents to report the greatest challenges associated with neuropsychological assessment of ethnic minorities (see Table 3.5). Respondents overwhelmingly endorsed the lack of appropriate norms as the greatest challenge, which is unsurprising in light of the previous discussion. Other commonly reported challenges, each endorsed by more than one third of participants, were the lack of appropriate tests and difficulties in finding referral sources or colleagues to consult, both previously discussed. Respondents spontaneously reported additional challenges that reflect unresolved issues without easy solutions such as lack of ecological validity of assessments (see Rabin et al., 2016, for discussion) and heterogeneity of members within ethnic minority designations leading to difficulty applying normative data (as previously discussed).

STUDY LIMITATIONS AND ADDITIONAL CHALLENGES

As noted previously, data presented in this chapter come from questionnaire items that were part of a larger study on neuropsychological assessment. The survey contained a limited number of questions about multicultural practice issues and did not solicit feedback about ways to enhance multicultural competence. Also, some respondents commented that certain response options were overly restrictive or offered ideals unattainable in practice. Future research might employ open-ended question formats or a qualitative investigatory approach to probe diversity issues more extensively. For example, one might inquire about the perceived quality of multicultural training experiences or about instruments utilized most frequently with specific cultural or linguistic subgroups and the supporting rationale for their application. Another area of inquiry could explore intracultural diversity (e.g., the idea that Whites as a cultural group have significant intracultural variability, and therefore it is problematic to treat all test data from non-Hispanic White individuals as the same). In addition to members of major neuropsychological organizations, future survey work should also target members of professional organizations with a cross-cultural emphasis (e.g.,

Hispanic Neuropsychological Society, Association of Black Psychologists), who may hold different experiences and perspectives.

Certain challenges that may arise during neuropsychological assessments of ethnic minorities were not addressed by our respondents. We discuss a few such challenges and provide references for those seeking more information.

- First, communication challenges may arise from cultural differences between the neuropsychologist and patient. For example, there can be significant differences in cultural proscriptions about when to talk, what to say, pacing and pausing within a conversation, listenership (e.g., eye contact norms), intonation and prosody, (in)directness, and formulaicity (i.e., use of aphorisms or clichés; Tannen, 1984). Fujii (2017) offers detailed recommendations for overcoming these challenges during the clinical interview and formal evaluation. When working with individuals whose style of communication is indirect, a neuropsychologist should give thought to phrasing questions in a manner that would not be perceived as rude. Also, neuropsychologists should be sensitive to behavioral clues that suggest patient offense, discomfort, or disengagement from examination. For additional discussion, the reader is referred to Fujii and Davis and D'Amato (2014).
- Second, it is not always clear when and how to adjust assessments when working with ethnic minorities (see Fujii, 2017; Hernández-Cardenache et al., 2016; Verney, Bennett, & Hamilton, 2016). For example, given cultural norms regarding participation of relatives in medical treatment, the neuropsychologist may invite key family members to attend clinical interview and feedback sessions. Also, the neuropsychologist may wish to conduct more extensive interviews with the patient than is typical to gather critical information about the patient's life experiences, familial roles, and level of acculturation. These issues highlight the need for neuropsychologists to enhance their cultural awareness, sensitivity, and preparation when working with ethnically diverse individuals. Moreover, neuropsychologists should document potential threats to data validity that occur as a result of changes made to assessment protocols (e.g., modifications to test administration, use of an interpreter).
- Third, in our discussion on language, we addressed the use of interpreters and translations and the need to ensure their responsible and effective use. Respondents did not report other challenges related to language such as how to select measures of language proficiency and level of acculturation, which are essential measures at the beginning of an evaluation (see Fujii, 2017; Guo & Uhm, 2014; Riccio et al., 2014; Salinas et al., 2016). Also, bilingualism can present unique challenges insofar as bilingual patients may have greater dominance in one language or equal mastery of both languages, with implications for neuropsychological test performance (Hernández-Cardenache et al., 2016; Paradis, 2008; Salinas et al., 2016).

- Fourth, the neuropsychological literature pertaining to ethnic minority children is sparse, and certain pediatric-specific issues can pose challenges. For example, because bilingual children are at risk for misidentification as having language disorders or learning disabilities (Salinas et al., 2016), neuropsychologists must be mindful of possible biases related to referral procedures. Also, before proceeding with learning disability or language evaluations, the neuropsychologist must be prepared to assess key characteristics of the child's learning environment (e.g., instructional setting, curriculum, accommodations) as well as the child's proficiency in dominant and non-dominant languages, rate of academic progress, and acculturation (Pham, Goforth, Oganes, Medina-Pekofsy, & Fine, 2016). For information on conducting neuropsychological assessments with ethnic minority children, the reader is referred to recent work by Pham et al. (2016) and Thames, Karimian, and Steiner (2016).

CONCLUSIONS AND PRACTICAL RECOMMENDATIONS

Shifting demographics both locally and globally create challenges for neuropsychologists in terms of delivering culturally competent services. Overall, our findings confirm what others have observed in recent decades: Ethnic minorities remain grossly underrepresented in neuropsychology and practitioners perceive the lack of appropriate norms, tests, and referral sources as challenges strongly associated with the assessment of ethnic minorities. Although the practice of neuropsychology is rooted in psychometric assessment and appropriate use of normative information, test translation and validation for various groups have been slow to emerge. Continued viability of neuropsychology depends on the ability to assess and treat diverse individuals in a sensitive, responsible, and valid manner. We offer the following considerations:

- Even with advances in cross-cultural test validation and the publication of compendia of measures that include diverse groups in their standardization samples, only a small percentage of tests meet all recommended criteria—that is, follow APA and International Test Commission guidelines for test development, validation, and/or adaptation; are readily available in the patient's language; and contain appropriate normative data for the patient (American Educational Research Association et al., 2014; Geisinger, 2003; International Test Commission, 2010). Fujii (2017) and Robbins et al. (2016) offer useful recommendations for test selection and interpretation along with illustrative examples of how these recommendations are applied. They stress the need to review each case carefully to develop a plan for the evaluation process that considers the patient's unique sociodemographics as well as the referral questions and sources, available measures and norms, and other contextual variables. Overall, we feel that it is important for the neuropsychological report to contain all relevant caveats related to

cultural issues such as limitations of measures used, any departures from standardization, limitations of the normative data set, and possible bias resulting from the presence of an interpreter.
- Challenges in providing services to diverse individuals go beyond battery construction and interpretive decisions, as neuropsychologists who identify as ethnic minorities are as underrepresented as they were decades ago. Important goals for the field include early exposure to neuropsychology (e.g., introducing neuropsychologists to diverse high school and college students), recruitment efforts that target underrepresented individuals for both doctoral and faculty positions, greater financial incentives for students entering doctoral programs, increased exposure to diverse faculty and peer mentors, and diversity-focused research and clinical training opportunities. It is important to keep in mind, however, that ethnic minority identification does not guarantee multicultural competence and sensitivity. Regardless of demographic background, level of training, or specific area of expertise, all neuropsychologists must strive for cultural competency through education (preferably at the doctoral and postdoctoral levels), participation in courses and workshops, and peer consultation or supervision. It might be useful to highlight or reward specific organizations and training programs that demonstrate excellence in multicultural teaching, practice, and research. There is also a need for increased discussion about and delineation of the core education and skills requirements for practicing culturally competent neuropsychology (Salinas et al., 2016) and enforcement of these requirements by administrative structures (Wong et al., 2000). Although the APA requires multicultural training, neuropsychology as a subfield is arguably behind the curve in this area.
- Finally, it is imperative that neuropsychologists recognize that cultural competence is a lifelong process that requires ongoing self-analysis of how one's own attitudes, values, and biases influence perceptions of and interactions with diverse patients (Dugbartey, 2014). In addition to enhancing knowledge and competency skills through educational opportunities, several recently published books and articles are devoted to the conduct of culturally informed neuropsychological evaluations and offer opportunities for self-directed study. Of particular relevance are the comprehensive case studies and chapters devoted to working with specific ethnic/racial groups (e.g., American Indians, Alaska Natives, Asian Americans, multiracial individuals), multicultural patient and diagnostic populations (e.g., dementia in Latino elderly populations, ethnic minority children with learning disabilities), and special topics (e.g., assessment of malingering in ethnic minority populations). Additionally, these books provide links to professional organizations, census data on ethnic minorities, data on mental health disparities, and general information to facilitate understanding of specific cultural subgroups.

ACKNOWLEDGMENTS

The authors express their gratitude to the many INS and NAN members who completed and returned the survey. This project was supported, in part, by a PSC-CUNY Award #63160-00 41. Data and other information included in this chapter were utilized from M. M. Elbulok-Charcape, L. A. Rabin, A. T. Spadaccini, & W. B. Barr, 2014, Trends in the neuropsychological assessment of ethnic/racial minorities: A survey of clinical neuropsychologists in the United States and Canada. *Cultural Diversity & Ethnic Minority Psychology, 20*, 353–361. ©2014. Used with permission from the American Psychology Association.

RECOMMENDED READING

Davis, J. M., & D'Amato, R. C. (Eds.). (2014). *Neuropsychology of Asians and Asian Americans: Practical and theoretical considerations*. New York, NY: Springer.

Ferraro, F. R. (Ed.). (2016). *Minority and cross-cultural aspects of neuropsychological assessment: Enduring and emerging trends* (2nd ed.). New York, NY: Taylor & Francis.

Fujii, D. (2017). *Conducting a culturally informed neuropsychological evaluation*. Washington, DC: American Psychological Association.

REFERENCES

Abedi, J. (2002). Standardized achievement tests and English language learners: Psychometric issues. *Educational Assessment, 8*, 231–257. https://doi.org/10.1207/S15326977EA0803_02

Allison, K. W., Echemendia, R. J., Crawford, I., & Robinson, W. L. (1996). Predicting cultural competence: Implications for practice and training. *Professional Psychology: Research and Practice, 27*, 386–393. https://doi.org/10.1037/0735-7028.27.4.386

American Educational Research Association, American Psychological Association, & National Council on Measurement in Education. (2014). *Standards for educational and psychological testing*. Washington, DC: Author.

American Psychological Association. (2003). Guidelines on multicultural education, training, research, practice, and organizational change for psychologists. *American Psychologist, 58*, 377–402. https://doi.org/10.1037/0003-066X.58.5.377

American Psychological Association. (2010). Ethical principles of psychologists and code of conduct (2002, Amended June 1, 2010). Retrieved from http://www.apa.org/ethics/code/principles.pdf

Ardila, A., Bertolucci, P. H., Braga, L. W., Castro-Caldas, A., Judd, T., Kosmodis, M. H., . . . Roselli, M. (2010). Illiteracy: The neuropsychology of cognition without reading. *Archives of Clinical Neuropsychology, 25*, 689–712. https://doi.org/10.1093/arclin/acq079

Bender H., García A., & Barr W. B. (2010). An interdisciplinary approach to neuropsychological test construction: Perspectives from translation studies. *Journal of the International Neuropsychological Society, 16*, 227–232. https://doi.org/10.1017/S1355617709991378

Brickman, A. M., Cabo, R., & Manly, J. J. (2006). Ethical issues in cross-cultural neuropsychology. *Applied Neuropsychology, 13*, 91–100. https://doi.org/10.1207/s15324826an1302_4

Byrd, D., Razani, J., Suarez, P., LaFosse, J. M., Manly, J., & Attix, D. K. (2010). Diversity Summit 2008: Challenges in the recruitment and retention of ethnic minorities in neuropsychology. *The Clinical Neuropsychologist, 24,* 1279–1291. https://doi.org/10.1080/13854046.2010.521769

Casals-Coll, M., Sánchez-Benavides, G., Quintana, M., Manero, R. M., Rognoni, T., Calvo, L., . . . Peña-Dasanova, J. (2013). Spanish normative studies in young adults (NEURONORMA young adults project): Norms for verbal fluency tests. *Neurología (English Edition), 28,* 33–40. https://doi.org/10.1016/j.nrleng.2012.02.003

Casas, R., GuzmánVélez, E., CardonaRodriguez, J., Rodriguez, N., Quiñones, G., Izaguirre, B., & Tranel, D. (2012). Interpretermediated neuropsychological testing of monolingual Spanish speakers. *The Clinical Neuropsychologist, 26,* 88–101. https://doi.org/10.1080/13854046.2011.640641

Castaño, M., Biever, J., Gonzalez, C., & Anderson, K. (2007). Challenges of providing mental health services in Spanish. *Professional Psychology: Research and Practice, 38,* 667–673. https://doi.org/10.1037/0735-7028.38.6.667

Chan, A. S., Shum, D., & Cheung, R. W. (2003). Recent development of cognitive and neuropsychological assessment in Asian countries. *Psychological Assessment, 15,* 257–267. https://doi.org/10.1037/1040-3590.15.3.257

Colby, S. L. & Ortman, J. M. (2015). Projections of the size and composition of the U.S. population: 2014 to 2060. *Current Population Reports,* 25-1143, U.S. Census Bureau, Washington, DC, 2014. Retrieved from http://www.census.gov/content/dam/Census/library/publications/2015/demo/p25-1143.pdf

Cole, M. (2008). *The illusion of culture-free intelligence testing.* Unpublished manuscript. The Laboratory of Comparative Human Cognition, University of California, San Diego. Retrieved from http://lchc.ucsd.edu/mca/Paper/Cole/iq.html

Curiel, R. E., Hernández-Cardenache, R., Giraldo, N., Rosado, M., Restrepo, L., Raffo, A., . . . Whitt, N. (2016). A compendium of neuropsychological measures for Hispanics in the United States. In F. R. Ferraro (Ed.), *Minority and cross-cultural aspects of neuropsychological assessment: Enduring and emerging trends* (2nd ed., pp. 471–513). New York, NY: Taylor & Francis.

Davis, J. M. & D'Amato, R. C. (Eds.). (2014). *Neuropsychology of Asians and Asian Americans: Practical and theoretical considerations.* New York, NY: Springer.

Díaz-Santos, M., & Hough, S. (2016). Cultural competence guidelines for neuropsychology trainees and professionals: Working with ethnically diverse individuals. In F. R. Ferraro (Ed.), *Minority and cross-cultural aspects of neuropsychological assessment: Enduring and emerging trends* (2nd ed., pp. 11–33). New York, NY: Taylor & Francis.

Duff, K., Pattern, D., Schoenberg, M. R., Mold, J., Scott, J. G., & Adams, R. L. (2003). Age- and education-corrected independent normative data for the RBANS in a community dwelling elderly sample. *The Clinical Neuropsychologist, 17,* 351–366. https://doi.org/10.1076/clin.17.3.351.18082

Dugbartey, A. T. (2014). Ethical considerations in neuropsychological assessment of Asian heritage clients. In J. M. Davis & R. C. D'Amato (Eds.), *Neuropsychology of Asians and Asian Americans: Practical and theoretical considerations* (pp. 17–25). New York, NY: Springer.

Elbulok-Charcape, M. M., Rabin, L. A., Spadaccini, A. T., & Barr, W. B. (2014). Trends in the neuropsychological assessment of ethnic/racial minorities: A survey of clinical neuropsychologists in the U.S. and Canada. *Cultural Diversity & Ethnic Minority Psychology, 20,* 353–361. https://doi.org/10.1037/a0035023

Fujii, D. (2017). *Conducting a culturally informed neuropsychological evaluation.* Washington, DC: American Psychological Association.

Gasquoine, P. G. (2009). Race-norming of neuropsychological tests. *Neuropsychology Review, 19*, 250–262. https://doi.org/10.1007/s11065-009-9090-5

Geisinger, K. F. (2003). Testing and assessment in cross-cultural psychology. In I. B. Weiner & J. R. Graham (Eds.), *Handbook of psychology: Vol. 10. Assessment psychology* (pp. 95–117). Hoboken, NJ: Wiley.

Greenfield, P. M. (1997). You can't take it with you. Why ability assessments don't cross cultures. *American Psychologist, 52*, 1115–1124. https://doi.org/10.1037/0003-066X.52.10.1115

Guàrdia-Olmos, J., Peró-Cebollero, M., Rivera, D., & Arango-Lasprilla, J. (2015). Methodology for the development of normative data for ten Spanish-language neuropsychological tests in eleven Latin American countries. *NeuroRehabilitation, 37*, 93–499. https://doi.org/10.3233/NRE-151277

Guo, T., & Uhm, S. Y. (2014). Society and acculturation in Asian American communities. In J. M. Davis & R. C. D'Amato (Eds.), *Neuropsychology of Asians and Asian Americans: Practical and theoretical considerations* (pp. 55–76). New York, NY: Springer.

Heaton, R. K., Avitable, N., Grant, I., & Matthews, C. G. (1999). Further cross validation of regression-based neuropsychological norms with an update for the Boston Naming Test. *Journal of Clinical and Experimental Neuropsychology, 21*, 572–582. https://doi.org/10.1076/jcen.21.4.572.882

Heaton, R. K., Ryan, L., & Grant, I. (2009). Demographic influences and use of demographically corrected norms in neuropsychological assessment. In I. Grant & K. M. Adams (Eds.), *Neuropsychological assessment of neuropsychiatric and neuromedical disorders* (3rd ed., pp. 127–155). New York, NY: Oxford University Press.

Hernández-Cardenache, R., Curiel, R. E., Raffo, A., Kitaigorodsky, M., & Burguera, L. (2016). Current trends in neuropsychological assessment with Hispanic/Latinos. In F. R. Ferraro (Ed.), *Minority and cross-cultural aspects of neuropsychological assessment: Enduring and emerging trends* (2nd ed., pp. 259–278). New York, NY: Taylor & Francis.

Horton, A. M., Carrington, C. H., & Lewis-Jack, O. (2001). Neuropsychological assessment in a multicultural context. In L. A. Suzuki, J. G. Ponterotto, & P. J. Meller (Eds.), *Handbook of multicultural assessment: Clinical, psychological, and educational applications* (2nd ed., pp. 433–460). San Francisco, CA: Jossey-Bass.

Howe, L. L. S., & McCaffrey, R. J. (2010). Third party observation during neuropsychological evaluation: An update on the literature, practical advice for practitioners, and future directions. *The Clinical Neuropsychologist, 24*, 518–537. https://doi.org/10.1080/13854041003775347

Howieson, D. B., Loring, D. W., & Hannay, H. J. (2004). Neurobehavioral variables and diagnostic issues. In M. D. Lezak, D. B. Howieson, & D. W. Loring (Eds.), *Neuropsychological Assessment* (4th ed., pp. 286–336). New York, NY: Oxford University Press.

International Test Commission. (2010). Guidelines for translating and adapting tests. Retrieved from http://www.intestcom.org/

Judd, T., Capetillo, D., Carrión-Baralt, J., Marmol, L. M., Miguel-Montes, L., Navarrete, M. G., . . . Valdés, J. (2009). Professional considerations for improving the neuropsychological evaluation of Hispanics: A National Academy of Neuropsychology education paper. *Archives of Clinical Neuropsychology, 24*, 127–135. https://doi.org/10.1093/arclin/acp016

Lau, E. Y. Y. (2014). Clinical interviewing and qualitative assessment with Asian heritage clients. In J. M. Davis & R. C. D'Amato (Eds.), *Neuropsychology of Asians and*

Asian Americans: Practical and theoretical considerations (pp. 135–149). New York, NY: Springer.

Lim, M., Collinson, S. L., Feng, L., & Ng, T. P. (2010). Cross-cultural application of the Repeatable Battery for the Assessment of Neuropsychological Status (RBANS): Performance of elderly Chinese Singaporeans. *The Clinical Neuropsychologist, 24,* 811–826. https://doi.org/10.1080/13854046.2010.490789

Malec, J. F., Ivnik, R. J., Smith, G. E., Tangalos, E. G., Petersen, R. C., Kokmen, E., & Kurland, L. T. (1992). Mayo's older Americans normative studies: Utility of corrections for age and education for the WAIS-R. *The Clinical Neuropsychologist,* 6(Suppl), 31–47. https://doi.org/10.1080/13854049208401878

Manly, J. J. (2006). Cultural issues. In D. K. Attix & K. A. Welsh-Bohmer (Eds.), *Geriatric neuropsychology: Assessment and intervention* (pp. 198–222). New York, NY: Guilford.

Manly, J. J., Jacobs, D. M., Touradji, P., Small, S. A., & Stern, J. (2002). Reading level attenuates differences in neuropsychological test performance between African American and White elders. *Journal of the International Neuropsychological Society, 8,* 341–348. https://doi.org/10.1017/S1355617702813157

Nell, V. (2004). Translation and test administration techniques to meet the assessment needs of ethnic minorities, migrants, and refugees. In G. Goldstein, S. R. Beers, & M. Hersen (Eds.), *Comprehensive handbook of psychological assessment: Vol. 1. Intellectual and neuropsychological assessment* (pp. 333–338). Hoboken, NJ: Wiley.

Nelson, A. P., Roper, B. L., Slomine, B. S., Morrison, C., Greher, M. R., Janusz, J., . . . Wodushek, T. R. (2015). Official position of the American Academy of Clinical Neuropsychology (AACN): Guidelines for practicum training in clinical neuropsychology. *The Clinical Neuropsychologist, 29,* 879–904. https://doi.org/10.1080/13854046.2015.1117658

Paradis, M. (2008). Bilingualism and neuropsychiatric disorders. *Journal of Neurolinguistics, 21,* 199–230. https://doi.org/10.1016/j.jneuroling.2007.09.002

Pedraza, O., & Mungas, D. (2008). Measurement in cross-cultural neuropsychology. *Neuropsychology Review, 18,* 184–193. https://doi.org/10.1007/s11065-008-9067-9

Perez-Arce, P., & Puente, A. E. (1996). Neuropsychological assessment of ethnic-minorities: The case of assessing Hispanics living in North America. In R. J. Sbordone & C. J. Long (Eds.), *Ecological validity of neuropsychological testing* (pp. 283–300). Boca Raton, FL: St. Lucie.

Pham, A. V., Goforth, A. N., Oganes, M., Medina-Pekofsky, E., & Fine, J. G. (2016). Nondiscriminatory neuropsychological assessment of children with learning disabilities. In F. R. Ferraro (Ed.), *Minority and cross-cultural aspects of neuropsychological assessment: Enduring and emerging trends* (2nd ed., pp. 359–378). New York, NY: Taylor & Francis.

Proctor, S. L., & Simpson, C. (2016). Improving service delivery to ethnic and racial minority students though multicultural program training. In S. L. Graves Jr. & J. J. Blake (Eds.), *Psychoeducational assessment and intervention for ethnic minority children: Evidence-based approaches* (pp. 251–265). Washington, DC: American Psychology Association.

Proctor, S. L., Simpson, C. M., Levin, J., & Hackimer, L. (2014). Recruitment of diverse students in school psychology programs: Direction for future research and practice. *Contemporary School Psychology, 18,* 117–126. https://doi.org/10.1007/s406880140012z

Puente, A. E., & Perez-Garcia, M. P. (2000). Psychological assessment of ethnic minorities. In G. Goldstein & M. Hersen (Eds.), *Handbook of psychological assessment* (3rd ed., pp. 527–551). Boston, MA: Allyn & Bacon.

Rabin, L. A., Barr, W. B., & Burton, L. A. (2005). Assessment practices of clinical neuropsychologists in the United States and Canada: A survey of INS, NAN, and APA Division 40 members. *Clinical Neuropsychology, 20,* 33–65. https://doi.org/10.1016/j.acn.2004.02.005

Rabin, L. A., Paolillo, E., & Barr, W. B. (2016). Stability in test-usage practices of clinical neuropsychologists in the United States and Canada over a 10-year period: A follow-up survey of INS and NAN members. *Archives of Clinical Neuropsychology, 31,* 206–230. https://doi.org/10.1093/arclin/acw007

Rey-Casserly, C., Roper, B. L., & Bauer, R. M. (2012). Application of a competency model to clinical neuropsychology. *Professional Psychology: Research and Practice, 43,* 422–431. https://doi.org/10.1037/a0028721

Riccio, C. A., Yoon, H., & McCormick, A. S. (2014). Neuropsychological test selection. In J. M. Davis & R. C. D'Amato (Eds.), *Neuropsychology with Asian Americans* (pp. 151–174). New York, NY: Springer.

Rivera Mindt, M., Byrd, D., Saez, P., & Manly, J. (2010). Increasing culturally competent neuropsychological services for ethnic minority populations: A call to action. *The Clinical Neuropsychologist, 24,* 429–453. https://doi.org/10.1080/13854040903058960

Robbins, R. N., Schuler, M., Ferrett, H. L., & Strutt, A. M. (2016). Operationalizing a standard and ethical approach to neuropsychological assessment across diverse cultures. In F. R. Ferraro (Ed.), *Minority and cross-cultural aspects of neuropsychological assessment: Enduring and emerging trends* (2nd ed., pp. 411–435). New York, NY: Taylor & Francis.

Rohit, M., Levine, A., Hinkin, C., Abramyan, S., Saxton, E., Valdes-Sueiras, M., & Singer, E. (2007). Education correction using years in school or reading grade-level equivalent? Comparing the accuracy of two methods in diagnosing HIV-associated neurocognitive impairment. *Journal of the International Neuropsychological Society, 13,* 462–470. https://doi.org/10.1017/S1355617707070506

Romero, H. R., Lageman, S. K., Kamath, V., Farzin, I., Sim, A., Suarez, P., & Summit Participants. (2009). Challenges in the neuropsychological assessment of ethnic minorities: Summit proceedings. *Clinical Neuropsychologist, 23,* 761–779. https://doi.org/10.1080/13854040902881958

Salinas, C. M., Bordes-Edgar, V., & Puente, A. E. (2016). Barriers and practical approaches to neuropsychological assessment of Spanish speakers. In F. R. Ferraro (Ed.), *Minority and cross-cultural aspects of neuropsychological assessment: Enduring and emerging trends* (2nd ed., pp. 229–257). New York, NY: Taylor & Francis.

Sayegh, P. (2016). Cross-cultural issues in the neuropsychological assessment of dementia. In F. R. Ferraro (Ed.), *Minority and cross-cultural aspects of neuropsychological assessment: Enduring and emerging trends* (2nd ed., pp. 54–71). New York, NY: Taylor & Francis.

Shuttleworth-Edwards, A. B., & Van der Merwe, A. S. (2016). WAIS-III and WISC-IV South African cross-cultural normative data stratified for quality of education. In F. R. Ferraro (Ed.), *Minority and cross-cultural aspects of neuropsychological assessment: Enduring and emerging trends* (2nd ed., pp. 72–96). New York, NY: Taylor & Francis.

Strutt, A. M., Burton, V. J., Resendiz, C. V., & Peery, S. (2016). Neurocognitive assessment of Hispanic individuals residing in the United States. In F. R. Ferraro (Ed.), *Minority and cross-cultural aspects of neuropsychological assessment: Enduring and emerging trends* (2nd ed., pp. 201–228). New York, NY: Taylor & Francis.

Sunderaraman, P., Love, C. E., & Madore, M. (2016). Utility of using webinars as an educational tool in neuropsychology: A preliminary study using pre- and post-webinar surveys. *National Academy of Neuropsychology Bulletin, 29*(2), 21–23.

Sweet, J. J., Meyer, D. G., Nelson, N. W., & Moberg, P. J. (2011). The TCN/AACN 2010 "Salary Survey": Professional practices, beliefs, and incomes of U.S. neuropsychologists. *The Clinical Neuropsychologist, 25,* 12–61. https://doi.org/10.1080/13854046.2010.544165

Sweet, J. J., Nelson, N. W., & Moberg, P. J. (2006). The TCN/AACN 2005 "Salary Survey": Professional practices, beliefs, and incomes of U.S. neuropsychologists. *The Clinical Neuropsychologist, 20,* 325–364. https://doi.org/10.1080/13854040600760488

Tannen, D. (1984). The pragmatics of cross-cultural communication. *Applied Linguistics, 5,* 189–195.

Testa, S. M., Winicki, J. M., Pearlson, G. D., Gordon, B., & Schretlen, D. J. (2009). Accounting for estimated IQ in neuropsychological test performance with regression-based techniques. *Journal of International Neuropsychological Society, 15,* 1012–1022. https://doi.org/10.1017/S1355617709990713

Thames, A. D., Karimian, A., & Steiner, A. J. (2016). Neuropsychological assessment of ethnic minority children. In S. L. Graves Jr. & J. J. Blake (Eds.), *Psychoeducational assessment and intervention for ethnic minority children: Evidence-based approaches* (pp. 133–161). Washington, DC: American Psychology Association.

U.S. Census Bureau. (2011). Statistical abstract of the United States 2011: Table 6. Resident population by sex, race, and Hispanic-origin status: 2000 to 2009. Retrieved from https://www2.census.gov/library/publications/2010/compendia/statab/130ed/tables/11s0006.pdf

U.S. Census Bureau. (2017). Quick facts: United States. Retrieved from https://www.census.gov/quickfacts/table/PST045216/00

Van Breukelen, G. J. P., & Vlaeyen, J. W. S. (2005). Norming clinical questionnaires with multiple regression: The Pain Cognition List. *Psychological Assessment, 17,* 336–344. https://doi.org/10.1037/1040-3590.17.3.336

Van der Elst, W., Van Boxtel, M. P. J., Van Breukelen, G. J. P., & Jolles, J. (2006). The Stroop Color–Word Test: Influence of age, sex, and education; and normative data for a large sample across the adult age range. *Assessment, 13,* 62–79. https://doi.org/10.1177/1073191105283427

Van Gorp, W. G., Myers, H. F., & Drake, E. B. (2000). Neuropsychology training: Ethnocultural considerations in the context of general competency training. In E. Fletcher-Janzen, T. L. Strickland, & C. R. Reynolds (Eds.), *Handbook of cross-cultural neuropsychology* (pp. 19–27). Dordrecht, the Netherlands: Kluwer Academic.

Verney, S. P., Bennett, J., & Hamilton, J. M. (2016). Cultural considerations in the neuropsychological assessment of American Indians/Alaska Natives. In F. R. Ferraro (Ed.), *Minority and cross-cultural aspects of neuropsychological assessment: Enduring and emerging trends* (2nd ed., pp. 115–158). New York, NY: Taylor & Francis.

Vilar-López, R., & Puente, A. E. (2010). Forensic neuropsychological assessment of members of minority groups: The case for assessing Hispanics. In A. M. Horton & L. C. Hartlange (Eds.), *Handbook of forensic neuropsychology* (2nd ed., pp. 309–331). New York, NY: Springer.

Wong, T., & Fujii, D. E. (2004). Neuropsychological assessment of Asian Americans: Demographic factors, cultural diversity, and practical guidelines. *Applied Neuropsychology, 11,* 23–36. https://doi.org/10.1207/s15324826an1101_4

Wong, T. M., Strickland, T. L., Fletcher-Janzen, E., Ardila, A., & Reynolds, C. A. (2000). Theoretical and practical issues in the neuropsychological assessment and treatment of culturally dissimilar patients. In E. Fletcher-Janzen, T. L. Strickland, & C. R. Reynolds (Eds.), *Handbook of cross-cultural neuropsychology* (pp. 3–18). New York, NY: Springer.

APPENDIX: SURVEY ITEMS RELATED TO THE ASSESSMENT OF ETHNIC MINORITIES

The following 11 survey questions originate from the 2011 Neuropsychological Assessment Survey (available in full from http://rabin.us/nas-2011.pdf).

39. What is your race/ethnicity? (✓ <u>any and all that</u> apply)

☐ White
☐ Black or African American
☐ Latino or Hispanic
☐ Asian
☐ Native American or American Indian
☐ Native Hawaiian or Other Pacific Islander

☐ other (*specify*) _____

64. What percentage of your professional time is spent with the following populations? (must <u>sum</u> to 100%)

White	_____ %
Black or African American	_____ %
Latino or Hispanic	_____ %
Asian	_____ %
Native American or American Indian	_____ %
Native Hawaiian or Other Pacific Islander	_____ %
other (*specify*) _____	_____ %

65. Have you received any training in neuropsychological assessment of ethnic/racial minority populations or cross-cultural neuropsychology?

☐ yes   ☐ no

66. <u>If yes [to Q65]</u>, at which level did you receive this training? (e.g., graduate school, postdoctoral training)

_____

67. <u>If yes [to Q65]</u>, what did this training primarily consist of? (✓ <u>no more than two</u> boxes)

☐ formal didactics
☐ assigned readings
☐ conference presentations
☐ other (*specify*) _____
☐ group discussion
☐ case conferences
☐ consultation with colleagues

68. Do you conduct assessments in languages other than English?

☐ yes   ☐ no

69. If yes [to Q65], which language(s) other than English at what level of fluency? (list no more than two languages)

_____
☐ limited working proficiency
☐ full professional proficiency
☐ native or bilingual proficiency

_____
☐ limited working proficiency
☐ full professional proficiency
☐ native or bilingual proficiency

70. If a patient seeking neuropsychological assessment is not fluent in English, what typically follows in your practice? (✓ no more than two boxes)

☐ referral to a neuropsychologist fluent in the patient's language
☐ use of a bilingual technician/psychometrist
☐ administer a translated version of the test(s)
☐ administer tests designed to be culturally unbiased, such as non-verbal tests
☐ use of an interpreter
☐ other (*specify*) _____

71. Do you use special approaches to interpreting cognitive scores of ethnic/racial minorities?

☐ yes   ☐ no

72. If yes [to Q71], what approach(es) do you regularly use? (✓ no more than two boxes)

☐ separate norms based on a normative group that most closely matches the patient's race/ethnicity or cultural experience (in the absence of appropriate norms)
☐ education-corrected norms (e.g., adjust cognitive test scores for differences in years of education)
☐ adjusted cognitive test scores for differences is educational experience (e.g., using estimates of quality of education such as reading recognition
☐ regression-based normative techniques
☐ other (*specify*) _____

73. What do you view as the greatest challenge(s) associated with neuropsychological assessment of ethnic/racial minority populations? (✓ <u>no more than two</u> boxes)

- ☐ lack of appropriate tests
- ☐ lack of appropriate norms
- ☐ lack of training opportunities
- ☐ lack of trained neuropsychologists or psychometrists
- ☐ difficulty finding a colleague to whom the patient can be referred or who can be consulted
- ☐ other (*specify*) _____

4

Culture and Memory

LIXIA YANG, BRENDA WONG, AND LINGQIAN LI

INTRODUCTION

According to the brain–computer analogy in cognitive psychology, all human brains operate based on identical mechanisms, just like hard-wired computers (Block, 1995). This model assumes that most basic cognitive functions, such as attention and memory, are universal across cultures (e.g., Chomsky, 2002). However, in challenge of this universality view, a growing body of cross-cultural research suggests that culture shapes information processing styles. Most of these cross-cultural studies focus on the comparison between Western (e.g., Britain, Western Europe, United States, Canada, and Australia) and Eastern (e.g., China, Japan, and Korea) cultures. Overall, the results show that accumulated cultural experience (e.g., socialization and parent–child interactions) guides our attention to and subsequent memory for some aspects of information over others (Fernald & Morikawa, 1993; Nisbett & Masuda, 2003). Typically, individuals from Western cultures tend to think analytically, focus more on object-based or feature-specific aspects of information whereas those from East Asian cultures are inclined to think holistically and attend more to contextual and thematically associated information (Gutchess & Indeck, 2009; Kitayama & Uskul, 2011; Nisbett & Masuda, 2003; Nisbett & Miyamoto, 2005; Park & Gutchess, 2002; 2006; Park, Nisbett, & Hedden, 1999). For example, Choi, Koo, and Choi (2007) found that East Asians scored higher (e.g., suggesting a more holistic thinking style) than Western participants on the Analysis-Holism Scale (AHS), which scores were, in turn, positively correlated with the tendency to prioritize contextual information in cognitive tasks (Choi et al., 2007). The cultural differences have been demonstrated in various cognitive domains, such as perception, attention, and memory (Buchtel & Norenzayan, 2009; Han et al., 2013; Nisbett & Norenzayan, 2002; Norenzayan & Heine, 2005; Yama, Nishioka, Horishita, Kawasaki, & Taniguchi, 2007).

With regards to perception/attention, behavioral, eye movement, and neuroimaging studies consistently show that East Asians attend more to peripheral contextual information whereas Westerners focus more on focal information (e.g., Kitayama, Duffy, Kawamera, & Larsen, 2003). Using a standard Framed-Line Task, Kitayama et al. (2003) found that Japanese performed better on the relative task (i.e., drawing a line in reference to the relationship between the line and the frame) than the absolute task (i.e., drawing a line to reproduce a given line, regardless of the size of the frame) whereas Americans showed the reversed pattern, suggesting that Japanese are more field-dependent and attend more to the context (i.e., the frame) whereas Americans are more field-independent and attend more to focal information (i.e., the line). Similarly, Kuhnen et al. (2001) found that Malaysian and Russian participants (i.e., representing Eastern culture) detected fewer individual figures embedded in geometrical patterns than Western participants (e.g., Americans and Germans). Furthermore, using eye movement tracking technique, Chua, Boland, and Nisbett (2005) found that when looking at photographs that contained a focal object presented against a background scene, American participants fixated more on the focal objects whereas their Chinese counterparts showed a more balanced fixation on the objects and the backgrounds. Cultural differences in attention have also been evidenced in neuroimaging studies. For example, Jenkins Yang, Goh, Hong, and Park (2010) found that Chinese participants, but not their American counterparts, showed greater activation in the lateral occipital complex when viewing objects repeatedly presented against incongruent relative to congruent backgrounds, suggesting a more context-dependent attention.

Going beyond the aforementioned attention/perception differences across cultures, accumulated research also demonstrated clear cultural differences in memory (e.g., Gutchess & Indeck, 2009). In the following section, we will zoom in to review studies on cultural differences in memory processes (e.g., memory organization and specificity) and various types of memory (e.g., memory for items and contexts, emotional memory, self-reference memory, and autobiographical memory). The chapter will then conclude with a section on mechanisms underlying these cultural effects.

CULTURE AND MEMORY

Memory Organization

It has been revealed in literature, with both children (Chiu, 1972) and young adults (Ji, Zhang, & Nisbett, 2004), that Western and East Asian individuals differ in the way of information organization (Nisbett & Masuda, 2003; Norenzayan, Smith, Kim, Nisbett, 2002; Unsworth, Sears, & Pexman, 2005). Specifically, Westerners prefer to engage rule-based categorical classification and tend to sort information based on taxonomic categorization (i.e., based on shared features) whereas East Asians are more likely to engage experience-based intuitive relational classification of information (i.e., based on thematic and functional relationship). For example, it has been shown that Chinese classify objects in a more relational (e.g., to group monkey and banana together because monkeys eat bananas) and less categorical (e.g., to group monkey and panda together because they are both animals) way relative to Americans (Ji et al., 2004).

The cultural differences in information organization have also been found in memory (e.g., Gutchess, Yoon et al., 2006; Yang, Chen, Ng, & Fu, 2013). For instance, Gutchess, Yoon et al. (2006) found that Americans were more likely than Chinese, in older adults particularly, to cluster information based on their categorical groups (e.g., types of fruits) in their free recall, despite culturally equivalent memory performance. As shown in figure 4.1, Yang, Chen et al. (2013) found that Canadians showed a memory advantage over Chinese when the to-be-remembered information was encoded in a categorical manner, such as when faces were sorted into "good" or "evil" categories according to externally provided cues or words that were learned either with an accompanying image ("SEEN" category) or a mentally generated image ("IMAGINED" category). Interestingly, in both of these studies, cultural differences were more pronounced in older than young adults, suggesting that cultural differences in memory organization magnify with age. This might be because young adults are more flexible to use both culturally preferred and non-preferred strategies, but using culturally non-preferred strategies becomes more demanding and difficult for older adults due to normal aging-related cognitive declines (Park et al., 1999). As a result, East Asian older adults particularly have trouble using culturally nonpreferred taxonomic categorization strategies to study information during encoding (Yang, Chen et al., 2013) or to organize information at retrieval (Gutchess, Yoon, et al., 2006).

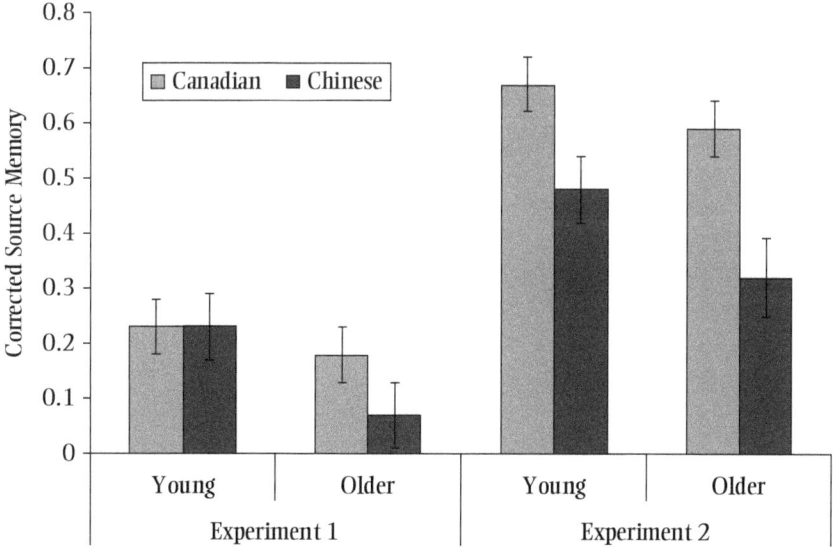

Figure 4.1. Corrected source memory scores (hit rate – false alarm rate) across age groups in the two experiments (Yang, Chen, et al., 2013).

Cultural variations in memory organization also have been demonstrated in memory errors. In a cued-recall memory paradigm, American and Turkish participants studied categorically related word pairs and later completed a cued-recall task (Schwartz, Boduroglu, & Gutchess, 2014). Americans were more likely than Turks to make an error by responding with an unstudied word that was in the same taxonomic category as the cue (e.g., mistakenly recall "banana" for the cue "pear"), suggesting that Americans (representing Western cultures) have a greater reliance on categorical memory organization than Turks, who live in a blend of Western and Eastern cultures.

In summary, research generally suggests that relative to East Asians, Westerners are more likely to and better able to organize information categorically in their memory. This effect has been demonstrated in both memory encoding and retrieval, and the effect tends to magnify with aging. It also makes Westerners especially vulnerable to categorically related memory errors.

Memory Specificity

Research suggests that cultural background not only affects the *quantity* of memory (i.e., how much information one can remember) but also the *quality* of memory. For example, Millar, Serbun, Vadalia, and Gutchess (2013) found that due to Western individuals' analytic information processing preference, they remember more specific details of the learned information than East Asians who prefer a holistic processing style. In this study, American and East Asian participants viewed pictures of objects (Experiment 1) or pictures with both objects and backgrounds (Experiment 2) during encoding. They then received a surprising recognition test in which they were asked to indicate whether a picture was exactly the same as the picture they had seen, similar to the encoded picture, or a new picture. The results showed culturally equivalent general memory performance (i.e., recognizing encoded pictures either as "same" or "similar"). However, in comparison to East Asians, American participants showed more accurate and specific memory (i.e., recognizing encoded pictures as "same" pictures).

Similar cultural differences in memory specificity also have been revealed in autobiographical memory (e.g., Wang, 2016). In a study by Wang (2009), Asian and European American college students were asked to segment a narrative text into discrete events. Driven by the East–West cultural differences in holistic versus analytic processing, Americans perceived, encoded, and subsequently recalled a larger number of discrete event episodes than Asians. Similarly, Wang (2013) found that in comparison to Asians who moved to the United States at an older age (less exposure to Western culture), those who moved at a younger age (more exposure to Western culture) perceived and subsequently recalled more specific events as reflected in a series of randomly sampled 20-minute autobiographical event recordings.

Taken together, memory tends to be more detailed and specific in Western relative to East Asian individuals, as demonstrated in memory for both specific pictures (Millar et al., 2013) and autobiographical event episodes (Wang, 2009, 2013).

Memory for Items and Their Contexts

Past research suggests that East Asians show a better performance on tasks that require context-oriented holistic processing, whereas Western individuals perform better on tasks that rely on feature-driven analytic or categorical processing (e.g., Nisbett & Miyamoto, 2005; Nisbett, Peng, Choi, & Norenzayan, 2001). In this context, it has been found that East Asians are more likely than Western individuals to pay attention and remember meaningful item-context associations, such as a focal object and its background scene in a picture (e.g., Masuda & Nisbett, 2001) or the context under which the information is learned (e.g., Yang, Li et al., 2013).

MEMORY FOR OBJECTS AND SCENES

A substantial number of studies have shown that East Asians are more likely than Western individuals to prioritize contextual information, pay attention to object–background relations, and holistically integrate these pictorial features together in their memory for complex scenes (Kitayama et al., 2003; Masuda & Nisbett, 2001). Eye-tracking data revealed that Western individuals focused their attention on objects in pictures sooner and longer than East Asians, whereas East Asians showed a more balanced fixation toward both objects and background scenes (Chua, Boland et al., 2005). In addition, East Asians were more likely than Western individuals to describe background scenes and the object–scene interactions in animated video clips (Masuda & Nisbett, 2001). In a functional magnetic resonance imaging study, Gutchess, Welsh, Boduroglu, and Park (2006) found that Americans showed greater activation than East Asians in brain regions (i.e., the bilateral middle temporal gyrus, the angular gyrus, and the right superior temporal/supramarginal and superior parietal gyrus) associated with object processing when viewing objects. Similarly, neuroimaging data showed that East Asians demonstrated less attention or sensitivity to objects relative to their Western counterparts (Goh et al., 2004).

Taken together, these findings suggest that East Asians may more readily bind items to associated contexts than Westerners, who, in contrast, tend to process items and contexts discretely. Driven by cultural differences in object–background attention orientation, memory for objects and scenes also tends to differ across the two cultures. For example, it has been shown that Japanese remembered more background information and their memory for objects was more disrupted when background was changed or removed relative to Americans (Masuda & Nisbett, 2001). On the other hand, Western individuals' memory for the objects was not affected by the manipulations of the background scenes (Chua, Boland et al., 2005; Masuda & Nisbett, 2001). A similar cultural effect has also been found in memory for scenes, in which East Asians remembered fewer scenes than Western individuals when the scenes were to be recognized in isolation from the objects (Wong, Yin, Yang, Li, & Spaniol, 2018), an unbinding condition that was particularly disruptive to the retrieval of holistically processed information in East Asians (e.g., Gutchess & Park, 2009).

However, it should be noted that there are inconsistent findings on cultural differences in object–scene processing. For example, Evans, Rotello, Li, and Rayner (2009) conducted a study to replicate Chua, Boland et al.'s (2005) eye-tracking study (Evans et al., 2009). To increase statistical power, Evans et al. (2009) expanded the testing stimuli by adding more pictures to the stimulus set used in Chua, Boland et al.'s study (2005). However, Evans et al. did not find any cultural differences. In fact, both American and Chinese participants fixated more and longer on focal objects than on background scenes. Moreover, both cultural groups showed better memory for objects when they were presented with their original backgrounds than with new backgrounds. Evans et al. suggested that cultural differences in object memory in Chua, Boland et al.'s study might be attributed to group differences in response bias. However, this does not easily explain cultural differences in other studies in which false alarm rates (Masuda & Nisbett, 2001) and response bias (Wong et al., 2018) were accounted for. We speculate that the discrepancies might be because Evans et al. recruited their Chinese participants in the United States, and thus their exposure to Western culture might have eliminated the cultural effects (Huff, Yoon, Lee, Mandadi, & Gutchess, 2013). Lack of cultural differences has also been demonstrated in neural activities featuring object–scene binding. Goh et al. (2007) tested young and older Singaporeans and Americans using a functional magnetic resonance adaptation (i.e., functional magnetic resonance adaptation) paradigm in which participants passively viewed a series of four scene pictures, with the focal object and/or background systematically repeated or varied. The attenuation of the blood oxygen level dependent (BOLD) signal in response to the repetition of objects, backgrounds, and object–background associations was used to index neural adaptation for object, scene, or object–scene binding, respectively. The results revealed little cultural differences in neural adaptation for object-scene binding.

Taken together, a main body of research suggests cultural differences in memory for objects and background scenes by showing that Westerners attend more to and remember better the focal objects whereas East Asians' memory of object/scene is more affected by the manipulations of the object-scene paring. Nevertheless, the research is still inconclusive in this area given the mixed results, calling for further studies to consolidate the conclusions.

SOURCE/CONTEXT MEMORY

Considering East Asians' sensitivity to contexts (Ji, Schwarz, & Nisbett, 2000; Masuda et al., 2008), it is reasonable to speculate that they would also show better source/context memory than Western individuals. Source/context memory is referred to as the memory for the contexts or conditions (e.g., when, where, and how) in which the information is learned (Pandey, 2011). Chua, Chen, and Park (2006) took an initiative to investigate cultural differences in source memory. In this study, young and older American and Chinese participants studied a series of facts spoken by one of four speakers. They then completed a source memory task in which they identified the correct speaker for each studied fact statement. The results did not show any cultural differences in source memory performance.

However, Chua et al. acknowledged that the speakers were arbitrarily assigned to the facts. It is possible that East Asians, relative to Westerners, may engage more elaborative processing when being required to make meaningful associations between items and the corresponding contexts, and this may benefit their subsequent memory for context/sources.

This speculation is supported by Yang, Li et al.'s (2013) study through an intentional source memory paradigm in which young and older Chinese and Canadians studied pictures of familiar objects under socially meaningful contexts and were later tested on their memory for the pictures and the associated encoding contexts (Figure 4.2). Specifically, participants studied the images in two blocks, a "relational" and an "independent" block. In the relational block, each object was rated either by its usefulness in fostering relationship with others in a new city or its typicality in everyday life. In the independent block, each object was rated either by its usefulness in helping one live independently in a new city or its typicality in everyday life. Context memory was defined as the ability to recognize how the object was originally studied (e.g., "independent/relational" vs. "daily living") during encoding within each block. It was predicted that the relational context would be consistent with East Asians' value on collectivism and social relationships, whereas the independent context would be more aligned with Western cultures' value on individualism and independence (Nisbett et al., 2001). Overall, Chinese participants showed better context memory than their Canadian counterparts in both blocks, regardless of whether the encoded context aligned

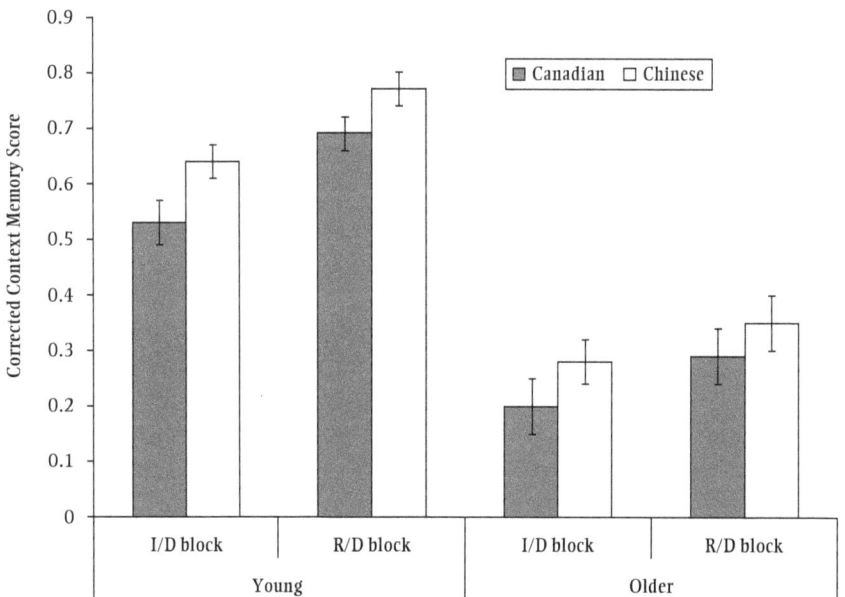

Figure 4.2. Corrected context memory scores across two memory blocks for the four age by culture groups (Yang, Li, et al., 2013). I/D = independent vs. daily life block; R/D = relational vs. daily life block.

with the theoretical cultural values. These findings suggest that socially meaningful item–context association processing in general promotes context memory in East Asians, even when the context is hypothetically more consistent with Western cultural values.

Taken together, previous cross-cultural studies provided mixed support for cultural differences in item–context binding. Overall, memory for arbitrary item–context/source perceptual associations appears to be universal across cultures (Chua et al., 2006) whereas memory for socially meaningful item–context tends to vary with cultures (Yang, Li et al., 2013). This might be because memory for arbitrary perceptual associations relies primarily on basic neurocognitive resources that tends to be culturally invariant (Baltes, 1993). In contrast, forming socially meaningful item–context associations may recruit culturally relevant knowledge or experience (Baltes, 1993; Yoon, Feinberg, & Gutchess, 2006), which, in turn, influences the subsequent memory for contexts/sources.

Emotional Memory and Positivity Effect

Research suggests that our memory can be influenced by the emotional valence of the to-be-remembered information (e.g., Reisberg & Hertel, 2004). It has been demonstrated that emotional values affect memory across cultures. Specifically, information (e.g., emotional experience) consistent with cultural values is likely to be maintained in memory for longer (Oishi et al., 2007). In a free recall test of narratives or videos about social interactions, it has been reported that Americans recalled more information about the main character and reported less emotional content than Taiwanese (Chua, Leu, & Nisbett, 2005). Similarly, the perception of the emotional expression of a person was more affected by conflicting social contexts in Japanese than in Westerners (Masuda et al., 2008).

In the aging literature with Western participants, it has been found that older adults tend to remember more positive and/or less negative information than young adults, a phenomenon labeled as the positivity effect (Carstensen, 1993). It is speculated that East Asian older adults should show even stronger positivity bias because of their differentially more positive views of aging (Chung & Lin, 2012). In a study by Kwon, Scheibe, Samanez-Larkin, Tsai, and Carstensen (2009), young and older Koreans viewed emotionally positive, negative, and neutral images and then recalled and recognized these images, as well as rated the emotional valence of each image. Although Korean young adults showed similar emotional valence ratings as the norms based on American samples, Korean older adults were more likely to rate neutral pictures (e.g., a cup, mushrooms) as more positive and positive pictures as less positive than the American norms. Despite these cultural differences in subjective valence perception, older adults in this study in general remembered more positive and fewer negative images than did young adults, showing a positivity effect that was comparable to Western samples (Kwon, Scheibe, Samanez-Larkin, Tsai, & Carstensen, 2009). This universality of age-related positivity effect has also been demonstrated in another study by Ko, Lee, Yoon, Kwon, and Mather (2011). In this study, young and older Koreans and Americans rated the emotionality of faces presented against emotional background pictures and later recalled the backgrounds. The results showed that in both cultures more negative backgrounds were recalled than positive backgrounds, a bias that was more pronounced in young than older adults. Furthermore, Chung and Lin (2012) conducted a study in which young and older Chinese and American participants viewed and subsequently recalled emotional and neutral pictures. Older adults across both cultures recalled more positive and fewer negative images than young adults, supporting a universal positivity effect in older adults. Nevertheless, there was a trend for Chinese older adults to remember fewer negative images than Western older adults, thus providing some preliminary evidence that the magnitude of the positivity effect might differ across cultures.

Overall, research generally suggests a universal positivity effect in older adults across cultures, with some preliminary evidence that the effect tends to be stronger for East Asians relative to Americans.

Self-Representation and Self-Reference Memory

Culture experience also shapes the representation of self, as evidenced in both behavioral (e.g., Markus & Kitayama, 1991; Ng & Lai, 2009) and neuroimaging studies (e.g., Sui & Han, 2007; Zhu, Zhang, Fan, & Han, 2007). Typically, East Asian societies endorse a more interdependent self-construal and view self as more socially connected with others, particularly close others such as family. In contrast, Western societies adopt a more independent self-construal and view self as an autonomous and unique identity, distinct and different from others (Markus & Kitayama, 1991; Ng & Lai, 2009). In line with this dimension, it has been shown that when primed with Western culture, individuals with bicultural identity (e.g., Chinese Canadians) or even mono-cultural identity (e.g., Beijing Chinese) tended to use fewer interdependent and more independent self-statements to describe themselves (Sui, Zhu, & Chiu, 2007; Ross, Xun, & Wilson, 2002). This supports a dynamic constructivist model for culture and cognition (Hong, Morris, Chiu, & Benet-Martinez, 2000). In a neuroimaging study (Sui & Han, 2007), Chinese participants identified the orientation of their own face or other familiar faces in photos under different self-construal priming conditions. The results showed that BOLD signals in the right frontal cortex differentiated self and other faces after independent self-construal priming, but this difference was not visible following interdependent self-construal priming.

Considering that self serves as a fundamental schema to organize and regulate mental processes and behavior (Markus & Kitayama, 1991) and the aforementioned research evidence for cultural differences in self representation, it has been suggested that cultural experience shapes self-related cognition, such as self-reference and autobiographical memory systems (e.g., Conway & Pleydell-Pearce, 2000; Sparks, Cunningham, & Kritikos, 2016). The self-reference effect refers to the mnemonic advantage for information encoded in reference to the self over that encoded in reference to a nonself or with any other encoding strategies (Rogers, Kuiper, & Kirker, 1997). Although both East Asians and Westerners demonstrated self-reference effect in memory (Symons & Johnson, 1997), the specific pattern of this effect appears to differ between the two cultures. In a 40-minute delayed recognition task (Wagar & Cohen, 2003), Asian Canadians showed no difference in recognition reaction time to words between self-reference and other-reference conditions; whereas European Canadians responded faster to self-referent words relative to other-referent words. In addition, only Asian Canadians responded faster to words representing collective self (e.g., friend, student, colleague) relative to those indexing personal self (i.e., happy and smart). Similarly, driven by their highly interdependent self-construal that emphasizes connection with others, East Asians would allocate similar attentional resources to close others, such as one's mother as to the self. Therefore, information associated with one's mother would elicit deep and elaborated processing similar to that associated with oneself and thus result in efficient memorization of mother-relevant information (Sui et al., 2007). In support of this, it has been found that although Western participants consistently demonstrated memory advantage for self-referent over other-referent information, Chinese participants showed equivalent memory for self-referent and mother-referent information, as both were remembered better than information processed in reference to a familiar public figure (Zhu & Zhang, 2002). Similarly, in cultural priming studies, this memory improvement for mother-relevant words was only observed in Chinese participants when they were primed with East Asian culture prior to encoding, but not when they were primed with Western culture (Ng & Lai, 2009; Sui et al., 2007). Neuroimaging data also support these cultural differences. For example, American participants showed differentiated BOLD signal changes in the ventromedial prefrontal cortex (vmMPFC) between self and mother trait judgements whereas Chinese participants showed equivalent BOLD signal changes between the two judgement conditions (Zhu et al., 2007).

The cultural differences in self-reference over mother-reference memory effect has also been extended to nonidentified but well-connected others, friends, or even strangers. For example, Ng and Lai (2009) assessed the self-reference effects in bicultural Chinese under East Asian or Western cultural priming conditions. The results showed a clear memory advantage for self-referent information relative to that processed in reference to mother, a nonidentified person (NIP), or font under Western cultural priming. The self–NIP and self–mother distinctions disappeared under Chinese culture priming in which only the self–font distinction remained. The results suggest that cultural priming shapes self-reference effect in memory, with a higher interpersonal connectedness (e.g., self–NIP and self–mother) under Chinese priming than under Western priming in bicultural Chinese individuals. Furthermore, Sparks et al. (2016) examined this phenomenon using an ownership paradigm. In this study, East Asian and Western participants were asked to sort items as either belonging to themselves or others (stranger/close friend/mother) and later asked to identify these items. Westerners showed better recognition of self-owned over other-owned items whereas East Asians showed a better recognition of mother-owned than self-owned items.

However, this self–mother equivalent memory effect observed in East Asian participants was wiped out in the sample of bicultural Asian Americans, who instead showed the typical Western self- over mother-reference memory benefit (Huff et al., 2013). The authors attributed this to the exposure to another culture or a multicultural context, as their participants were recruited in the United States, which has minimized or even eliminated the cultural differences in the self–mother reference memory effect.

Taken together, both behavioral and neuroimaging research supports clear cultural differences in self-reference memory, with a much reduced or even eliminated self-reference effect when comparing with encoding in reference to a close other (e.g., mother, close friend, or ownership encoding instruction) in East Asians relative to Westerners. This effect, however, could be minimized by exposure to Western culture (Huff et al., 2013) or being primed by Western culture (Ng & Lai, 2009; Siu et al., 2007).

Autobiographical Memory

Western and East Asian individuals have been found to differ in autobiographical memory or memory for their past life events (Nelson & Fivush, 2004). The general finding is that individuals with stronger independent values (e.g., Americans) are more likely to recall more individual memories whereas individuals with stronger interdependency values (e.g., Asians) tend to report more social interactions and involve more people in their autobiographical memory (Conway, Wang, Hanyu, & Haque, 2005; Jobson & O'Kearney, 2008; Wang & Conway, 2004; Wang & Rose, 2005). These cultural differences in autobiographical memory start to emerge as early as 3 to 4 years of age (Wang, 2004). Furthermore, it has also been found that Chinese participants are more attentive to observable behaviors and show more accurate memory for these behaviors than Americans (Ji et al., 2000) because attention to these behaviors promote smooth social relationships with others. Wang (2008a) further showed that the East–West cultural differences in the emphasis of self versus social relationships were driven by their perception of self. In this study, Asian Americans were primed to focus on their American or Asian self before their recall of personal memories. When their American self was activated, they recalled more self-focused memory episodes. However, when their Asian self was activated, they recalled more socially oriented content.

Knowledge about emotions, as well as emotional content of memory episodes, was also found to interact with the cultural effects on autobiographical memory. Wang (2008b) found that American children, driven by their cultural emphasis on individual feelings and internal states, showed earlier development in emotional knowledge (i.e., the understanding of specific situations that provoke different types of emotions) relative to Chinese children. This difference, in turn, mediated the effect of culture on autobiographical memory, such that American children's reports of personal memory were more likely to include features of internal states in comparison to those provided by Chinese children. In another study, Zaragoza Scherman, Salgado, Shao, and Berntsen (2015) asked middle-aged participants from Mexico, Greenland, China, and Denmark to recall their most stressful or positive personal life events. It was found that participants in general recalled more positive events from their 20s than from their 40s. However, there was an exception in that Chinese participants recalled as many positive events from these two stages of life, and they recalled significantly more positive events from their 40s relative to participants from other cultures. Interestingly, many of the positive memories recalled by Chinese were related to their children but not directly experienced by the participants themselves. These results once again support that in autobiographical memory, social relationships are highlighted in East Asians whereas self-focused is focused in Western individuals.

Overall, in autobiographical memory, Western individuals report more self-focused information whereas East Asians report more observable behaviors and events that focus on social relations (e.g., Wang, 2016). In addition, Western individuals' autobiographical report tends to involve more information on internal state and emotional feelings relative to East Asians (Wang, 2008b). Finally, Chinese tend to report more positive memories, particularly from their 40s, relative to individuals from other cultures (Zaragoza Scherman et al., 2015).

CULTURE AND MEMORY: MECHANISM

Cultural differences in memory could be driven by a variety of factors. First, there have been established cultural differences in historically and socially adopted values between Western and Eastern cultures. In ancient Greek philosophy, which is dominant in Western cultures, every individual object or person is viewed as a distinct entity separate from each other. In contrast, according to Confucianism and Taoism, two predominant schools of thoughts in ancient East Asia, the universe is considered as a network of interdependent substances (Hansen, 1983). These philosophical differences between the two cultures might contribute to the cultural differences in processing and memory for objects versus context, source/context memory, and memory specificity.

Meanwhile, the two cultures differ vastly in social and cultural orientation. East Asian cultures highly value interpersonal social relationships and group harmony, so it is crucial for them to attend and respond appropriately to social cues. This may shape their context-oriented holistic information processing style and thus influence their attention to and memory for context, relationship, or item–context associations (Masuda et al., 2008; Nisbett et al., 2001).

In contrast, Western individuals value autonomy and freedom (Nisbett et al., 2001), so they are comparatively more likely to engage rule-based object-oriented analytic information processing style. As a consequence, they are inclined to attend to and remember self-related information and individual objects, as well as categorically processed information (Symons & Johnson, 1997; Yang, Chen et al., 2013).

In addition, individualism–collectivism cultural orientation differences could also shape their information processing and memory. Driven by the individualism, independent and personal goals without being overly constrained by social demand are prioritized in Western society. In the context of a collectivism culture,

East Asians are more dependent on each other and work as a whole to serve the interest of the society (Triandis, 1995). These cultural differences have been captured in the social-cognitive framework (Nisbett & Masuda, 2003; Nisbett & Miyamoto, 2005). According to this framework, the specific value system a person holds will preferentially direct his or her attention to some aspects of the information over others, which subsequently will shape his or her culturally congruent memory processing (Nisbett et al., 2001). These cultural values and practice may affect memory performance such as autobiographical memory and self-reference memory. The influence may occur through various routes including language, education system, child-rearing practice, and other societal values (Hedden et al., 2002).

Finally, an emerging body of literature also suggests cultural differences in functions of various brain regions (e.g., hippocampus and medial temporal lobes, medial prefrontal context, frontal-parietal network) that are involved in different memory systems (Han & Northoff, 2008; Gutchess & Indeck, 2009). These findings identified neural mechanisms for cultural effects in memory. Specifically, cultural experience shapes our brain and, consequently, influences our cognitive functions such as memory.

CONCLUSION AND OUTLOOK

Overall, it has been well-documented that cultural values and practice affect information processing styles, attention, and memory performance (e.g., Hedden et al., 2002). Based on the review of accumulated research in culture and memory, the following general conclusions could be drawn:

1. Relative to East Asians, Westerners are more likely to and better able to organize information categorically in their memory (Gutchess, Yoon et al., 2006; Yang, Chen et al., 2013).
2. Memory tends to be more detailed and specific in Western relative to East Asian individuals (Millar et al., 2013; Wang, 2009).
3. Memory for meaningful object–scene associations tends to vary with cultures (Yang, Li et al., 2013), despite the universality in memory for arbitrary item–context/source perceptual associations (Chua et al., 2006).
4. There is an universal positivity effect in older adults across cultures, with some preliminary evidence that the effect tends to be stronger for East Asians relative to Americans (Kwon et al., 2009; Ko et al., 2011; Chung & Lin, 2012).
5. Self-reference effect in memory tends to be reduced or even eliminated when comparing self-reference encoding and the encoding in reference to a close other (e.g., mother, close friend, or ownership encoding instruction) in East Asians relative to Westerners (Ng & Lai, 2009; Sui et al., 2007; Zhu & Zhang, 2002).
6. Western individuals' autographical memory tends to be self-focused, whereas East Asians are more likely to recall personal events that involve other individuals (Conway et al., 2005; Wang, 2008a).

Despite the dominant research evidence for cultural differences in memory, it should be noted that there are also some mixed findings (Evans et al., 2009; Huff et al., 2013). The discrepancy in these studies might be due to the multicultural exposure that is prevalent in modern Western societies. Future research might explore further how acculturalization and multicultural exposure in immigrant populations may shift an individual's perspective and guide his or her information processing style and memory performance under different culture priming conditions. It also should be noted that the existing research is predominantly behavioral; our understanding of the neural and bio-physiological mechanisms associated with cultural effects on memory is still very limited. More research is required to further explore how accumulated cultural experience may shape the brain (structure and function) and how this culture-specific brain plasticity affects information processing and memory.

REFERENCES

Baltes, P. B. (1993). The aging mind: Potential and limits. *Gerontologist, 33,* 580–594. https://doi.org/10.1093/geront/33.5.580

Block, N. (1995). The mind as the software of the brain. In E. E. Smith & D. N. Osherson (Eds.), *Thinking: An invitation to the cognitive science* (pp. 377–425). Cambridge, MA: MIT Press.

Buchtel, E. E., & Norenzayan, A. (2009). Thinking across cultures: Implications for dual processes. In J. Evans & K. Frankish, (Eds.), *In two minds: Dual processes and beyond.* (pp. 217–238). Oxford, England: Oxford University Press.

Carstensen, L. L. (1993). Motivation for social contact across the life span: A theory of socioemotional selectivity. In J. Jacobs (Ed.), *Nebraska Symposium on Motivation: Developmental perspectives on motivation* (Vol. 40, pp. 209–254). Lincoln, NE: University of Nebraska Press.

Chiu, L. (1972). A cross-cultural comparison of cognitive styles in Chinese and American children. *International Journal of Psychology, 7,* 235–242. https://doi.org/10.1080/00207597208246604

Choi, I., Koo, M., & Choi, J. A. (2007). Individual differences in analytic versus holistic thinking. *Personality and Social Psychology Bulletin, 33,* 691–705. https://doi.org/10.1177/0146167206298568

Chomsky, N. (2002). *On nature and language.* Cambridge, England: Cambridge University Press

Chua, H. F., Boland, J. E., & Nisbett, R. E. (2005). Cultural variation in eye movements during scene perception. *Proceedings of the National Academy of Sciences of the United States of America, 102,* 12629–12633. https://doi.org/10.1073/pnas.0506162102

Chua, H. F., Chen, W., & Park, D. C. (2006). Source memory, aging, and culture. *Gerontology, 52,* 306–313. https://doi.org/10.1159/000094612

Chua, H. F., Leu, J., & Nisbett, R. E. (2005). Culture and diverging views of social events. *Personality and Social Psychology Bulletin, 31,* 925–934. https://doi.org/10.1177/0146167204272166

Chung, C., & Lin, Z. (2012). A cross-cultural examination of the positivity effect in memory: United States vs. China. *International Journal of Aging and Human Development, 75,* 31–44. https://doi.org/10.2190/AG.75.1.d

Conway, M. A., & Pleydell-Pearce, C. W. (2000). The construction of autobiographical memories in the self-memory system. *Psychological Review, 107,* 261–288. https://doi.org/10.1037//0033-295X.107.2.261

Conway, M. A., Wang, Q., Hanyu, K., & Haque, S. (2005). A cross-cultural investigation of autobiographical memory: On the universality and cultural variation of the reminiscence bump. *Journal of Cross-Cultural Psychology, 36,* 739–749. https://doi.org/10.1177/0022022105280512

Evans, K., Rotello, C. M., Li, X., & Rayner, K. (2009). Scene perception and memory revealed by eye movements and receiver-operating characteristic analyses: Does a cultural difference truly exist? *The Quarterly Journal of Experimental Psychology, 62,* 276–285. https://doi.org/10.1080/17470210802373720

Fernald, A., & Morikawa, H. (1993). Common themes and cultural variations in Japanese and American mothers' speech to infants. *Child Development, 64,* 637–656. https://doi.org/10.1111/j.1467-8624.1993.tb02933.x

Goh, J. O., Chee, M. W., Tan, J. C., Venkatraman, V., Hebrank, A., Leshikar, E. D., . . . Park, D. C. (2007). Age and culture modulate object processing and object-scene binding in the ventral visual area. *Cognitive, Affective, & Behavioural Neuroscience, 7,* 44–52. https://doi.org/10.3758/CABN.7.1.44

Goh, J. O. S., Siong, S. C., Park, D., Gutchess, A. H., Hebrank, A., & Chee, M. W. L. (2004). Cortical areas involved in object, background, and object-background processing revealed with functional magnetic resonance adaptation. *Journal of Neuroscience, 24,* 10223–10228. https://doi.org/10.1523/JNEUROSCI.3373-04.2004

Gutchess, A. H., & Indeck, A. (2009). Cultural influences on memory. *Progress in Brain Research, 178,* 137–150. https://doi.org/10.1016/S0079-6123(09)17809-3

Gutchess, A. H., & Park, D. C. (2009). Effects of ageing on associative memory for related and unrelated pictures. *European Journal of Cognitive Psychology, 21,* 235–254. https://doi.org/10.1080/09541440802257274

Gutchess, A. H., Welsh, R. C., Boduroglu, A., & Park, D. C. (2006). Cultural differences in neural function associated with object processing. *Cognitive, Affective, & Behavioural Neuroscience, 6,* 102–109. https://doi.org/10.3758/CABN.6.2.102

Gutchess, A. H., Yoon, C., Luo, T., Feinberg, F., Hedden, T., Jing, Q., Nisbett, R. E., & Park, D. C. (2006). Categorical organization in free recall across culture and age. *Gerontology, 52,* 314–323. https://doi.org/10.1159/000094613

Han, S., & Northoff, G. (2008). Culture-sensitive neural substrates of human cognition: A transcultural neuroimaging approach. *Nature Reviews Neuroscience, 9,* 646–654. https://doi.org/10.1038/nrn2456

Han, S., Northoff, G., Vogeley, K., Wexler, B. E., Kitayama, S., & Varnum, M. E. W. (2013). A cultural neuroscience approach to the biosocial nature of the human brain. *Annual Review of Psychology, 64,* 335–359. https://doi.org/10.1146/annurev-psych-071112-054629

Hansen, C. (1983). *Language and logic in ancient China.* Ann Arbor, MI: University of Michigan Press.

Hedden, T., Park, D. C., Nisbett, R., Ji, L., Jing, Q., & Jiao, S. (2002). Cultural variation in verbal versus spatial neuropsychological function across the lifespan. *Neuropsychology, 16,* 65–73. https://doi.org/10.1037//0894-4105.16.1.65

Hong, Y-y., Morris, M. W., Chiu, C-y., & Benet-Martinez, V. (2000). Multicultural minds: A dynamic constructivist approach to culture and cognition. *American Psychologist, 55,* 709–720. https://doi.org/10.1037/0003-066X.55.7.709

Huff, S., Yoon, C., Lee, F., Mandadi, A., & Gutchess, A. H. (2013). Self-referential processing and encoding in bicultural individuals. *Culture and Brain, 1,* 16–33. https://doi.org/10.1007/s40167-013-0005-1

Jenkins, L. J., Yang, Y-J., Goh, J., Hong, Y-Y., & Park, D. C. (2010). Cultural differences in the lateral occipital complex while viewing incongruent scenes. *Social Cognitive and Affective Neuroscience, 5,* 236–241. https://doi.org/10.1093/scan/nsp056

Ji, L-J., Schwarz, N., & Nisbett, R. E. (2000). Culture, autobiographical memory, and behavioural frequency reports: Measurement issues in cross-cultural studies. *Personality and Social Psychology Bulletin, 26,* 585–593. https://doi.org/10.1177/0146167200267006

Ji, L-J., Zhang, Z., & Nisbett, R. E. (2004). Is it culture of is it language?: Examination of language effects in cross-cultural research on categorization. *Journal of Personality and Social Psychology, 87,* 57–65. https://doi.org/10.1037/0022-3514.87.1.57

Jobson, L., & O'Kearney, R. (2008). Cultural differences in personal identity in post-traumatic stress disorder. *The British Journal of Clinical Psychology, 47,* 95–109. https://doi.org/10.1348/014466507X235953

Kitayama, S., & Uskul, A. K. (2011). Culture, mind, and the brain: Current evidence and future directions. *Annual Review of Psychology, 62,* 419–449. https://doi.org/10.1146/annurev-psych-120709-145357

Kitayama, S., Duffy, S., Kawamura, T., & Larsen, J. T. (2003). Perceiving an object and its context in different cultures: A cultural look at new look. *Psychological Science, 14,* 201–206. https://doi.org/10.1111/1467-9280.02432

Ko, S-G., Lee, T-H., Yoon, H-Y., Kwon, J-H., & Mather, M. (2011). How does context affect assessments of facial emotion? The role of culture and age. *Psychology and Aging, 26,* 48–59. https://doi.org/10.1037/a0020222

Kuhnen, U., Hannover, B., Roeder, U., Shah, A. A., Schubert, Upmeyer, A., & Zakaria, S. (2001). Cross cultural variations in identifying embedded figures: Comparisons from the United States, Germany, Russia, and Malaysia. *Journal of Cross-Cultural Psychology, 32,* 366–372. https://doi.org/10.1177/0022022101032003007

Kwon, Y., Scheibe, S., Samanez-Larkin, G. R., Tsai, J. L., & Carstensen, L. L. (2009). Replicating the positivity effect in picture memory in Koreans: Evidence for cross-cultural generalizability. *Psychology and Aging, 24,* 748–754. https://doi.org/10.1037/a0016054

Markus, H. R., & Kitayama, S. (1991). Culture and the Self: Implications for cognition, emotion, and motivation. *Psychological Review, 98,* 224–253. https://doi.org/10.1037/0033-295X.98.2.224

Masuda, T., & Nisbett, R. E. (2001). Attending holistically versus analytically: Comparing the context sensitivity of Japanese and Americans. *Journal of Personality and Social Psychology, 81,* 922–934. https://doi.org/10.1037//0022-3514.81.5.922

Masuda, T., Ellsworth, P. C., Mesquita, B., Leu, J., Tanida, S., & Van de Veerdonk, E. (2008). Placing the face in context: Cultural differences in the perception of facial emotion. *Attitudes and Social Cognition, 94,* 365–381. https://doi.org/10.1037/0022-3514.94.3.365

Millar, P. R., Serbun, S. J., Vadalia, A., & Gutchess, A. H. (2013). Cross-cultural differences in memory specificity. *Culture and Brain, 1,* 138–157. https://doi.org/10.1007/s40167-013-0011-3

Nelson, K., & Fivush, R. (2004). The emergence of autobiographical memory: A social cultural developmental theory. *Psychological Review, 111,* 486–511. https://doi.org/10.1037/0033- 295X.111.2.486

Ng, S. H., & Lai, J. C. L. (2009). Effects of culture priming on the social connectedness of the bicultural self: A self-reference effect approach. *Journal of Cross-Country Psychology, 40,* 170–186. https://doi.org/10.1177/0022022108328818

Nisbett, R. E., & Masuda, T. (2003). Culture and point of view. *Proceedings of the National Academy of Sciences of the United States of America, 100,* 11163–11170. https://doi.org/10.1073/pnas.1934527100

Nisbett, R. E., & Miyamoto, Y. (2005). The influence of culture: Holistic versus analytic perception. *Trends in Cognitive Sciences, 9,* 467–473. https://doi.org/10.1016/j.tics.2005.08.004

Nisbett, R. E., & Norenzayan, A. (2002). Culture and cognition. In H. Pashler & D. L. Medinq (Eds.), *Stevens handbook of experimental psychology: Cognition* (3rd Ed., Vol. 2, pp. 561–597). New York, NY: Wiley.

Nisbett, R. E., Peng, K., Choi, I., & Norenzayan, A. (2001). Culture and systems of thought: Holistic versus analytic cognition. *Psychological Review, 108,* 291–310. https://doi.org/10.1037//0033-295X.108.2.291

Norenzayan, A., & Heine, S. J. (2005). Psychological universals: What are they and how can we know? *Psychological Bulletin, 131,* 763–784. https://doi.org/10.1037/0033-2909.131.5.763

Norenzayan, A., Smith, E. E., Kim, B. J., & Nisbett, R. E. (2002). Cultural preferences for formal versus intuitive reasoning. *Cognitive Science, 26,* 653–684. https://doi.org/10.1207/s15516709cog2605_4

Oishi, S., Schimmack, U., Diener, E., Kim-Prieto, C., Scollon, C. N., & Choi, D. W. (2007). The value-congruence model of memory for emotional experiences: An explanation for cultural differences in emotional self-reports. *Journal of Personality and Social Psychology, 93,* 897–905. https://doi.org/10.1037/0022-3514.93.5.897

Pandey, J. (2011). Source memory. In J. Kreutzer, J. DeLuca, & B. Caplan (Eds.), *Encyclopedia of clinical neuropsychology.* (Vol. 1, pp. 2325–2326). New York, NY: Springer-Verlag.

Park, D. C., & Gutchess, A. H. (2002). Aging, cognition, and culture: A neuroscientific perspective. *Neuroscience and Biobehavioural Reviews, 26,* 859–867. https://doi.org/10.1016/S0149-7634(02)00072-6

Park, D. C., & Gutchess, A. H. (2006). The cognitive neuroscience of aging and culture. *Current Directions in Psychological Science, 15,* 105–108. https://doi.org/10.1111/j.0963-7214.2006.00416.x

Park, D. C., Nisbett, R., & Hedden, T. (1999). Aging, culture, and cognition. *The Journals of Gerontology, 54,* 75–84. https://doi.org/10.1093/geronb/54B.2.P75

Reisberg, D., & Hertel, P. (Eds.). (2004). *Memory and emotion.* New York, NY: Oxford University Press.

Rogers, T. B., Kuiper, N. A., & Kirker, W. S. (1977). Self-reference and the encoding of personal information. *Journal of Personality and Social Psychology, 35,* 677–688. https://doi.org/10.1037//0022-3514.35.9.677

Ross, M., Xun, W., & Wilson, A. (2002). Language and the bicultural self. *Personality and Social Psychology Bulletin, 28,* 1040–1050. https://doi.org/10.1177/01461672022811003

Schwartz, A. J., Boduroglu, A., & Gutchess, A. H. (2014). Cross-cultural differences in categorical memory errors. *Cognitive Science, 38,* 997–1007. https://doi.org/10.1111/cogs.12109

Sparks, S., Cunningham, S. J., & Kritikos, A. (2016). Culture modulates implicit ownership induced self-bias in memory. *Cognition, 153,* 89–98. https://doi.org/10.1016/j.cognition.2016.05.003

Sui, J., & Han, S. (2007). Self-construal priming modulates neural substrates of self-awareness. *Psychological Science, 18,* 861–866. https://doi.org/10.1111/j.1467-9280.2007.01992.x

Sui, J., Zhu, Y., & Chiu, C-y. (2007). Bicultural mind, self-construal, and self- and mother reference effects: Consequences of cultural priming on recognition memory. *Journal of Experimental Social Psychology, 43,* 818–824. https://doi.org/10.1016/j.jesp.2006.08.005

Symons, C. S., & Johnson, B. T. (1997). The self-reference effect in memory: A meta-analysis. *Psychological Bulletin, 121,* 371–394. https://doi.org/10.1037/0033-2909.121.3.371

Triandis, H. C. (1995). *Individualism and collectivism.* Boulder, CO: Westview.

Unsworth, S. J., Sears, C. R., Pexman, P. M. (2005). Cultural influences on categorization processes. *Journal of Cross-Cultural Psychology, 36,* 662–688. https://doi.org/10.1177/0022022105280509

Wagar, B. M., & Cohen, D. (2003). Culture, memory, and the self: An analysis of the personal and collective self in long-term memory. *Journal of Experimental Social Psychology, 39,* 468–475. https://doi.org/10.1016/S0022-1031(03)00021-0

Wang, Q. (2004). The emergence of cultural self-constructs: Autobiographical memory and self- description in European American and Chinese children. *Developmental Psychology, 40,* 3–15. https://doi.org/10.1037/0012-1649.40.1.3

Wang, Q. (2008a). Being American, being Asian: The bicultural self and autobiographical memory in Asian Americans. *Cognition, 107,* 743–751. https://doi.org/10.1016/j.cognition.2007.08.005

Wang, Q. (2008b). Emotion knowledge and autobiographical memory across the preschool years: A cross-cultural longitudinal investigation. *Cognition, 108,* 117–135. https://doi.org/10.1016/j.cognition.2008.02.002

Wang, Q. (2009). Are Asians forgetful? Perception, retention, and recall in episodic remembering. *Cognition, 111,* 123–131. https://doi.org/10.1016/j.cognition.2009.01.004

Wang, Q. (2013). Gender and emotion in everyday event memory. *Memory, 21,* 503–511. https://doi.org/10.1080/09658211.2012.743568

Wang, Q. (2016). Remembering the self in cultural contexts: A cultural dynamic theory of autobiographical memory. *Memory Studies, 9,* 295–304. https://doi.org/10.1177/1750698016645238

Wang, Q., & Conway, M. A. (2004). The stories we keep: Autobiographical memory in American and Chinese middle-aged adults. *Journal of Personality, 72,* 911–938. https://doi.org/10.1111/j.0022-3506.2004.00285.x

Wang, Q., & Ross, M. (2005). What we remember and what we tell: The effects of culture and self-priming on memory representations and narratives. *Memory, 13,* 594–606. https://doi.org/10.1080/09658210444000223

Wong, B. I., Yin, S., Yang, L., Li, J., & Spaniol, J. (2018). Cultural differences in attentional allocation and memory for pictures. *Journal of Cross-Cultural Psychology, 49,* 404–417. https://doi.org/10.1177/0022022117748763

Yama, H., Nishioka, M., Horishita, T., Kawasaki, Y., & Taniguchi, J. (2007). A dual process model for cultural differences in thought. *Mind & Society, 6,* 143–172. https://doi.org/10.1007/s11299-007-0028-4

Yang, L., Chen, W., Ng, A. H., & Fu, X. (2013). Aging, culture, and memory for categorically processed information. *The Journals of Gerontology, 68,* 872–881. https://doi.org/10.1093/geronb/gbt006

Yang, L., Li, J., Spaniol, J., Hasher, L., Wilkinson, A. J., Yu, J., & Niu, Y. (2013). Aging, culture, and memory for socially meaningful item-context associations: An East–West cross-cultural comparison study. *PLoS One, 8*(4), 1–7. https://doi.org/10.1371/journal.pone.0060703

Yoon, C., Feinberg, F., & Gutchess, A. H. (2006). Pictorial naming specificity across ages and cultures: A latent class analysis of picture norms for younger and older Americans and Chinese. *Gerontology, 52*, 295–305. https://doi.org/10.1159/000094611

Zaragoza Scherman, A., Salgado, S., Shao, Z., & Berntsen, D. (2015). Life span distribution and content of positive and negative autobiographical memories across cultures. *Psychology of Consciousness: Theory, Research, and Practice, 2*, 475–489. https://doi.org/10.1037/cns0000070

Zhu, Y., & Zhang, L. (2002). An experimental study on the self-reference effect. *Science in China Series C: Life Sciences, 45*, 120–128. https://doi.org/10.1360/02yc9014

Zhu, Y., Zhang, L., Fan, J., & Han, S. (2007). Neural basis of cultural influence on self-representation. *Neuroimage, 34*, 1310–1316. https://doi.org/10.1016/j.neuroimage.2006.08.047

5

Assessment of Mood Disorders in Ethnic Minorities

VONETTA M. DOTSON, SHELLIE-ANNE LEVY, DEIRDRE M. O'SHEA, MOLLY E. MCLAREN, AND SARAH M. SZYMKOWICZ

INTRODUCTION

Mood disorders are common in the United Sates (Huang, Chung, Kroenke, & Spitzer, 2006; Jonas, Brody, Roper, & Narrow, 2003; Kessler, Chiu, Demler, & Walters, 2005). Multiple studies have reported 12-month prevalence rates near 10% (Bromet et al., 2011; Kessler et al., 2005), and lifetime prevalence estimates range from 11.5% to as high as nearly 20% (Bromet et al., 2011; Jonas et al., 2003). True prevalence rates may be higher given the evidence that only half of individuals with a mood disorder seek treatment (Kessler et al., 1999, 2001). Mood disorders have significant public health and personal costs related to loss of productivity and use of healthcare services (Huang et al., 2006). Major depression is the most common of the mood disorders, with 12-month and lifetime prevalence rates estimated at 8.6% and 29.9%, respectively (Kessler, Petukhova, Sampson, Zaslavsky, & Wittchen, 2012). Depression is also the leading cause of disability for both men and women aged 15 to 44 years (Mathers, 2008). However, less is known about whether prevalence rates of mood disorders differ by race/ethnicity in the United States. Understanding disparities in mood disorders is essential to increasing public awareness so that public health policies can better target at-risk groups.

Simning et al. (2011) reported a 5.25% 12-month prevalence and a 10.34% lifetime prevalence of depression in community-dwelling African Americans who took part in the National Survey of American Life study. A few studies have reported depression prevalence by ethnicity and are summarized in Table 5.1. Some studies show similar prevalence rates across ethnic groups, while others show ethnic differences, without a consensus about lower or higher depression prevalence in ethnic groups compared to non-Hispanic White individuals. It is

Table 5.1. SUMMARY OF STUDIES REPORTING ETHNIC DIFFERENCES IN MOOD DISORDER PREVALENCE

	Sample	European American	African American	Hispanic/Latino
Smith et al. (2006)[a]	Community sample	MDD = 7.38% Dysthymia = 1.88% Mania = 1.66%	MDD = 6.36% Dysthymia = 1.86% Mania =1.87%	MDD = 5.67% Dysthymia = 1.56% Mania = 1.54%
Huang, Chung, Kroenke, Delucchi, and Spitzer (2006)	Outpatients in medical clinics	MDD = 32.5%	MDD = 30.4%	MDD = 30.0%
Minsky et al. (2003)	Inpatients and outpatients	MDD = 18.33% Other depression = 17.86% Bipolar = 7.08%	MDD = 19.22% Other depression = 15.7% Bipolar = 3.9%	MDD = 28.4% Other depression = 16% Bipolar = 3.25%
Williams et al. (2003)[b]	Nationally representative community sample	MDD = 17%	MDD = 10.4%	N/A

Note. MDD = major depressive disorder.

[a]In Asian patients, MDD = 4.78%, dysthymia = 1.20%, and mania = 0.95%. In Native American patients, MDD = 12.38%, dysthymia = 2.99%, mania = 2.54%
[b]In Caribbean Blacks, MDD = 12.9%

apparent from these studies that mood disorder prevalence rates based on patient information or healthcare utilization data are higher than rates based on epidemiologic data. While this is of no surprise, the more severe and distressing the disorder, the more likely an individual will seek out health care (Huang et al., 2006). The assumption is that all ethnic groups have similar perceptions of the healthcare system and seek out services in accordance with the severity of the mood disorder. However, accumulating evidence shows that there is much variability in service utilization for mental health across ethnic groups. Specifically, all ethnic minorities showed significant lower use of mental health services compared to Whites (Jimenez, Cook, Bartels, & Alegría, 2013; Ojeda & McGuire, 2006; Wang et al., 2005). This is an important consideration with respect to understanding true prevalence rates of mood disorders in the United States.

ETHNIC DIFFERENCES IN CLINICAL PRESENTATION AND SYMPTOM PROFILES

The social concept of *culture* is broad and dynamic and involves interactions between individuals, communities, institutions, and ideologies (Bibeau, 1997; Eagleton, 2000; Kirmayer, 2001). Ethnicity refers to a shared common heritage within a group (Zenner, 1996). One's ethnicity and culture are widely accepted to influence variations in the expression of mood disorders, though there is consistency in the core features of depression across ethnic groups (Ballenger et al., 2001; Myers et al., 2002; Weissman et al., 1996). The link between ethnicity/culture and specific mood symptoms is not always simple given the variability in other factors, such as race, age, gender, education, and socioeconomic (Blazer, Landerman, Hays, Simonsick, & Saunders, 1998; U.S. Department of Health and Human Services [HHS], 2001). Other salient factors, such as stigma, mistrust, perceived discrimination, level of acculturation, and cultural assumptions of clinicians can also influence minority symptom mood profiles (HHS, 2001). In light of these confounds, we present a brief review of the literature regarding ethnic and cultural presentations of mood symptoms, including idioms of distress and culture-bound syndromes, with the caveat that clinicians must remain cautious about overgeneralizations and stereotypes due to the heterogeneity within ethnic and cultural groups.

African Americans

Early studies have found that African Americans are just as likely to experience depression as their White counterparts when adjusting for age, sex, education, and socioeconomic status (Comstock & Helsing, 1976; Eaton & Kessler, 1981; Roberts, Stevenson, & Breslow, 1981; Williams, 1986). Although depression rates are similar among Blacks and Whites, symptom presentation may differ. Indeed, in comparison to Whites, African Americans have been shown to be misdiagnosed and overdiagnosed with schizophrenia relative to affective disorders (Adebimpe, 1984; Bell & Mehta, 1980; Jones & Gray, 1986; Jones, Gray, & Parson, 1981; Raskin,

Crook, & Herman, 1975), which could be attributed to different symptom profiles of depression (Baker, Velli, Friedman, & Wiley, 1995; Carter, 1974; Scott & Gaitz, 1975) and/or diagnostic bias by the treating clinicians (Adebimpe, Klein, & Fried, 1981; Bell & Mehta, 1980; Jones & Gray, 1986).

Studies have shown that African Americans present with more somatic and vegetative symptoms of depression (e.g., weight loss, sleep disturbance, fatigue, appetite changes, etc.) rather than low mood, guilt, or suicidal thoughts (Carter, 1974; Fabrega, Mezzich, & Ulrich, 1988). However, these findings are not always consistent across studies and may be due to sample and methodological differences (Blazer et al., 1998; Gallo, Cooper-Patrick, & Lesikar, 1998; Wohl, Lesser, & Smith, 1997).

In a homogenous sample of clinically depressed individuals without comorbid psychiatric or medical illness, Wohl et al. (1997) compared the symptom presentation of age- and gender-matched African American and Whites. When assessed with the Hamilton Depression Rating Scale (Hamilton, 1960; Rhoades, 1983), results showed that African Americans were more likely to experience a morning diurnal variation, although there were no significant differences in other somatic or vegetative symptoms relative to Whites. In contrast, White participants reported more articulated severe depressed mood and agitation/anxiety.

Following the assessment of African American elders in San Antonio, Texas, and clinical interviews with other African American adults and elderly in Baltimore Maryland, Baker (2001) confirmed a developing clinical impression that depressed African Americans did not spontaneously report depressed mood. Thus, he proposed three alternative presentations of depression in African Americans that did not reflect standard diagnostic criteria, the stoic believer; the angry, "evil" one with a personality change; and the John Henry doer. The stoic believer denied feeling sad, blue, down, or helpless in the presence of significant health and social stressors, but believed his or her faith kept him or her going. The angry, "evil" one with a personality change presented to family members as angry, irritable, and prone to giving abrupt responses. Lastly, the John Henry doer continues to take on multiple tasks without regard for his or her health and often expresses that "they can get it done right" or complains that "others do not pull their weight." The John Henry doer was named after the legend of an African American who was famous for his ability to drive steel pins into railroad ties, but died from exhaustion after he challenged a newly invented steam-driven machine and won. In the course of his interviews, Baker made sure to follow up with questions about neurovegetative symptoms, changes in energy level, and continued pleasurable activities to confirm a diagnosis of a mood disorder.

Hispanic Americans

Hispanic Americans represent the largest minority group in the United States with the largest subgroups including Mexican Americans, Puerto Ricans, and Cuban Americans (Ramirez & de La Cruz, 2002). Correctly diagnosing mood disorders in Hispanic Americans is complicated by barriers, such as limited English proficiency in some Hispanic Americans, reduced health literacy, tendency toward somatization, and use of cultural idioms of distress (Lewis-Fernandez, Guarnaccia, Patel, Lizardi, & Diaz, 2005).

Early studies reveal that somatic presentations of depression are common in Hispanic subgroups (Canino, Rubio-Stipec, Canino, & Escobar, 1992; Escobar, Gomez, & Tuason, 1983; Kolody, Vega, Meinhardt, & Bensussen, 1986; Mezzich & Raab, 1980), but evidence suggesting that Hispanics are more likely to report more somatic complaints than Caucasians is inconsistent (Compton & Jones, 1991; Escobar, Rubio-Stipec, Canino, & Karno, 1989; Kirmayer & Young, 1998; Roberts, 1992). Many Hispanics express their somatization through cultural idioms of distress—linguistic and behavioral styles of expressing and experiencing psychosocial distress or illness that may not correlate with traditional diagnostic categories (Guarnaccia, Lewis-Fernandez, & Marano, 2003; Lewis-Fernandez et al., 2005). Latinos in the United States may describe their symptoms as *nervios* (nerves). *Nervios* can be a general state of vulnerability or a syndrome in response to significant life stressors; it may include headaches, *dolor de cerebro* (brain ache), irritability, stomach problems, sleep disturbance, tingling sensations, poor concentration, and/or *mareos* (dizziness). Another frequent idiom of distress among Latin American groups is *ataque de nervios* (attacks of nerves) that refers to discrete episodes of labile mood in response to stress (Lewis-Fernandez et al., 2005). These episodes may include shouting, crying, trembling, bouts of verbal and physical aggression, or a general sense of being out of control. *Ataques* may represent a normal expression of distress following an acute stressor, but can also rise to the level of symptoms reflecting a mood disorder.

Asian Americans

Asian Americans are a heterogeneous group representing more than 28 Asian groups with their own culture, language, and immigration history (Lin & Cheung, 1999). Despite the diversity, many Asian groups have been influenced by major traditions and belief systems, such as Buddhism and Confucianism, that have influenced conceptualizations of the self, health beliefs without a distinction between mind and body, and symptom presentation (Marsella, DeVos, & Hsu, 1985).

Although research is limited, a review of mental health issues for Asian Americans (Lin & Cheung, 1999) revealed that there is a long-standing view that Asians typically present their distress in the form of somatic symptoms. Indeed, somatization may represent a culturally approved "idiom of distress" or prescribed mode of communication with the majority of patients fully aware of their emotional problems and stressors; when directly queried, they can identify and report psychological symptoms (Lin & Cheung, 1999). Indeed, a study examining illness beliefs of depressed Chinese Americans (Yeung, Chang, Gresham, Nierenberg, & Fava, 2004) found that none spontaneously complained of depressed mood but reported psychological symptoms when asked directly and endorsed depressed mood on the Chinese Beck Depression Inventory. Additionally, the majority of the depressed Chinese Americans in the study sample (76%) reported chief complaints of somatic symptoms involving fatigue, insomnia, headache, cough, pain, dizziness, cervical dysfunction, and sexual dysfunction. Notably, the majority of the patients were unaware that they were suffering from depression, and some never heard of the diagnosis of major depression; one patient labeled his illness as post-traumatic stress and two patient labeled their symptoms as *shen jing bing* (insanity).

American Indians and Alaska Natives

American Indians and Alaska Natives are markedly heterogeneous group with 561 federally recognized tribes and over 200 indigenous languages spoken (Fleming, 1992). Among the various tribes, the concept of mental illness is viewed as supernatural possession, disharmony between inner and outer natural forces of the earth, expression of a special gift, or as the terminal phase of an illness (Kunitz, 1983; Thompson, Walker, & Silk-Walker, 1993). Categories of illness recognized in this group do not correspond well to traditional diagnostic criteria of mood disorders. Indeed, words such as *depressed* and *anxious* do not exist in some American Indian and Alaska Native languages (Manson, Shore, & Bloom, 1985). Manson et al. (1985) described two prominent culture bound syndromes *ghost sickness* and *heart-break syndrome*. Individuals with ghost sickness are believed to be possessed by the deceased and may have symptoms of fatigue, loss of appetite, dizziness, and recurrent nightmares. Heart-break syndrome is typically brought about by the death of a loved one and may cause death. Research on the symptom presentation of mood disorders in this minority group is scarce. However, screening of mood in two studies using the Center for Epidemiological Studies Depression Scale (CES-D) revealed no differential experience of somatic versus psychological symptoms of depression (Manson, Ackerson, Dick, Baron, & Fleming, 1990; Somervell et al., 1992).

ASSESSMENT AND INTERVIEW TOOLS

The accurate assessment of mood disorders in ethnic minorities requires consideration of ethnic differences in symptom presentation, as described in the previous section, as well as the impact of ethnicity and culture on the interview process (Aklin & Turner, 2006). Cultural biases might impact the clinician's diagnosis of psychiatric disorders in ethnic minorities. Moreover, the current diagnostic systems and assessment instruments were primarily developed in individuals from the majority culture (Adebimpe, 1981; Choi et al., 2012; Neighbors, Trierweiler, Ford, & Muroff, 2003). Nonetheless, some efforts have been made to determine the appropriateness of available measures for use in ethnic minorities.

Structured and Semi-Structured Interviews

Both clinicians and researchers use structured and semi-structured interviews to diagnose psychological disorders, including depression. When assessing ethnic minorities, these measures are considered to have the advantage of minimizing variability between patients and reducing the tendency to misdiagnose disorders due to clinician biases or a priori hypotheses (Zink, Lee, & Allen, 2015). There is evidence that, compared to open clinical interviews, structured and semi-structured interviews have better reliability and validity in ethnic minorities, but cross-cultural research on these interviews is limited.

The Mini International Neuropsychiatric Interview (MINI; Sheehan et al., 1998) and the Diagnostic Interview Schedule (DIS; Robins, Cottler, Bucholz, & Compton, 1995) are fully structured interviews that can be used in the assessment of depression. The MINI has the advantage of a short administration time—15 minutes compared to other interviews that take upwards of 2 hours to administer. Both interviews have been shown to have adequate validity and reliability (e.g., Sheehan et al., 1998; Wells, Burnam, Leake, & Robins, 1988). The MINI was found to accurately classify depression in low-income African Americans with HIV (Himelhoch et al., 2011) and the third edition of the *Diagnostic and Statistical Manual of Mental Disorders* (DSM-III; American Psychiatric Association, 1980) version of the DIS was found to have adequate validity in African Americans (Hendricks et al., 1983). Nonetheless, in a sample of predominantly African American homeless individuals presenting to mental health clinics, North et al. (1997) found that compared to an open clinical interview in two mental health clinics for homeless people, the DIS overdiagnosed depression. Both interviews have been translated into Spanish (Bobes, 1998; Bravo, Canino, Rubio-Stipec, & Woodbury-Farina, 1991). The MINI is available in multiple other languages, including French, Japanese, and Kurdish (Durieux-Paillard, Whitaker-Clinch, Bovier, & Eytan, 2006; Mitchell et al., 2011; Otsubo et al., 2005).

The Structured Clinical Interview for DSM disorders (First, Williams, Karg, & Spitzer, 2015) is perhaps the most widely used semi-structured interview for psychiatric disorders. The most recent edition, designed for the fifth edition of the DSM (DSM-V; American Psychiatric Association, 2013), includes research, clinician, and clinical trials versions. Other available semi-structured interviews include the World Health Organization's Schedules for Clinical Assessment in Neuropsychiatry (Schutzwohl, Kallert, & Jurjanz, 2007), which is the successor to the Present State Examination (Wing, Birley, Cooper, Graham, & Isaacs, 1967), and the Schedule for Affective Disorders and Schizophrenia (SADS; Endicott & Spitzer, 1978). There is a striking dearth of cross-cultural studies using these interviews; however, one study reported that suicidality is rated lower by the SADS for African Americans compared to Caucasians (Dilsaver, Chen, Swann, Shoaib, & Krajewski, 1994). The interviews have been translated into multiple languages and are used in research studies outside of the United States (Hodiamont, Peer, & Syben, 1987; Katz et al., 1988; Krisanaprakornkit, Paholpak, & Piyavhatkul, 2006; Torrens, Serrano, Astals, Perez-Dominguez, & Martin-Santos, 2004).

Depression Questionnaires

BECK DEPRESSION INVENTORY, SECOND EDITION

The Beck Depression Inventory–second edition (BDI-II) is a widely used 21-item self-report questionnaire assessing the type and severity of depressive symptoms over the 2 weeks prior to the date of assessment (Beck, Steer, & Brown, 1996). The BDI-II includes substantial revisions from the original BDI and focuses on symptoms of depression as they are outlined in the DSM (fourth edition, text revision; DSM-IV-TR; American Psychiatric Association, 2000). In the original standardization sample of mostly Caucasian psychiatric outpatients and college students (Beck et al., 1996), the BDI-II was shown to have high internal consistency ($\alpha = 0.91$) and high test-retest reliability ($r = 0.93$). A two-factor structure, consisting of cognitive-affective and somatic dimensions, emerged from the data.

In cross-cultural studies, the psychometric properties of the BDI-II have been most investigated in subgroups of African Americans. The BDI-II has been found to be a valid and reliable measure in African American primary care patients (Dutton et al., 2004), low-income medical outpatients (Grothe et al., 2005), recent suicide attempters (Joe, Woolley, Brown, Ghahramanlou-Holloway, & Beck, 2008), and rural women from north-central Florida (Gary & Yarandi, 2004). Several studies have found no ethnic differences in BDI-II scores across diverse college samples (Carmody, 2005; Hambrick et al., 2010; Sashidharan, Pawlow, & Pettibone, 2012; Whisman, Judd, Whiteford, & Gelhorn, 2013), including Caucasians, African Americans, Hispanics, and Asian Americans. Within all of these samples, various two- and three-factor structures have been described.

A Spanish version of the BDI-II has been developed (Beck et al., 1996). In a sample of Hispanic patients with end-stage renal disease (Penley, Wiebe, & Nwosu, 2003), the Spanish BDI-II was shown to have good internal consistency ($\alpha = 0.92$) and a two-factor structure (cognitive and somatic-affective; similar to the BDI-II factors found in a medical sample by Arnau, Meagher, Norris, & Bramson, 2001). A strong correlation ($r = 0.70$) and no differences in total scores were found when comparing the English and Spanish versions of the BDI-II. Moreover, comparable reliability and validity between the English and Spanish versions of the BDI-II have been reported in a nonclinical college sample (Wiebe & Penley, 2005), and the two-factor structure (cognitive and affective-somatic) was comparable to the factors found by Beck et al. (1996).

Taken together, the BDI-II is a widely used measure of depressive symptoms and has been validated for use in African American samples and in Hispanic samples using the Spanish BDI-II. No ethnic differences have been found for total depression scores; however, different factor structures have been described in different groups, suggesting that the heterogeneity of these subscale scores across diverse samples should be interpreted with caution and warrant further investigation. In addition, there is little to no research on the psychometric properties of the BDI-II in Asian Americans, indicating that use of this measure should be interpreted with caution until further research is conducted.

CENTER FOR EPIDEMIOLOGICAL STUDIES DEPRESSION SCALE

The CES-D is a 20-item self-report questionnaire assessing the frequency and severity of depressive symptoms over the previous week (Radloff, 1977). The original standardization sample for this scale was mostly Caucasians but did include a subsample of African Americans. Radloff (1977) reported an internal consistency reliability of α = 0.85 in the general population and α = 0.90 in a patient sample. The CES-D was shown to comprise four distinct factors: depressed affect, lack of positive affect, somatic and retarded activity, and interpersonal issues.

The psychometric properties and factor structure of the CES-D have been widely investigated in various ethnic and cultural groups. In African Americans, the CES-D has been shown to be a valid and reliable measure in low-income individuals (Rozario & Menon, 2010), middle-class women (Williams et al., 2007), single mothers (Atkins, 2014), and women with and without a history of cancer (Makambi, Williams, Taylor, Rosenberg, & Adams-Campbell, 2009). Several studies have replicated Radloff's (1977) original four-factor structure (Blazer et al., 1998; Makambi et al., 2009; Nguyen, Kitner-Triolo, Evans, & Zonderman, 2004; Rozario & Menon, 2010; Williams et al., 2007). However, other studies have found two-, three-, and alternative four-factor structures in different subgroups of African Americans (Atkins, 2014; McCallion & Kolomer, 2000; Rozario & Menon, 2010). A study by Callahan and Wolinsky (1994) found significant race/gender disparities in the CES-D factor structure in African-American and Caucasian older adults. These disparities were eliminated when five items from the CES-D were removed, suggesting that sociocultural differences in response styles may influence the expression of depressive symptoms on the CES-D.

Hispanic/Latino ethnic and cultural groups are quite diverse, and this reflects in the research on CES-D factor structures. In U.S.-born and Mexico-born Mexican Americans and U.S.-born non-Hispanic Whites, a four-factor model, similar to that reported by Radloff (1977), fit the data; however, the underlying construct of depression, as measured by the CES-D, was not identical across groups. Specifically, different somatic symptoms loaded onto different factors (Golding & Aneshensel, 1989). In large, diverse Hispanic samples, a three-factor structure fits best (Guarnaccia, Angel, & Worobey, 1989), while Mexican American and Puerto Rican women from single and couple-headed homes showed significant variability in their factor structures depending on their group status (Stroup-Benham, Lawrence, & Trevino, 1992). Radloff's (1977) four-factor structure was confirmed in urban Latina women when age and acculturation were included in the models; however, it was not confirmed in Latino men (Posner, Stewart, Marin, & Perez-Stable, 2001). Interestingly, there is diverging evidence with respect to the CES-D factor structure in English and Spanish speakers, with two- and three-factor models best fitting some data (Leykin, Torres, Aguilera, & Munoz, 2011) and Radloff's (1977) four-factor model fitting other data (Roberts, 1980).

With respect to Asian Americans, the CES-D was found to have good internal consistency (α = 0.86), sensitivity of 100%, and specificity of 76% in Chinese American women (Li & Hicks, 2010). Factor analytic studies in Asian American samples have confirmed that Radloff's (1977) four-factor structure best fits the data (Kuo, 1984; Ying, Lee, Tsai, Yeh, & Huang, 2000). Factor analyses of the CES-D have also been conducted in second generation Arab Americans (Amer, Awad, & Hovey, 2014) and American Indian college students (Beals, Manson, Keane, & Dick, 1991), with results suggesting that three-factor structures fit best.

Overall, the CES-D appears to be a valid and reliable measure of depression in various ethnic and cultural groups. However, the factor structure of the CES-D varies widely between groups, suggesting that the CES-D may not be measuring the same depression construct across these groups.

DEPRESSION ANXIETY STRESS SCALES

The Depression Anxiety Stress Scales (DASS) is a 42-item self-report measure assessing the frequency or severity of negative emotional symptoms over the past week (Lovibond & Lovibond, 1995). The DASS includes three subscales: depression, anxiety, and stress. A short-version of the DASS, containing only 21 items, is also available (DASS-21; Antony, Bieling, Cox, Enns, & Swinson, 1998). The original standardization sample consisted of first-year undergraduates at an Australian university (Lovibond & Lovibond, 1995). It was then normed on a sample aged 17 to 69 years from varying backgrounds. Scores were then checked for validity in various outpatient groups suffering from mood disorders. All three subscales were found to be reliable (Depression: $\alpha = 0.91$, Anxiety: $\alpha = 0.84$, Stress: $\alpha = 0.90$). Antony et al. (1998) reported comparable factor structures and scale content for the DASS and DASS-21. The internal consistency for each scale was high (DASS αs ≥ 0.92; DASS-21 αs ≥ 0.87). DASS and DASS-21 scales were found to have adequate concurrent validity, as they correlated moderately high with other anxiety and depression measures.

The psychometric properties of the DASS-21 have been investigated in a diverse sample of undergraduates (Norton, 2007). It was found to have acceptable internal consistency across Caucasian, African American, Hispanic, and Asian American individuals (αs for each subscale ≥ 0.74 across groups). The three-factor structure held across all four racial groups; however, the factor covariances were not invariant across groups, suggesting that the depression, anxiety, and stress constructs may be differentially expressed in each ethnic group. Taken together, the use of the DASS-21 in ethnic minorities should be taken with caution until further research is conducted, as the constructs that the subscales measure may not be identical across groups.

GERIATRIC DEPRESSION SCALE

The Geriatric Depression Scale (GDS) is a 30-item self-report or orally administered measure that assesses behavioral and affective symptoms of depression in the elderly using a yes/no format (Yesavage et al., 1982). A 15-item short-form (GDS-SF) has been developed and is an adequate substitute (Lesher & Berryhill, 1994). The original standardization sample consisted of community-dwelling older adults and depressed patients (Yesavage et al., 1982). The GDS was found to be a reliable measure with high internal consistency ($\alpha = 0.94$) and high test–retest reliability at a 1-week interval ($\alpha = 0.85$). The GDS-SF has been shown to be highly correlated with the GDS (rs = 0.89 and 0.84, respectively; Lesher & Berryhill, 1994; Sheikh & Yesavage, 1986) and has similar sensitivity and specificity.

In African American older adults (Pedraza, Dotson, Willis, Graff-Radford, & Lucas, 2009), the GDS-SF was found to have good internal consistency ($\alpha = 0.71$)

and adequate test–retest reliability after 15-month interval ($\alpha = 0.68$). In Japanese American older adults (Iwamasa, Hilliard, & Kost, 1998; Mui & Shibusawa, 2003), the GDS was found to be a reliable measure with high internal consistency (αs = 0.92 and 0.86, respectively). More research is needed, particularly with Hispanic and other Asian American older adults, to determine the utility of the GDS in cross-cultural clinical populations and in research.

PATIENT HEALTH QUESTIONNAIRE-9

The Patient Health Questionnaire-9 (PHQ-9) is a 9-item self-report measure that assesses the frequency of depressive symptoms over the previous 2 weeks (Kroenke, Spitzer, & Williams, 2001). It was designed for use in primary care settings and is based on the mood module from the Primary Care Evaluation of Mental Disorders questionnaire (PRIME-MD; Spitzer, Kroenke, & Williams, 1999). In the original Primary Care and Obstetrics-Gynecology Studies (Kroenke et al., 2001), a mostly Caucasian and female sample was used, though African Americans, Hispanics, and men were represented. Internal reliability estimates of $\alpha = 0.89$ and $\alpha = 0.86$ were reported, with excellent test–retest reliability.

In a diverse sample of primary care patients (Huang, Chung, Kroenke, Delucchi, & Spitzer, 2006), the PHQ-9 was found to have good internal consistency across Caucasian, African American, Chinese American, and Latino participants (αs ranging from 0.79 to 0.89 across groups). An exploratory principal components factor analysis found that a single factor that included all 9 items was derived for all four ethnic groups, although there were minor differences in the expression of individual symptoms across ethnicities. Another study found that in Latinas, both English and Spanish versions of the PHQ-9 showed good internal consistency ($\alpha = 0.84$ and 0.85, respectively; Merz, Malcarne, Roesch, Riley, & Sadler, 2011), and a single factor that included all 9 items was derived for both groups.

Taken together, these studies suggest that the PHQ-9 is an appropriate measure of depression for these groups; however, more research in diverse subgroups (e.g., Japanese, Korean, or Mexican Americans, Puerto Ricans) is needed to be certain of the psychometric properties of the PHQ-9 in these groups. Moreover, research on the sensitivity and specificity of the PHQ-9 in diverse clinical populations is needed.

Comments and Recommendations

There are several additional depression questionnaires that are used both clinically and in research. We were unable to find any literature examining the psychometric properties of the following scales in ethnic minorities: Clinically Useful Depression Outcome Scale (CUDOS; Zimmerman, Chelminski, McGlinchey, & Posternak, 2008), Hospital Anxiety and Depression Scale (HADS; Zigmond & Snaith, 1983), Hamilton Depression Rating Scale (HDRS or HAM-D; Hamilton, 1960), Inventory for Depressive Symptomatology (IDS; Rush et al., 1986; Rush, Gullion, Basco, Jarrett, & Trivedi, 1996), Montgomery-Asberg Depression Rating Scale (MADRS; Montgomery & Asberg, 1979), Quick Inventory of Depressive

Symptomatology (QIDS; Rush et al., 2003), and Zung Self-Rating Depression Scale (Zung SDS; Zung, 1965). While some of these scales have been translated into other languages and investigated in various ethnic and cultural groups outside of the United States, their use in diverse American samples has not been examined. The BDI-II and CES-D have received the most psychometric scrutiny, although many other scales are used in research and practice with diverse populations and have not been validated. More research on cut-off scores and sensitivity and specificity for specific ethnic and cultural groups would be beneficial for both physicians and researchers.

COGNITIVE AND NEUROIMAGING STUDIES OF DEPRESSION IN MINORITIES

Mood disorders and subthreshold mood symptoms have been associated with cognitive deficits in multiple domains, including processing speed, episodic memory, language, working memory, and executive functioning (Dotson et al., 2014; Sheline et al., 2006). Several studies have examined the impact of depressive symptoms on cognition in ethnic minorities, as summarized in the following discussion.

Cognitive Studies

Mood disorders and subthreshold mood symptoms have been associated with cognitive deficits in multiple domains, including processing speed, episodic memory, language, working memory, and executive functioning (Dotson et al., 2014; Sheline et al., 2006). Similar to the general population, depressive symptoms are associated with lower scores on memory, language, attention, and processing speed measures in African Americans (Hamilton et al., 2014). There is some evidence to suggest African Americans may be more susceptible to depression-related executive dysfunction and memory deficits compared to non-Hispanic Whites: A recent study reported that increased depressive symptoms were associated with worse performance on task-switching, inhibition, and episodic memory in African Americans, while in non-Hispanic Whites, depressive symptoms were only associated with reduced processing speed (Zahodne, Nowinski, Gershon, & Manly, 2014). Longitudinal studies in African Americans have shown that depression increases the odds of older African Americans developing mild cognitive impairment over 5 years (Unverzagt et al., 2011). Another study found that reduced positive affect was associated with declining global cognitive functioning including episodic memory and perceptual speed, and total depressive symptoms was associated with faster rates of decline in semantic and working memory in older African Americans (Turner, Capuano, Wilson, & Barnes, 2015). In contrast, another longitudinal study examining the impact of depressive symptoms on cognition over time did not find depression significantly predicted cognitive change over 2.5 years (Carmasin, Mast, Allaire, & Whitfield, 2014). Instead, authors found scores on the digit symbol task had a small significant prediction value on future depressive scores, suggesting declining cognition may increase the risk for developing depressive symptoms.

In Hispanic individuals, increased depressive symptoms, and dysphoria in particular, has been associated with poorer performance on tasks of immediate memory, visuospatial skills, and language (O'Bryant et al., 2011). As with studies examining the impact of depression on cognition longitudinally in African Americans, some studies have found that depressive symptoms predict changes in cognitive functioning over time (Raji, Reyes-Ortiz, Kuo, Markides, & Ottenbacher, 2007), while others report that decreased cognitive functioning predict future depressive symptoms (Perreira, Deeb-Sossa, Harris, & Bollen, 2005).

In Asian individuals, the association between increased depressive symptoms and cognition has been somewhat mixed. One study, examining the impact of late-onset depression in Chinese older adults found depression to be associated with poorer working memory, episodic memory, and processing speed (Tam & Lam, 2012). Another study found no correlations between depressive symptoms and Mini-Mental State Exam (MMSE) scores in Chinese individuals, while a third study reported depression significantly correlated with cognitive functioning as measured by the MMSE in older Korean women (Ji-Rong et al., 2010; Kim, 2009).

One known study to date has examined the impact of depressive symptoms on cognition in older Native Americans: Those who endorsed depressive symptoms performed more poorly on the Mattis Dementia Rating Scale conceptualization subscale (Verney et al., 2008).

Overall, research examining the impact of depressive symptoms on cognitive functioning in older minority populations is rather limited. Evidence suggests cognitive deficits do exist in these populations, and the type of deficits observed may be distinct from cognitive deficits seen in non-Hispanic White individuals with depression.

Neuroimaging Studies

Depression, especially in older adults, is associated with structural and functional changes in grey matter in frontolimbic pathways and increased white matter hyperintensities (Kirton, Resnick, Davatzikos, Kraut, & Dotson, 2014; McLaren et al., 2016; Naismith, Norrie, Mowszowski, & Hickie, 2012). The literature is lacking in studies examining brain mechanisms of depression in ethnic minorities. Given the difference in clinical presentation of depression across ethnic minorities and evidence that different symptom dimensions of depression are associated with distinct neurobiological mechanisms (Kirton et al., 2014; McLaren et al., 2016), we might expect to see unique patterns of underlying brain changes in depressed individuals of different ethnicities. For example, evidence that the risk for depression associated with vascular factors is higher in older African Americans compared to Caucasians (Reinlieb et al., 2014) suggests that white matter changes in the brain might be more prominent in older African Americans.

SUMMARY AND CONCLUSIONS

There is a clear need for more clinical, cognitive, and neuroimaging studies of depression in ethnic minorities. Research to date suggests that accurate assessment of depression in ethnic minorities can be limited by both biases on the part of the clinician as well as ethnic differences in the expression of mood disorders. While efforts have been made to determine the psychometrics of some depression measures in diverse samples, many instruments have yet to be examined cross-culturally, and existing studies are often limited to one ethnic minority group. Cognitive neuroscience studies that include higher numbers of ethnic minorities and that specifically compare different ethnic groups are needed to clarify whether or not brain mechanisms of depression differ across groups and, if so, what impact these differences may have on cognition, treatment response, and long-term outcomes.

REFERENCES

Adebimpe, V. R. (1981). Overview: White norms and psychiatric diagnosis of Black patients. *American Journal of Psychiatry, 138,* 279–285. https://doi.org/10.1176/ajp.138.3.279

Adebimpe, V. R. (1984). American Blacks and psychiatry. *Transcultural Psychiatric Research, 21,* 83–111. https://doi.org/10.1177/136346158402100201

Adebimpe, V. R., Klein, H. E., & Fried, J. (1981). Hallucinations and delusions in Black psychiatric patients. *Journal of the National Medical Association, 73,* 517–520.

Aklin, W. M., & Turner, S. M. (2006). Toward understanding ethnic and cultural factors in the interviewing process. *Psychotherapy (Chicago), 43,* 50–64. https://doi.org/10.1037/0033-3204.43.1.50

Amer, M. M., Awad, G. H., & Hovey, J. d. (2014). Evaluation of the CES-D factor structure in a sample of second-generation Arab-Americans. *International Journal of Culture and Mental Health, 7,* 46–58. https://doi.org/10.1080/17542863.2012.693514

American Psychiatric Association. (1980). *Diagnostic and statistical manual of mental disorders* (3rd ed.). Washington, DC: Author.

American Psychiatric Association. (2000). *Diagnostic and statistical manual of mental disorders* (4th ed., Text rev, ed.). Washington, DC: Author.

American Psychiatric Association. (2000). *Diagnostic and statistical manual of mental disorders* (5th ed.). Washington, DC: Author.

Antony, M. M., Bieling, P. J., Cox, B. J., Enns, M. W., & Swinson, R. P. (1998). Psychometric properties of the 42-item and 21-item versions of the Depression Anxiety Stress Scales in clinical groups and a community sample. *Psychological Assessment, 10,* 176–181. https://doi.org/10.1037/1040-3590.10.2.176

Arnau, R. C., Meagher, M. W., Norris, M. P., & Bramson, R. (2001). Psychometric evaluation of the Beck Depression Inventory-II with primary care medical patients. *Health Psychology, 20,* 112–119. https://doi.org/10.1037/0278-6133.20.2.112

Atkins, R. (2014). Validation of the Center for Epidemiologic Studies Depression Scale in Black single mothers. *Journal of Nursing Measurement, 22,* 511–524. https://doi.org/10.1891/1061-3749.22.3.511

Baker, F. M. (2001). Diagnosing depression in African Americans. *Community Mental Health Journal, 37,* 31–38. https://doi.org/10.1023/a:1026540321366

Baker, F. M., Velli, S. A., Friedman, J., & Wiley, C. (1995). Screening tests for Depression in older Black vs. White patients. *American Journal of Geriatric Psychiatry, 3,* 43–51. https://doi.org/https://doi.org/10.1097/00019442-199524310-00006

Ballenger, J. C., Davidson, J. R., Lecrubier, Y., Nutt, D. J., Kirmayer, L. J., Lepine, J. P., ... Ono, Y. (2001). Consensus statement on transcultural issues in depression and anxiety from the International Consensus Group on Depression and Anxiety. *Journal of Clinical Psychiatry, 13,* 47–55.

Beals, J., Manson, S. M., Keane, E. M., & Dick, R. W. (1991). Factorial structure of the Center for the Epidemiologic Studies—Depression Scale among American Indian college students. *Psychological Assessment, 3,* 623–627. https://doi.org/10.1037/1040-3590.3.4.623

Beck, A. T., Steer, R. A., & Brown, G. K. (1996). *Manual for the Beck Depression Inventory-II.* San Antonio, TX: Psychological Corporation.

Bell, C. C., & Mehta, H. (1980). The misdiagnosis of Black patients with manic depressive illness. *Journal of the National Medical Association, 72,* 141–145.

Bibeau, G. (1997). Cultural psychiatry in a creolizing world: Questions for a new research agenda. *Transcultural Psychiatry, 34,* 9–41. https://doi.org/10.1177/136346159703400102

Blazer, D. G., Landerman, L. R., Hays, J. C., Simonsick, E. M., & Saunders, W. B. (1998). Symptoms of depression among community-dwelling elderly African-American and White older adults. *Psychological Medicine, 28,* 1311–1320. https://doi.org/10.1017/S0033291798007648

Bobes, J. (1998). A Spanish validation study of the mini international neuropsychiatric interview. *European Psychiatry, 13,* 198s–199s. https://doi.org/10.1016/S0924-9338(99)80240-5

Bravo, M., Canino, G. J., Rubio-Stipec, M., & Woodbury-Farina, M. (1991). A cross-cultural adaptation of a psychiatric epidemiologic instrument: The diagnostic interview schedule's adaptation in Puerto Rico. *Culture, Medicine and Psychiatry, 15,* 1–18. https://doi.org/10.1007/BF00050825

Bromet, E., Andrade, L. H., Hwang, I., Sampson, N. A., Alonso, J., de Girolamo, G., ... Kessler, R. C. (2011). Cross-national epidemiology of DSM-IV major depressive episode. *BMC Medicine, 9,* 90. https://doi.org/10.1186/1741-7015-9-90

Callahan, C. M., & Wolinsky, F. D. (1994). The effect of gender and race on the measurement properties of the CES-D in older adults. *Medical Care, 32,* 341–356. https://doi.org/10.1097/00005650-199404000-00003

Canino, I. A., Rubio-Stipec, M., Canino, G., & Escobar, J. I. (1992). Functional somatic symptoms: A cross-ethnic comparison. *American Journal of Orthopsychiatry, 62,* 605–612. https://doi.org/10.1037/h0079376

Carmasin, J. S., Mast, B. T., Allaire, J. C., & Whitfield, K. E. (2014). Vascular risk factors, depression, and cognitive change among African American older adults. *International Journal of Geriatric Psychiatry, 29,* 291–298. https://doi.org/10.1002/gps.4007

Carmody, D. P. (2005). Psychometric characteristics of the Beck Depression Inventory-II with college students of diverse ethnicity. *International Journal of Psychiatry in Clinical Practice, 9,* 22–28. https://doi.org/10.1080/13651500510014800

Carter, J. H. (1974). Recognizing psychiatric symptoms in Black Americans. *Geriatrics, 29,* 95–99.

Choi, M. R., Eun, H. J., Yoo, T. P., Yun, Y., Wood, C., Kase, M., ... Yang, J. C. (2012). The effects of sociodemographic factors on psychiatric diagnosis. *Psychiatry Investigation, 9,* 199–208. https://doi.org/10.4306/pi.2012.9.3.199

Compton, A. B., & Jones, F. D. (1991). Clinical features of young adult Hispanic psychiatric in-patients: The so-called "Puerto Rican syndrome." *Military Medicine, 156,* 351–354.

Comstock, G. W., & Helsing, K. J. (1976). Symptoms of depression in two communities. *Psychological Medicine, 6,* 551–563. https://doi.org/10.1017/S0033291700018171

Dilsaver, S. C., Chen, Y. W., Swann, A. C., Shoaib, A. M., & Krajewski, K. J. (1994). Suicidality in patients with pure and depressive mania. *American Journal of Psychiatry, 151,* 1312–1315. https://doi.org/10.1176/ajp.151.9.1312

Dotson, V. M., Szymkowicz, S. M., Kirton, J. W., McLaren, M. E., Green, M. L., & Rohani, J.Y. (2014). Unique and interactive effect of anxiety and depressive symptoms on cognitive and brain function in young and older adults. *Journal of Depression & Anxiety, Suppl 1,* 003. https://doi.org/10.4172/2167-1044.s1-003

Durieux-Paillard, S., Whitaker-Clinch, B., Bovier, P. A., & Eytan, A. (2006). Screening for major depression and posttraumatic stress disorder among asylum seekers: Adapting a standardized instrument to the social and cultural context. *Canadian Journal of Psychiatry, 51,* 587–597. https://doi.org/10.1177/070674370605100907

Dutton, G. R., Grothe, K. B., Jones, G. N., Whitehead, D., Kendra, K., & Brantley, P. J. (2004). Use of the Beck Depression Inventory-II with African American primary care patients. *General Hospital Psychiatry, 26,* 437–442. https://doi.org/10.1016/j.genhosppsych.2004.06.002

Eagleton, T. (2000). *The idea of culure.* Oxford, England: Blackwell.

Eaton, W. W., & Kessler, L. G. (1981). Rates of symptoms of depression in a national sample. *American Journal of Epidemiology, 114,* 528–538. https://doi.org/10.1093/oxfordjournals.aje.a113218

Endicott, J., & Spitzer, R. L. (1978). A diagnostic interview: The schedule for affective disorders and schizophrenia. *Archives Of General Psychiatry, 35,* 837–844. https://doi.org/10.1001/archpsyc.1978.01770310043002

Escobar J., Gomez, J., Tuason V. (1983). Depressive phenomenology in North and South American patients. *American Journal of Psychiatry, 140,* 47–51. https://doi.org/10.1176/ajp.140.1.47

Escobar, J. I., Rubio-Stipec, M., Canino, G., & Karno, M. (1989). Somatic symptom index (SSI): A new and abridged somatization construct. Prevalence and epidemiological correlates in two large community samples. *Journal of Nervous and Mental Disease, 177,* 140–146. https://doi.org/10.1097/00005053-198903000-00003

Fabrega, H., Jr., Mezzich, J., & Ulrich, R. F. (1988). Black–White differences in psychopathology in an urban psychiatric population. *Comprehensive Psychiatry, 29,* 285–297. https://doi.org/10.1016/0010-440X(88)90051-X

First, M. B., Williams, J. B. W., Karg, R. S., & Spitzer, R. L. (2015). *Structured clinical interview for DSM-5—Research version.* Arlington, VA: American Psychiatric Association.

Fleming, C. M. (1992). American Indians and Alaska Natives: Changing societies past and present. In M. A. Orlandi, R. Weston & L. G. Epstein (Eds.), *Cultural competence for evaluators: A guide for alcohol and other drug abuse prevention practitioners working with ethnic/racial communities* (pp. 147–171). Rockville, MD: U.S. Department of Health & Human Services.

Gallo, J. J., Cooper-Patrick, L., & Lesikar, S. (1998). Depressive symptoms of Whites and African Americans aged 60 years and older. *The Journals of Gerontology: Series B. Psychological Sciences and Social Sciences, 53,* 277–286. https://doi.org/10.1093/geronb/53B.5.P277

Gary, F. A., & Yarandi, H. N. (2004). Depression among southern rural African American women: A factor analysis of the Beck Depression Inventory-II. *Nursing Research, 53,* 251–259. https://doi.org/10.1097/00006199-200407000-00008

Golding, J. M., & Aneshensel, C. S. (1989). Factor structure of the Center for Epidemiologic Studies Depression Scale among Mexican Americans and non-Hispanic Whites. *Psychological Assessment, 1,* 163–168. https://doi.org/10.1037/1040-3590.1.3.163

Grothe, K. B., Dutton, G. R., Jones, G. N., Bodenlos, J., Ancona, M., & Brantley, P. J. (2005). Validation of the Beck Depression Inventory-II in a low-income African American sample of medical outpatients. *Psychological Assessment, 17,* 110–114. https://doi.org/10.1037/1040-3590.17.1.110

Guarnaccia, P. J., Angel, R., & Worobey, J. L. (1989). The factor structure of the CES-D in the Hispanic Health and Nutrition Examination Survey: The influences of ethnicity, gender and language. *Social Science & Medicine, 29,* 85–94. https://doi.org/10.1016/0277-9536(89)90131-7

Guarnaccia, P. J., Lewis-Fernandez, R., & Marano, M. R. (2003). Toward a Puerto Rican popular nosology: Nervios and ataque de nervios. *Culture, Medicine and Psychiatry, 27,* 339–366.

Hambrick, J. P., Rodebaugh, T. L., Balsis, S., Woods, C. M., Mendez, J. L., & Heimberg, R. G. (2010). Cross-ethnic measurement equivalence of measures of depression, social anxiety, and worry. *Assessment, 17,* 155–171. https://doi.org/10.1177/1073191109350158

Hamilton, J. L., Brickman, A. M., Lang, R., Byrd, G. S., Haines, J. L., Pericak-Vance, M. A., & Manly, J. J. (2014). Relationship between depressive symptoms and cognition in older, non-demented african americans. *Journal of the International Neuropsychological Society, 20,* 756–763. https://doi.org/10.1017/S1355617714000423

Hamilton, M. (1960). A rating scale for depression. *Journal of Neurology, Neurosurgery, and Psychiatry, 23,* 56–62. https://doi.org/10.1136/jnnp.23.1.56

Hendricks, L. E., Bayton, J. A., Collins, J. L., Mathura, C. B., McMillan, S. R., & Montgomery, T. A. (1983). The NIMH Diagnostic Interview Schedule: A test of its validity in a population of Black adults. *Journal of the National Medical Association, 75,* 667–671.

Himelhoch, S., Mohr, D., Maxfield, J., Clayton, S., Weber, E., Medoff, D., & Dixon, L. (2011). Feasibility of telephone-based cognitive behavioral therapy targeting major depression among urban dwelling African-American people with co-occurring HIV. *Psychology, Health & Medicine, 16,* 156–165. https://doi.org/10.1080/13548506.2010.534641

Hodiamont, P., Peer, N., & Syben, N. (1987). Epidemiological aspects of psychiatric disorder in a Dutch health area. *Psychological Medicine, 17,* 495–505. https://doi.org/10.1017/S0033291700025058

Huang, F. Y., Chung, H., Kroenke, K., Delucchi, K. L., & Spitzer, R. L. (2006). Using the Patient Health Questionnaire-9 to measure depression among racially and ethnically diverse primary care patients. *Journal of General Internal Medicine, 21,* 547–552. https://doi.org/10.1111/j.1525-1497.2006.00409.x

Huang, F. Y., Chung, H., Kroenke, K., & Spitzer, R. L. (2006). Racial and ethnic differences in the relationship between depression severity and functional status. *Psychiatr Services, 57,* 498–503. https://doi.org/10.1176/ps.2006.57.4.498

Iwamasa, G. Y., Hilliard, K. M., & Kost, C. R. (1998). The Geriatric Depression Scale and Japanese American older adults. *Clinical Gerontologist, 19,* 13–24. https://doi.org/10.1300/J018v19n03_03

Ji-Rong, Y. M. D., Bi-Rong, D. M. D., Chang-Quang, H. M. D., Hong-Mei, W. M. D., Yan-Ling, Z. M. D., Qing-Xiu, L. K. S., . . . Qi-Yuan, Y. M. S. (2010). Cognitive impairment and depression among Chinese nonagenarians/centenarians. *American Journal of Geriatric Psychiatry, 18,* 297–304. https://doi.org/10.1097/JGP.0b013e3181d143bc

Jimenez, D. E., Cook, B., Bartels, S. J., & Alegría, M. (2013). Disparities in mental health service use of racial and ethnic minority elderly adults. *Journal of the American Geriatrics Society, 61,* 18–25. https://doi.org/10.1111/jgs.12063

Joe, S., Woolley, M. E., Brown, G. K., Ghahramanlou-Holloway, M., & Beck, A. T. (2008). Psychometric properties of the Beck Depression Inventory-II in low-income, African American suicide attempters. *Journal of Personality Assessment, 90,* 521–523. https://doi.org/10.1080/00223890802248919

Jonas, B. S., Brody, D., Roper, M., & Narrow, W. E. (2003). Prevalence of mood disorders in a national sample of young American adults. *Social Psychiatry and Psychiatric Epidemiology, 38,* 618–624. https://doi.org/10.1007/s00127-003-0682-8

Jones, B. E., & Gray, B. A. (1986). Problems in diagnosing schizophrenia and affective disorders among Blacks. *Hospital & Community Psychiatry, 37,* 61–65. https://doi.org/10.1176/ps.37.1.61

Jones, B. E., Gray, B. A., Parson, E. B. (1981). Manic-depressive illness among poor urban Blacks. *American Journal of Psychiatry, 138,* 654–657. https://doi.org/10.1176/ajp.138.5.654

Katz, M. M., Marsella, A., Dube, K. C., Olatawura, M., Takahashi, R., Nakane, Y., . . . Jablensky, A. (1988). On the expression of psychosis in different cultures: Schizophrenia in an Indian and in a Nigerian community. *Culture, Medicine and Psychiatry, 12,* 331–355. https://doi.org/10.1007/BF00051973

Kessler, R. C., Berglund, P. A., Bruce, M. L., Koch, J. R., Laska, E. M., Leaf, P. J., . . . Wang, P. S. (2001). The prevalence and correlates of untreated serious mental illness. *Health Services Research, 36*(6 Pt 1), 987–1007.

Kessler, R. C., Chiu, W. T., Demler, O., & Walters, E. E. (2005). Prevalence, severity, and comorbidity of 12-month DSM-IV disorders in the National Comorbidity Survey Replication. *Archives Of General Psychiatry, 62,* 617–627. https://doi.org/10.1001/archpsyc.62.6.617

Kessler, R. C., Petukhova, M., Sampson, N. A., Zaslavsky, A. M., & Wittchen, H. U. (2012). Twelve-month and lifetime prevalence and lifetime morbid risk of anxiety and mood disorders in the United States. *International Journal of Methods in Psychiatric Research, 21,* 169–184. https://doi.org/10.1002/mpr.1359

Kessler, R. C., Zhao, S., Katz, S. J., Kouzis, A. C., Frank, R. G., Edlund, M., & Leaf, P. (1999). Past-year use of outpatient services for psychiatric problems in the National Comorbidity Survey. *American Journal of Psychiatry, 156,* 115–123. https://doi.org/10.1176/ajp.156.1.115

Kim, O. (2009). Cognitive impairment and depression in community-dwelling older Korean women. *Psychological Reports, 105,* 569–574. https://doi.org/10.2466/pr0.105.2.569-574

Kirmayer, L. J. (2001). Sapir's vision of culture and personality. *Psychiatry: Interpersonal and Biological Processes, 64,* 23–31. https://doi.org/10.1521/psyc.64.1.23.18233

Kirmayer, L. J., & Young, A. (1998). Culture and somatization: Clinical, epidemiological, and ethnographic perspectives. *Psychosomatic Medicine, 60,* 420–430. https://doi.org/10.1097/00006842-199807000-00006

Kirton, J. W., Resnick, S. M., Davatzikos, C., Kraut, M. A., & Dotson, V. M. (2014). Depressive symptoms, symptom dimensions, and White matter lesion volume in older adults: A longitudinal study. *American Journal of Geriatric Psychiatry, 22,* 1469–1477. https://doi.org/10.1016/j.jagp.2013.10.005

Kolody, B., Vega, W., Meinhardt, K., & Bensussen, G. (1986). The correspondence of health complaints and depressive symptoms among Anglos and Mexican-Americans. *Journal of Nervous and Mental Disease, 174,* 221–228. https://doi.org/10.1097/00005053-198604000-00005

Krisanaprakornkit, T., Paholpak, S., & Piyavhatkul, N. (2006). The validity and reliability of the WHO Schedules for Clinical Assessment in Neuropsychiatry (SCAN Thai Version): Mood disorders section. *Journal of the Medical Association of Thailand, 89,* 205–211.

Kroenke, K., Spitzer, R. L., & Williams, J. B. (2001). The PHQ-9: Validity of a brief depression severity measure. *Journal of General Internal Medicine, 16,* 606–613. https://doi.org/10.1046/j.1525-1497.2001.016009606.x

Kunitz, S. J. (1983). *Disease change and the role of medicine: The Navajo experience.* Berkeley, CA: University of California Press.

Kuo, W. H. (1984). Prevalence of depression among Asian-Americans. *Journal of Nervous and Mental Disease, 172,* 449–457. https://doi.org/10.1097/00005053-198408000-00002

Lesher, E. L., & Berryhill, J. S. (1994). Validation of the Geriatric Depression Scale—Short form among inpatients. *Journal of Clinical Psychology, 50,* 256–260. https://doi.org/10.1002/1097-4679(199403)50:2<256::AID-JCLP2270500218>3.0.CO;2-E

Lewis-Fernandez, R., Guarnaccia, P. J., Patel, S., Lizardi, D., & Diaz, N. (2005). Ataque de nervios: Anthropological, epidemiological, and clinical dimensions of a cultural syndrome. In A. M. Georgiopoulos & J. F. Rosenbaum (Eds.), *Perspectives in cross cultural psychiatry* (Vol. 63–85). Philadelphia, PA: Lippincott, Williams & Wilkins.

Leykin, Y., Torres, L. D., Aguilera, A., & Munoz, R. F. (2011). Factor structure of the CES-D in a sample of Spanish- and English-speaking smokers on the Internet. *Psychiatry Research, 185,* 269–274. https://doi.org/10.1016/j.psychres.2010.04.056

Li, Z., & Hicks, M. H. (2010). The CES-D in Chinese American women: Construct validity, diagnostic validity for major depression, and cultural response bias. *Psychiatry Research, 175,* 227–232. https://doi.org/10.1016/j.psychres.2009.03.007

Lin, K. M., & Cheung, F. (1999). Mental health issues for Asian Americans. *Psychiatric Services, 50,* 774–780. https://doi.org/10.1176/ps.50.6.774

Lovibond, S. H., & Lovibond, P. F. (1995). *Manual for the depression anxiety stress scales* (2nd ed.). Sydney, Australia: Psychology Foundation.

Makambi, K. H., Williams, C. D., Taylor, T. R., Rosenberg, L., & Adams-Campbell, L. L. (2009). An assessment of the CES-D scale factor structure in Black women: The Black Women's Health Study. *Psychiatry Research, 168,* 163–170. https://doi.org/10.1016/j.psychres.2008.04.022

Manson, S. M., Ackerson, L. M., Dick, R. W., Baron, A. E., & Fleming, C. M. (1990). Depressive symptoms among American Indian adolescents: Psychometric characteristics of the Center for Epidemiologic Studies Depression Scale (CES-D). *Psychological Assessment, 2,* 231–237. https://doi.org/10.1037/1040-3590.2.3.231

Manson, S. M., Shore, J. H., & Bloom, J. D. (1985). The depressive experience in American Indian communities: A challenge for psychiatric theory and diagnosis. In A. Kleinman & B. Good (Eds.), *Culture and depression* (pp. 331–368). Berkeley, CA: University of California Press.

Marsella, A. J., DeVos, G., & Hsu, F. (1985). *Culture and self: Asian and Western perspectives.* New York, NY: Tavistock.

Mathers, C., Fat, D. M., Boerma, J. T., & World Health Organization. (2008). *The global burden of disease: 2004 update.* Geneva, Switzerland: World Health Organization.

McCallion, P., & Kolomer, S. R. (2000). Depressive symptoms among African American caregiving grandmothers: The factor structure of the CES-D. *Journal of Mental Health & Aging, 6,* 325–338.

McLaren, M. E., Szymkowicz, S. M., O/'Shea, A., Woods, A. J., Anton, S. D., & Dotson, V. M. (2016). Dimensions of depressive symptoms and cingulate volumes in older adults. *Translational Psychiatry, 6,* e788. https://doi.org/10.1038/tp.2016.49

Merz, E. L., Malcarne, V. L., Roesch, S. C., Riley, N., & Sadler, G. R. (2011). A multigroup confirmatory factor analysis of the Patient Health Questionnaire-9 among English- and Spanish-speaking Latinas. *Cultural Diversity and Ethnic Minority Psychology, 17,* 309–316. https://doi.org/10.1037/a0023883

Mezzich, J. E., & Raab, E. S. (1980). Depressive symptomatology across the Americas. *Archives of General Psychiatry, 37,* 818–823. https://doi.org/10.1001/archpsyc.1980.01780200096012

Minsky, S., Vega, W., Miskimen, T., Gara, M., & Escobar, J. (2003). Diagnostic patterns in Latino, African American, and European American psychiatric patients. *Arch Gen Psychiatry, 60*(6), 637–644. doi:10.1001/archpsyc.60.6.637

Mitchell, A. J., Smith, A. B., Al-salihy, Z., Rahim, T. A., Mahmud, M. Q., & Muhyaldin, A. S. (2011). Redefining diagnostic symptoms of depression using Rasch analysis: Testing an item bank suitable for DSM-V and computer adaptive testing. *Australian and New Zealand Journal of Psychiatry, 45,* 846–852. https://doi.org/10.3109/00048674.2011.596477

Montgomery, S. A., & Asberg, M. (1979). A new depression scale designed to be sensitive to change. *British Journal of Psychiatry, 134,* 382–389. https://doi.org/10.1192/bjp.134.4.382

Mui, A. C., & Shibusawa, T. (2003). Japanese American elders and the Geriatric Depression Scale. *Clinical Gerontologist, 26*(3–4), 91–104. https://doi.org/10.1300/J018v26n03_08

Myers, H. F., Lesser, I., Rodriguez, N., Mira, C. B., Hwang, W. C., Camp, C., . . . Wohl, M. (2002). Ethnic differences in clinical presentation of depression in adult women. *Cultural Diversity and Ethnic Minority Psychology, 8,* 138–156. https://doi.org/10.1037/1099-9809.8.2.138

Naismith, S. L., Norrie, L. M., Mowszowski, L., & Hickie, I. B. (2012). The neurobiology of depression in later-life: Clinical, neuropsychological, neuroimaging and pathophysiological features. *Progress in Neurobiology, 98,* 99–143. https://doi.org/10.1016/j.pneurobio.2012.05.009

Neighbors, H. W., Trierweiler, S. J., Ford, B. C., & Muroff, J. R. (2003). Racial differences in DSM diagnosis using a semi-structured instrument: The importance of clinical judgment in the diagnosis of African Americans. *Journal of Health and Social Behavior, 44,* 237–256. https://doi.org/10.2307/1519777

Nguyen, H. T., Kitner-Triolo, M., Evans, M. K., & Zonderman, A. B. (2004). Factorial invariance of the CES-D in low socioeconomic status African Americans compared with a nationally representative sample. *Psychiatry Research, 126,* 177–187. https://doi.org/10.1016/j.psychres.2004.02.004

North, C. S., Pollio, D. E., Thompson, S. J., Ricci, D. A., Smith, E. M., & Spitznagel, E. L. (1997). A comparison of clinical and structured interview diagnoses in a homeless mental health clinic. *Community Mental Health Journal, 33,* 531–543. https://doi.org/10.1023/A:1025052720325

Norton, P. J. (2007). Depression Anxiety and Stress Scales (DASS-21): Psychometric analysis across four racial groups. *Anxiety, Stress & Coping, 20,* 253–265. https://doi.org/10.1080/10615800701309279

O'Bryant, S. E., Hall, J. R., Cukrowicz, K. C., Edwards, M., Johnson, L. A., Lefforge, D., ... Dentino, A. (2011). The differential impact of depressive symptom clusters on cognition in a rural multi-ethnic cohort: A Project FRONTIER study. *International Journal of Geriatric Psychiatry, 26,* 199–205. https://doi.org/10.1002/gps.2514

Ojeda, V. D., & McGuire, T. G. (2006). Gender and racial/ethnic differences in use of outpatient mental health and substance use services by depressed adults. *Psychiatric Quarterly, 77,* 211–222. https://doi.org/10.1007/s11126-006-9008-9

Otsubo, T., Tanaka, K., Koda, R., Shinoda, J., Sano, N., Tanaka, S., . . . Kamijima, K. (2005). Reliability and validity of Japanese version of the Mini-International Neuropsychiatric Interview. *Psychiatry and Clinical Neurosciences, 59,* 517–526. https://doi.org/10.1111/j.1440-1819.2005.01408.x

Pedraza, O., Dotson, V. M., Willis, F. B., Graff-Radford, N. R., & Lucas, J. A. (2009). Internal consistency and test-retest stability of the Geriatric Depression Scale–Short Form in African American older adults. *Journal of Psychopathology and Behavioral Assessment, 31,* 412–416. https://doi.org/10.1007/s10862-008-9123-z

Penley, J. A., Wiebe, J. S., & Nwosu, A. (2003). Psychometric properties of the Spanish Beck Depression Inventory-II in a medical sample. *Psychological Assessment, 15,* 569–577. https://doi.org/10.1037/1040-3590.15.4.569

Perreira, K. M., Deeb-Sossa, N., Harris, K. M., & Bollen, K. (2005). What are we measuring? An evaluation of the CES-D across race/ethnicity and immigrant generation. *Social Forces, 83,* 1567–1602. https://doi.org/10.1353/sof.2005.0077

Posner, S. F., Stewart, A. L., Marin, G., & Perez-Stable, E. J. (2001). Factor variability of the Center for Epidemiological Studies Depression Scale (CES-D) among urban Latinos. *Ethnicity & Health, 6,* 137–144. https://doi.org/10.1080/13557850120068469

Radloff, L. S. (1977). The CES-D scale: A self-report depression scale for research in the general population. *Applied Psychological Measurement, 1,* 385–401. https://doi.org/10.1177/014662167700100306

Raji, M. A., Reyes-Ortiz, C. A., Kuo, Y.-F., Markides, K. S., & Ottenbacher, K. J. (2007). Depressive Mexican Americans. *Journal of Geriatric Psychiatry* symptoms and cognitive change in older *and Neurology, 20,* 145–152. https://doi.org/10.1177/0891988707303604

Ramirez, R. R., & de La Cruz, G. P. (2002). The Hispanic population in the United States: Population characteristics: March 2002. *U.S. Census Bureau Current Population Reports.* Retrieved from https://www.census.gov/prod/2003pubs/p20-545.pdf

Raskin, A., Crook, T. H., & Herman, K. D. (1975). Psychiatric history and symptom differences in Black and White depressed patients. *Journal of Consulting and Clinical Psychology, 43,* 73–80. https://doi.org/10.1037/h0076322

Reinlieb, M. E., Persaud, A., Singh, D., Garcon, E., Rutherford, B. R., Pelton, G. H., ... Sneed, J. R. (2014). Vascular depression: Overrepresented among African Americans? *International Journal of Geriatric Psychiatry, 29,* 470–477. https://doi.org/10.1002/gps.4029

Rhoades, H. M., & Overall, J. E. (1983). The Hamilton Depression Scale: Factor scoring and profile classification. *Psychopharmacology Bulletin, 19,* 91–96.

Roberts, R. E. (1980). Reliability of the CES-D Scale in different ethnic contexts. *Psychiatry Research, 2,* 125–134. https://doi.org/10.1016/0165-1781(80)90069-4

Roberts, R. E. (1992). Manifestation of depressive symptoms among adolescents. A comparison of Mexican Americans with the majority and other minority populations. *Journal of Nervous and Mental Disease, 180,* 627–633. https://doi.org/10.1097/00005053-199210000-00003

Roberts, R. E., Stevenson, J. M., & Breslow, L. (1981). Symptoms of depression among Blacks and Whites in an urban community. *Journal of Nervous and Mental Disease, 169,* 774–779. https://doi.org/10.1097/00005053-198112000-00005

Robins, L., Cottler, L., Bucholz, K., & Compton, W. (1995). *Diagnostic interview schedule for DSM-IV (DIS-IV).* St. Louis, MO: Washington University, School of Medicine, Department of Psychiatry.

Rozario, P. A., & Menon, N. (2010). An examination of the measurement adequacy of the CES-D among African American women family caregivers. *Psychiatry Research, 179,* 107–112. https://doi.org/10.1016/j.psychres.2010.06.022

Rush, A. J., Giles, D. E., Schlesser, M. A., Fulton, C. L., Weissenburger, J., & Burns, C. (1986). The Inventory for Depressive Symptomatology (IDS): Preliminary findings. *Psychiatry Research, 18,* 65–87. https://doi.org/10.1016/0165-1781(86)90060-0

Rush, A. J., Gullion, C. M., Basco, M. R., Jarrett, R. B., & Trivedi, M. H. (1996). The Inventory of Depressive Symptomatology (IDS): Psychometric properties. *Psychological Medicine, 26,* 477–486. https://doi.org/10.1017/S0033291700035558

Rush, A. J., Trivedi, M. H., Ibrahim, H. M., Carmody, T. J., Arnow, B., Klein, D. N., ... Keller, M. B. (2003). The 16-Item Quick Inventory of Depressive Symptomatology (QIDS), clinician rating (QIDS-C), and self-report (QIDS-SR): A psychometric evaluation in patients with chronic major depression. *Biological Psychiatry, 54,* 573–583. https://doi.org/10.1016/S0006-3223(02)01866-8

Sashidharan, T., Pawlow, L. A., & Pettibone, J. C. (2012). An examination of racial bias in the Beck Depression Inventory-II. *Cultural Diversity and Ethnic Minority Psychology, 18,* 203–209. https://doi.org/10.1037/a0027689

Schutzwohl, M., Kallert, T., & Jurjanz, L. (2007). Using the Schedules for Clinical Assessment in Neuropsychiatry (SCAN 2.1) as a diagnostic interview providing dimensional measures: Cross-national findings on the psychometric properties of psychopathology scales. *European Psychiatry, 22,* 229–238. https://doi.org/10.1016/j.eurpsy.2006.10.005

Scott, J., & Gaitz, C. M. (1975). Ethnic and age differences in mental health measurements. *Diseases of the Nervous System, 36,* 389–393.

Sheehan, D. V., Lecrubier, Y., Sheehan, K. H., Amorim, P., Janavs, J., Weiller, E., ... Dunbar, G. C. (1998). The Mini-International Neuropsychiatric Interview (M.I.N.I.): The development and validation of a structured diagnostic psychiatric interview for DSM-IV and ICD-10. *Journal of Clinical Psychiatry, 59*(Suppl 20), 22–33; quiz 34–57.

Sheikh, V. I., & Yesavage, J. A. (1986). Geriatric Depression Scale (GDS): Recent evidence and development of a shorter version. In T. L. Brink (Ed.), *Clinical gerontology: A guide to assessment and intervention* (pp. 165–174). New York, NY: Haworth.

Sheline, Y. I., Barch, D. M., Garcia, K., Gersing, K., Pieper, C., Welsh-Bohmer, K., ... Doraiswamy, P. M. (2006). Cognitive function in late life depression: Relationships to Depression severity, cerebrovascular risk factors and processing speed. *Biological Psychiatry, 60,* 58–65. https://doi.org/10.1016/j.biopsych.2005.09.019

Simning, A., van Wijngaarden, E., & Conwell, Y. (2011). Anxiety, mood, and substance use disorders in United States African-American public housing residents. *Social Psychiatry and Psychiatric Epidemiology, 46,* 983–992. https://doi.org/10.1007/s00127-010-0267-2

Smith, S. M., Stinson, F. S., Dawson, D. A., Goldstein, R., Huang, B., & Grant, B. F. (2006). Race/ethnic differences in the prevalence and co-occurrence of substance use disorders and independent mood and anxiety disorders: Results from the National

Epidemiologic Survey on Alcohol and Related Conditions. *Psychol Med, 36*(7), 987–998. doi:10.1017/S0033291706007690

Somervell, P. D., Beals, J., Kinzie, J. D., Boehnlein, J., Leung, P., & Manson, S. M. (1992). Use of the CES-D in an American Indian village. *Culture, Medicine and Psychiatry, 16*, 503–517. https://doi.org/10.1007/BF00053590

Spitzer, R. L., Kroenke, K., & Williams, J. B. (1999). Validation and utility of a self-report version of PRIME-MD: The PHQ primary care study. *JAMA, 282*, 1737–1744. https://doi.org/10.1001/jama.282.18.1737

Stroup-Benham, C. A., Lawrence, R. H., & Trevino, F. M. (1992). CES-D factor structure among Mexican American and Puerto Rican women from single- and couple-headed households. *Hispanic Journal of Behavioral Sciences, 14*, 310–326. https://doi.org/10.1177/07399863920143002

Tam, C. W. C., & Lam, L. C. W. (2012). Cognitive function, functional performance and severity of depression in Chinese older persons with late-onset depression. *East Asian Archives of Psychiatry, 22*, 7–12.

Thompson, J. W., Walker, R. D., & Silk-Walker, P. (1993). Psychiatric care of American Indians and Alaska Natives. In A. C. Gaw (Ed.), *Culture, ethnicity and mental illness* (pp. 189–243). Washington, DC: American Psychiatric Press.

Torrens, M., Serrano, D., Astals, M., Perez-Dominguez, G., & Martin-Santos, R. (2004). Diagnosing comorbid psychiatric disorders in substance abusers: Validity of the Spanish versions of the Psychiatric Research Interview for Substance and Mental Disorders and the Structured Clinical Interview for DSM-IV. *American Journal of Psychiatry, 161*, 1231–1237. https://doi.org/10.1176/appi.ajp.161.7.1231

Turner, A. D., Capuano, A. W., Wilson, R. S., & Barnes, L. L. (2015). Depressive symptoms and cognitive decline in older African Americans: Two scales and their factors. *American Journal of Geriatric Psychiatry, 23*, 568–578. https://doi.org/10.1016/j.jagp.2014.08.003

Unverzagt, F. W., Ogunniyi, A., Taler, V., Gao, S., Lane, M. K. A., Baiyewu, O., . . . Hall, K. S. (2011). Incidence and risk factors for cognitive impairment no dementia and mild cognitive impairment in African Americans. *Alzheimer Disease and Associated Disorders, 25*, 4–10. https://doi.org/10.1097/WAD.0b013e3181f1c8b1

U.S. Department of Health and Human Services. (2001). *Mental health: Culture, race, and ethnicity—a supplement to mental health: A report of the surgeon general.* Washington, DC: U.S. Department of Health and Human Services, Substance Abuse and Mental Health Services Administration, Center for Mental Health Services.

Verney, S. P., Jervis, L. L., Fickenscher, A., Roubideaux, Y., Bogart, A., & Goldberg, J. (2008). Symptoms of depression and cognitive functioning in older American Indians. *Aging & Mental Health, 12*, 108–115. https://doi.org/10.1080/13607860701529957

Wang, P. S., Lane, M., Olfson, M., Pincus, H. A., Wells, K. B., & Kessler, R. C. (2005). Twelve-month use of mental health services in the United States: Results from the National Comorbidity Survey Replication. *Archives of General Psychiatry, 62*, 629–640. https://doi.org/10.1001/archpsyc.62.6.629

Weissman, M. M., Bland, R. C., Canino, G. J., Faravelli, C., Greenwald, S., Hwu, H. G., . . . Yeh, E. K. (1996). Cross-national epidemiology of major depression and bipolar disorder. *JAMA, 276*, 293–299. https://doi.org/10.1001/jama.1996.03540040037030

Wells, K. B., Burnam, M. A., Leake, B., & Robins, L. N. (1988). Agreement between face-to-face and telephone-administered versions of the depression section of the NIMH Diagnostic Interview Schedule. *Journal of Psychiatric Research, 22*, 207–220. https://doi.org/10.1016/0022-3956(88)90006-4

Whisman, M. A., Judd, C. M., Whiteford, N. T., & Gelhorn, H. L. (2013). Measurement invariance of the Beck Depression Inventory–Second Edition (BDI-II) across gender, race, and ethnicity in college students. *Assessment, 20,* 419–428. https://doi.org/10.1177/1073191112460273

Wiebe, J. S., & Penley, J. A. (2005). A psychometric comparison of the Beck Depression Inventory-II in English and Spanish. *Psychological Assessment, 17,* 481–485. https://doi.org/10.1037/1040-3590.17.4.481

Williams, C. D., Taylor, T. R., Makambi, K., Harrell, J., Palmer, J. R., Rosenberg, L., & Adams-Campbell, L. L. (2007). CES-D four-factor structure is confirmed, but not invariant, in a large cohort of African American women. *Psychiatry Research, 150,* 173–180. https://doi.org/10.1016/j.psychres.2006.02.007

Williams, D. H. (1986). The Epidemiology of Mental Illness in Afro-Americans. *Psychiatric Services, 37,* 42–49. https://doi.org/10.1176/ps.37.1.42

Williams, D. R., Neighbors, H. W., & Jackson, J. S. (2003). Racial/ethnic discrimination and health: findings from community studies. *Am J Public Health, 93*(2), 200–208. doi:10.2105/ajph.93.2.200

Wing, J. K., Birley, J. L., Cooper, J. E., Graham, P., & Isaacs, A. D. (1967). Reliability of a procedure for measuring and classifying "present psychiatric state". *British Journal of Psychiatry, 113,* 499–515. https://doi.org/10.1192/bjp.113.498.499

Wohl, M., Lesser, I., & Smith, M. (1997). Clinical presentations of depression in African American and White outpatients. *Cultural Diversity and Mental Health, 3,* 279–284.

Yesavage, J. A., Brink, T. L., Rose, T. L., Lum, O., Huang, V., Adey, M., & Leirer, V. O. (1982). Development and validation of a geriatric depression screening scale: A preliminary report. *Journal of Psychiatric Research, 17,* 37–49. https://doi.org/10.1016/0022-3956(82)90033-4

Yeung, A., Chang, D., Gresham, R. L., Jr., Nierenberg, A. A., & Fava, M. (2004). Illness beliefs of depressed Chinese American patients in primary care. *Journal of Nervous and Mental Disease, 192,* 324–327. https://doi.org/10.1097/01.nmd.0000120892.96624.00

Ying, Y. W., Lee, P. A., Tsai, J. L., Yeh, Y. Y., & Huang, J. S. (2000). The conception of depression in Chinese American college students. *Cultural Diversity and Ethnic Minority Psychology, 6,* 183–195. https://doi.org/10.1007/978-1-4615-0735-2_12

Zahodne, L. B., Nowinski, C. J., Gershon, R. C., & Manly, J. J. (2014). Depressive symptoms are more strongly related to executive functioning and episodic memory among African American compared with non-Hispanic White older adults. *Archives of Clinical Neuropsychology, 29,* 663–669. https://doi.org/10.1093/arclin/acu045

Zenner, W. (1996). Ethnicity. In D. Levinson & M. Ember (Eds.), *Encyclopedia of cultural anthropology* (pp. 393–395). New York, NY: Holt.

Zigmond, A. S., & Snaith, R. P. (1983). The hospital anxiety and depression scale. *Acta Psychiatrica Scandinavica, 67,* 361–370. https://doi.org/10.1111/j.1600-0447.1983.tb09716.x

Zimmerman, M., Chelminski, I., McGlinchey, J. B., & Posternak, M. A. (2008). A clinically useful depression outcome scale. *Comprehensive Psychiatry, 49,* 131–140. https://doi.org/10.1016/j.comppsych.2007.10.006

Zink, D., Lee, B., & Allen, D. (2015). Structured and semi-structured clinical interviews available for use among African American clients: Cultural considerations in the diagnostic interview process. In L. T. Benuto & B. D. Leany (Eds.), *Guide to psychological assessment with African Americans* (pp. 19–42). New York, NY: Springer-Verlag.

Zung, W. W. (1965). A self-rating depression scale. *Archives of General Psychiatry, 12,* 63–70. https://doi.org/10.1001/archpsyc.1965.01720310065008

6

Visual Cognition and Culture

JOSHUA O. S. GOH, CHUN-YIH LI, YU-ZHEN TU, AND CAROLINE DALLAIRE-THÉROUX

INTRODUCTION

Accurate neural representation of visual information is critical for individuals to effectively navigate and interact with the environment and society. A reasonable expectation that follows is that the human brain should objectively encode information about physical quantities or events in a similar and universal manner. However, evidence shows that there are significant and systematic differences in visual representations across individuals related to their cultural preferences. For instance, when judging the physical length of lines embedded in a square frame, Westerners performed better at absolute length judgments whereas East Asians performed better at relative judgments involving the association between line length and contextual frame area (Kitayama, Duffy, Kawamura, & Larsen, 2003). This remarkable finding is consistent with the emphasis on analytic and holistic cognitive styles in Western and East Asian cultures, respectively. Moreover, such culture-related differences in the visual perception of the same physical quantity has been replicated, extended, and even associated with neural processes across many studies (Goh & Huang, 2012; Goh & Park, 2009).

We considered that such fundamental culture-related differences in visual perception between human beings, in fact, arise from the very neural impetus to accurately represent the environment. To achieve this aim, the brain must encode stimulus information in relation to the context so that neural representations are essentially bound contextualized experiences of objects and events (Tulving, 2002). Thus, while neural function should respond robustly and universally to environmental stimuli, some processes might also maintain tempered sensitivity to contextual contingencies that influence even perceptions—even those as fundamental as line length. Using the previous example, the processing of line lengths might occur in contexts that emphasize the features of lines as distinct units of information or in contexts in which the relationship of lines to other episodic elements form the base unit of information instead. If a particular context is regularly experienced over time, neural representations of line lengths should then gravitate to, or be organized in a manner that is consistent with, its emphases so that subsequent processing in future encounters will be facilitated. Indeed, the analytic and holistic dichotomy of visual cognition styles, suggested as a central dimension that dissociates Westerner from East Asian perception and attention processing, is thought to stem from their cultural contexts that emphasize independent and interdependent values, respectively (Nisbett, 2003).

Practically, apart from instrument comparability issues, it has been difficult to conduct adequately powered and well-controlled neural-based investigations that involve direct manipulation of cultural values in human participants with their heterogeneous and complex life experiences. Thus, at present the specific neural mechanisms underlying cultural differences in visual perception remain speculative. We suggest that one critical knowledge gap is whether a common process or distinct processes explain culture-related influences on perception of different types of visual stimuli (e.g., objects, scenes, and faces; see the following discussion for details) or during different tasks (e.g., passive viewing vs. reporting features vs. using feature information toward a goal). Also, rigorous frameworks based on neurobiological and neurocomputational perspectives are needed so that the scope and condition under which cultural forces might be particularly influential so that cultural biases are acquired or not can be more precisely specified.

Here, we suggest that culture-related biases, vis-à-vis visual perception, are acquired via behavioral reinforcement from reward and suppression after detecting error as with many other learned behaviors. We describe a framework to understand culture-related differences in visual processing that involves the notion that the brain is geared toward veridical representation of the environment and that the integration of physical and valuative (reward and error) signals from the environment are paramount to this goal. With respect to cultural differences in visual perception, valuative signals likely primarily come from social congruence signals, as suggested in recent studies (Hitokoto, Glazer, & Kitayama, 2016; Mu, Kitayama, Han, & Gelfand, 2015). Invoking the line example again, a child might be learning to draw a line while interacting with caretakers who give feedback or focus attention on the aesthetics of the line's features while neglecting its relationship with other visual items. Similarly, caretaker feedback or attention might highlight the meaningfulness of the role of the line with respect to other visual elements while glossing over featural details. Further, certain aspects of how the child draws the line, such as holding the pencil inappropriately or drawing out of the boundaries of the paper, might be discouraged. Such social interactions with caretakers may take several forms including but not limited to direct instructions, indirect comments, emotional expressions, or gestures. These social signals are necessarily encoded as part of the child's episodic representation of visual lines perhaps at first as relatively conscious traces that then become more automatized processes, or biases, with chronic experience. Further, these social-environmental signals might be integrated in neural representations to the extent that should resulting biased behaviors lead to incongruent feedback, the feedback may also be construed to be erroneous instead of the bias being inaccurate. Under this circumstance, more neural resources are needed to overcome the bias and integrate the new incongruent feedback.

To develop this framework, we first highlight current findings already extensively reviewed elsewhere (Goh & Huang, 2012; Goh & Park, 2009; Nisbett, 2003) that characterize the nature of cultural differences in behavior, cognition, and neural processing, particularly in relation to visual perception. Over the course of the review of findings, we progressively consider evidence showing that differences in visual processing might stem from culture-related differences in emphasis on socio-affective information, focusing on facial expression as one specific source as well as possible contributions from formal educational styles and language. Following this, we consider more specific aspects of the role of reinforcement learning in the acquisition of culture-related biases that implicates the dopaminergic reward system and its involvement in the formation of habits. With this framework established, we finally consider how understanding on culture-related differences in visual processing might be applied in normative as well as clinical settings.

BEHAVIORAL AND NEURAL EVIDENCE FOR CULTURAL INFLUENCES ON HUMAN VISUAL PERCEPTION

Early behavioral approaches on culture-related differences in visual perception were largely focused on differences between Westerners and East Asians, although it is well noted that culture is a far more heterogeneous concept than this simple dichotomy suggests. The standing notion is that Western culture is more individualistic and independent and associated with an emphasis on analytic visual perception and attention. This visual processing style tends to prioritize separating visual elements and focuses on their individual features. By contrast, East Asian culture is more collectivistic and interdependent and is associated with an emphasis on holistic visual perception and attention. Such a visual processing style prioritizes the relationships between visual elements, binding visual elements together more tightly within a context. Later in vivo functional neuroimaging techniques afforded further validation of these distinctive visual styles in Westerners and East Asians characterized by differential engagement of neural processing in visual as well as higher-level processing brain areas. Here, we briefly cover the behavioral and neuroimaging findings that characterize differences between individual human visual perception and attention related to cultural differences.

Behavioral and Eye-Movement Responses Associated with Analytic Versus Holistic Visual Processing of Objects and Backgrounds in Westerners and East Asians

The analytic–holistic dichotomy of cultural differences in visual processing has been reported in many behavioral studies involving drawing relative versus absolute line lengths (framed-line test; Kitayama et al., 2003), detection of changes in foreground or background elements in scenes (Boduroglu, Shah, & Nisbett, 2009; Masuda & Nisbett, 2006), recall of visual content in complex scenes (Masuda & Nisbett, 2001), and photographic portraits consisting of faces and backgrounds (Masuda et al., 2008; Figure 6.1). Across these studies, Westerners consistently show either better performance in task conditions that favor attention toward central objects or object features, or greater tendencies to focus on objects and their features rather than the backgrounds. By contrast, East Asians perform better in task conditions that favor the binding of objects to the backgrounds and

(a) Stimulus for the Frame-Line Task

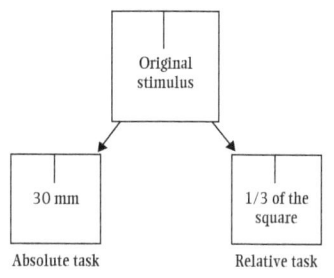

(b) Frame-Line Task behavioral performance

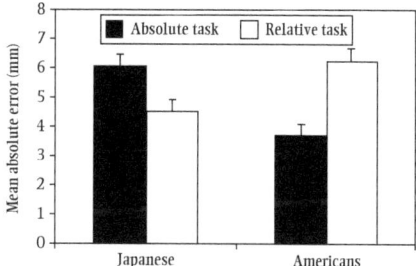

(c) Stimuli for the Change Detection Task

(d) Change Detection Task behavioral performance

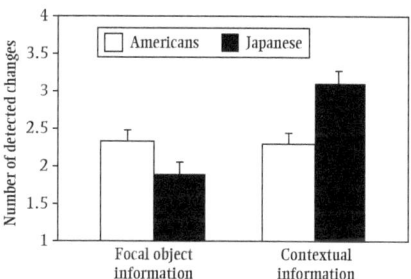

Figure 6.1. (a) Sample stimulus in the Frame-Line Task. Participants first encoded the original stimulus then drew the vertical line in a different sized square that was either relative or absolute length. (b) Behavioral performances in the frame-line test showing Americans making smaller errors when drawing absolute line lengths and Japanese making smaller errors when drawing line lengths in relation to the area of a contextual square frame. Adapted and reproduced with permission from S. Kitayama, S. Duffy, T. Kawamura, & J. T. Larsen, 2003, Perceiving an object and its context in different cultures: A cultural look at new look. *Psychological Science, 14,* 201–206. (c) Sample stimuli in the Change Detection Task showing subtle changes in foreground and background portions of the scene. (d) Behavioral performances in the change detection task showing Americans detected more changes occurring in focal objects whereas Japanese detected more changes occurring in contextual backgrounds. Adapted and reproduced with permission from T. Masuda & R. E. Nisbett, 2006, Culture and change blindness. *Cognitive Science, 30,* 381–399. Copyright © 2006 Cognitive Science Society, Inc. All Rights Reserved.

show tendency to focus on associations between objects or between objects and the backgrounds.

Supportive evidence also comes from eye-tracking studies that directly observe cultural differences between Westerner and East Asian eye movements as a surrogate of attention to objects and backgrounds. Chua, Boland, and Nisbett (2005) found that when viewing scenes containing central objects embedded against background contexts, Westerners showed faster onsets of the first fixation to objects with longer object fixation durations compared to East Asians, who had shorter fixation durations and more background fixations. In extension, Goh, Tan, and Park (2009) demonstrated that when objects and backgrounds were selectively changed, Westerner eye movements again showed longer fixation durations in general that were, additionally, were more sensitive to novel objects than background contexts, consistent with a focus on object featural details. By contrast, East Asian eye movements were characterized by shorter fixation durations that alternated more between objects and backgrounds characteristic of a focus on binding objects to their contexts (Figure 6.2). Jointly, these findings strongly support the analytic-holistic dichotomy of visual perception and attention between Westerners and East Asians.

Interestingly, Knox and Wolohan (2014) suggested that a more biological factor might also account for eye movement differences between Westerners and East Asians apart from culture-related influences. Caucasians and Chinese who were born and educated in the United Kingdom and Chinese born and educated in China underwent an eye-tracking study on reflexive saccades. Participants fixated on central targets and then made saccades to peripheral targets when they appeared. Importantly, introducing a temporal gap between the offset of the central target and the onset of the peripheral target induces a reflexive express saccade response in which the saccade latencies are reduced relative to when there is no temporal gap. The authors found that China Chinese had faster express saccade responses than UK Caucasians, consistent with previous findings of shorter fixation durations in East Asians than Westerners. However, UK Chinese, who had more similar cultural values to the UK Caucasians than China Chinese, also showed faster express saccade responses than UK Caucasians, suggesting a more biological rather than cultural basis for shorter fixation durations in East Asians than Westerners. Nevertheless, it should be noted that express saccades are very specific low-level eye movement responses to target offsets and onsets that rely on distinct neural mechanisms from other, more higher-level processes involved in studies such as in Chua et al. (2005) and Goh et al. (2009) who used more complex and semantically rich stimuli. Thus, the extent to which more biological differences interact with cognitive culture-related mechanisms remains an avenue for future research (Han et al., 2013; Kim & Sasaki, 2014).

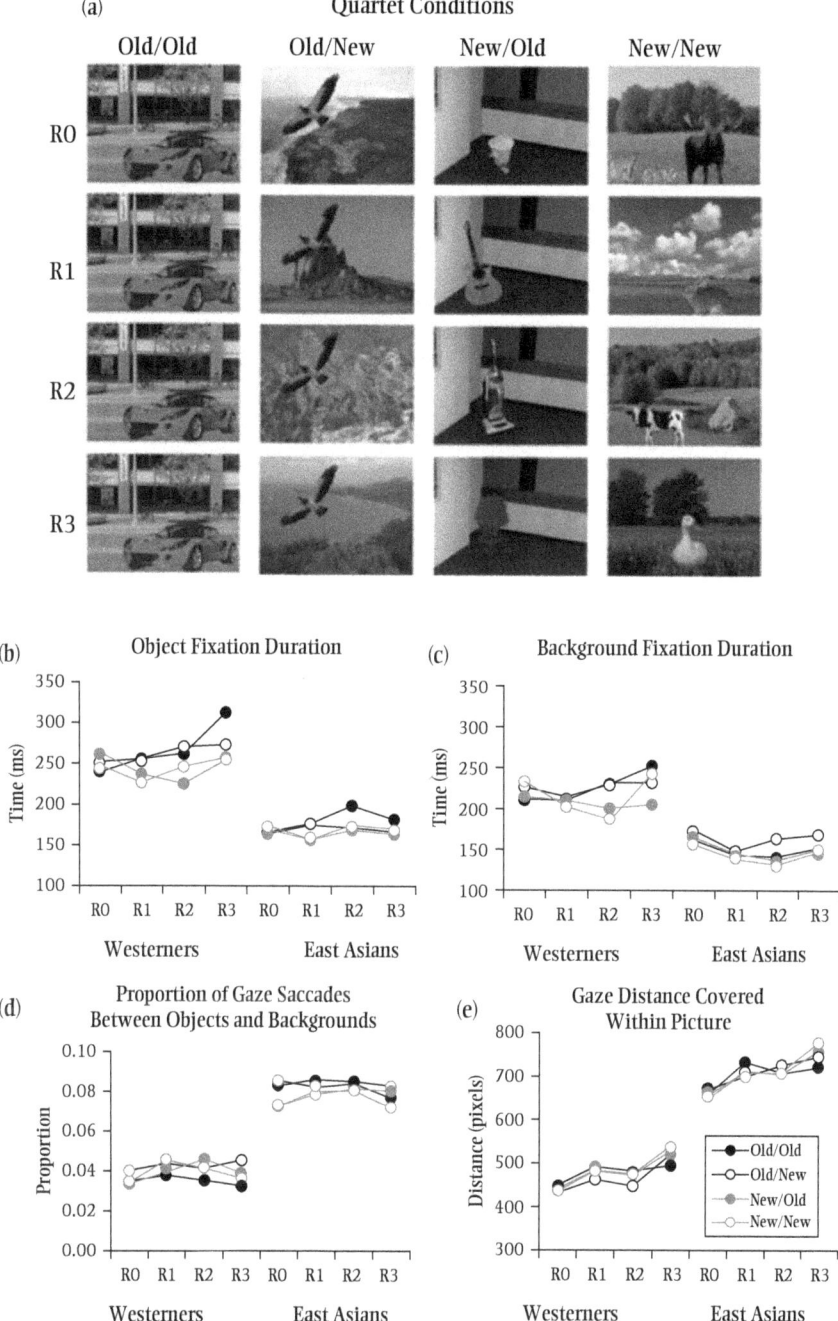

Figure 6.2. Eye-movement responses in Westerners and East Asians associated with passive viewing of complex pictures consisting of central objects embedded amidst background scenes. (a) Sample pictures presented as quartets (serially from R0 to R3) such that objects and backgrounds were repeated (Old/Old), objects were repeated against backgrounds that changed (Old/New), objects were changed against repeated

Neurophysiological Responses Associated With Analytic Versus Holistic Visual Processing of Objects and Backgrounds in Westerners and East Asians

Functional magnetic resonance imaging (fMRI) and electroencephalographic event-related potential (ERP) measures of neural processing of objects and backgrounds also support the analytic–holistic dichotomy between Westerners and East Asians. Using the framed-line test, Hedden, Ketay, Aron, Markus, and Gabrieli (2008) reported higher activity in precentral and parietal brain areas when Westerners and East Asians were engaged in nonpreferred than preferred modes of visuospatial judgments. Whereas East Asians had higher neural responses in these brain areas during absolute (nonpreferred) than relative (preferred) judgments of line lengths in square frame contexts, Westerners showed higher neural responses during relative than absolute judgments. Similarly, Goh et al. (2013) used a visuospatial task that favored holistic strategies (judging distances between a dot and a horizontal line; Baciu et al., 1999) and found higher frontoparietal responses and greater default-mode network suppression in Westerners than East Asians (Figure 6.3). Also, compared to Westerners, East Asians engaged greater neural processing of scenes containing incongruent than congruent object–background pairings (e.g., a cow in a restaurant vs. a cow in a farm) suggesting they were more sensitive to semantic or associative violations (Jenkins, Yang, Goh, Hong, & Park, 2010). Moreover, the key loci for such greater neural engagement in East Asians during incongruent scenes was the lateral occipital complex (which includes the inferior occipital, fusiform, and posterior inferior temporal gyri), suggesting that East Asians considered the objects as the incongruent elements of the pictures rather than the backgrounds. Interestingly, Wang, Umla-Runge, Hofmann, Ferdinand, and Chan (2014) used ERP methodology and also found greater P3 component responses to visual oddball targets and novel stimuli relative to neutral stimuli in Chinese (more holistic) compared to German (more analytic) participants. This result is consistent with stronger binding of object–background visual relations in East Asians than Westerners, such that East Asians engage more neural processing in the form of the accentuated ERP response to irregular occurrences of objects. That these differences were observed in the P3 responses (250–500 ms from stimulus onset) suggests that the influence of culture occurs relatively early on in the visual processing time course.

---

backgrounds (New/Old), or both objects and backgrounds were changed (New/New). (b) East Asians showed shorter objects fixation durations than Westerners who also increased object fixation durations as objects were repeated against repeating or changed backgrounds. Also, across all quartet conditions, East Asians showed (c) shorter background fixation durations than Westerners, (d) made more saccades between objects and backgrounds, and (e) covered greater gaze distances within pictures reflecting a more expansive visual scanning strategy. Adapted and reproduced with permission from J. O. S. Goh, J. C. Tan, & D. C. Park, 2009, Culture modulates eye-movements to visual novelty. *PloS One*, 4(12), e8238.

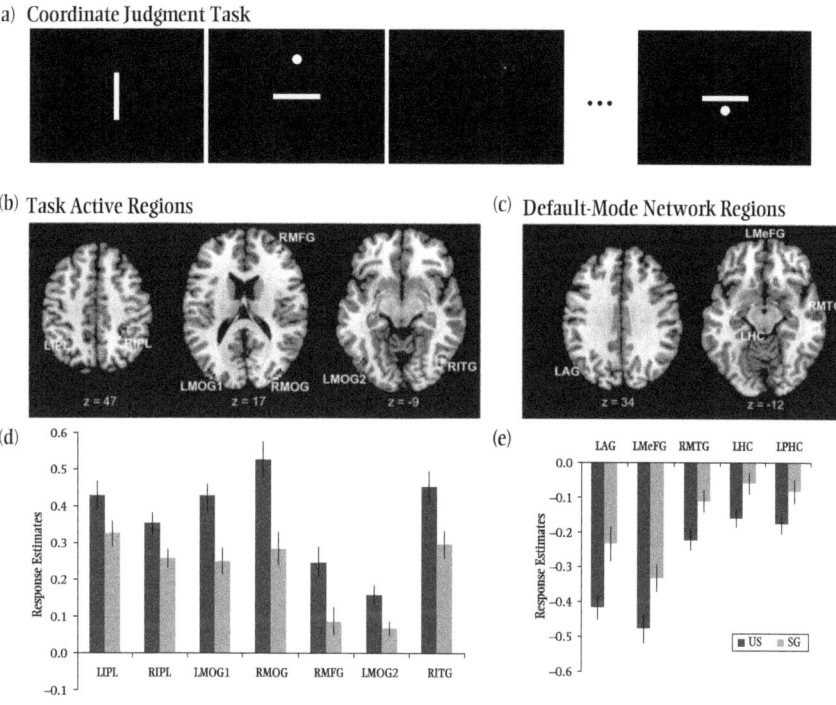

Figure 6.3. Functional neural responses in American (US) Westerners and Singaporean (SG) East Asians when making relative visuospatial judgments. (a) Sample stimuli for the visuospatial Coordinate Judgment Task in which participants were to encode the length of the target vertical bar stimuli (leftmost slide) and decide in subsequent trials whether a dot was farther or nearer from a horizontal bar relative to the length of the vertical bar (right slides; slides were blank during rest trials). (b) Task active and (c) default-mode network regions that significantly modulated their functional responses during the coordinate judgment task across Westerners and East Asians. L = left; R = right; IPL = inferior parietal lobule; MOG = middle occipital gyrus; MFG = Middle Frontal Gyrus; ITG = inferior temporal gyrus, AG = angular gyrus, HC = hippocampus; MeFG = medial frontal gyrus; MTG = middle temporal gyrus. (d) Westerners showed higher functional neural responses than East Asians in task active regions. (e) Westerners showed more suppressed functional neural responses than East Asians in default-mode network regions. (See color plate) Adapted and reproduced with permission from Joshua O. S. Goh et al. Culture differences in neural processing of faces and houses in the ventral visual cortex. J. O.S. Goh, E. D. Leshikar, B. P. Sutton, J. C. Tan, S. K. Y. Sim, A. C. Hebrank, & D. C. Park, 2010, Culture differences in neural processing of faces and houses in the ventral visual cortex. *Social Cognitive and Affective Neuroscience, 5*, 227–235. © The Author (2010). Published by Oxford University Press.

In a series of three studies, Goh and colleagues evaluated functional neural correlates of the above culture-related differences in object–background processing in a sample of Westerners and East Asians that included young as well as older adults (Chee et al., 2006; Goh et al., 2004, 2007; Figure 6.4). Using passive

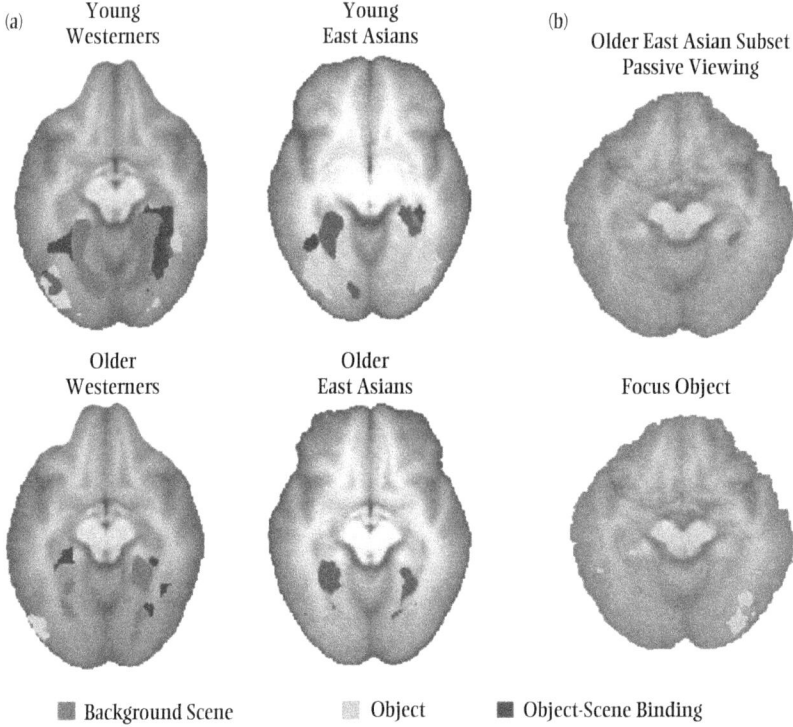

Figure 6.4. (a) Object, background scene, and object-background binding processing brain regions involved during passive viewing of picture quartets (see Figure 6.2). Compared to young Westerners, young East Asians, and older Westerners, older East Asians showed reduced involvement of object processing in the lateral occipital areas and binding processing in the medial temporal areas. (b) Object processing in the lateral occipital areas in a subset of older East Asians whose passive viewing response is shown above is rescued when instructed to focus on the central objects in pictures. (See color plate) Adapted and reproduced with permission from M. W. Chee, J. O. S. Goh, V. Venkatraman, J. C. Tan, A. Gutchess, B. Sutton, ... D. Park, Agerelated changes in object processing and contextual binding revealed using fMR adaptation, *Journal of Cognitive Neuroscience, 18,* 495–507. © 2006 by the Massachusetts Institute of Technology; J. O. S. Goh, S. C. Siong, D. Park, A. Gutchess, A. Hebrank, & M. W. L. Chee, 2004, Cortical areas involved in object, background, and object-background processing revealed with functional magnetic resonance adaptation. *Journal of Neuroscience, 24,* 10223–10228; and J. O.S. Goh, M. W. Chee, J C. Tan, V. Venkatraman, A. Hebrank, E. D. Leshikar, ... D. C. Park, 2007, Age and culture modulate object processing and object–scene binding in the ventral visual area. *Cognitive, Affective, & Behavioral Neuroscience, 7,* 44–52. Copyright © 2007 Springer US. All Rights Reserved.

viewing of scenes with selectively repeated objects and backgrounds in an fMRI experiment, they distinguished the lateral occipital complex, parahippocampal place area (PPA; part of the parahippocampal gyrus), and medial temporal lobe as brain regions involved in processing objects, background contexts, and the binding of objects and backgrounds, respectively. Critically, whereas young East Asians, young Westerners, and older Westerners clearly showed the object, background, and binding processing regions, older East Asians only showed background-processing responses in the PPA. Further, when the older East Asians underwent the experiment again but with instruction to focus on objects, lateral occipital complex object-processing responses were rescued but at the expense of PPA background processing responses, and medial temporal lobe binding responses remained absent. These findings suggest that when neural resources are attenuated with age, cultural biases might become more evident such that older adults default to their preferred visual processing style more strongly. Thus, whereas older Westerners maintained dissociated neural areas processing objects and backgrounds in accordance with more analytic visual processing, older East Asians may have treated the components more holistically as one contextual scene and only segregated the objects under instruction.

Integrating across these evaluations of Western and East Asian visual processing, we first see that the analytic-holistic visual styles can be construed as visual strategies that are adopted in a default manner when individuals process visual stimuli. Further, individuals can adopt nondefault visual strategies although greater difficulty is experienced, and more neural effort is required. Overall, culturally specific styles are an innate aspect of visual perception and attention of stimuli that range from simple line drawings to complex scenes and affects even judgments about physical properties regarding these visual elements. Next, we consider cultural influences that also bias processing of visual facial features and thus encompasses social processing as well.

Face Processing Differences Associated with Social Norms in Westerners and East Asians

Being a social animal, one of the most critical of human visual processing abilities is the parsing and interpretation of facial features and expressions. Accurate representation of visual facial features is necessary to correctly understand the intentions and states of others and generate appropriate behaviors. Note, however, that compared to objects and scenes, variations of visual information in faces are drastically reduced and generally limited to configurations of the eyes and mouth, as well as the nose, hair, and facial skin. Because of these, visual processing of faces arguably must rely much more on contextual information than is the case with object–scene processing. Consequently, cultural influences should be evident here such that the same facial stimuli should be evaluated differently and associated with different interpretations in line with culture-specific social norms other than but not excluding the analytic–holistic dichotomy.

Using eye-tracking methodology, Blais, Jack, Scheepers, Fiset, and Caldara (2008) found that whereas Westerners fixated on the eyes and mouths of face stimuli, East Asians fixated around the center of the faces. This eye-movement pattern was observed during face encoding, recognition, and categorization and applied regardless of the ethnicity of the face stimuli, which were counterbalanced across both Caucasian and Asian faces Figure 6.5). The central face fixation adopted by East Asians was not due to them utilizing different facial features for cognitive processes compared to Westerners (e.g., extracting information from the nose rather than the eyes) (Caldara, Zhou, & Miellet, 2010). Rather, the eye-movement patterns appeared to be strategically controlled and possibly linked to the East Asian social practice of not looking others directly in the eyes, which might be construed as aggression (Akechi et al., 2013). Based on their findings, Caldara et al. (2010) concluded that under natural viewing conditions East Asians still utilize similar facial features for recognition and expression processing as Westerners, but these might be based on extrafoveal input due to socially acquired eye-movement habits.

A functional neuroimaging study further extended eye-movement differences to faces between Westerners and East Asians to differences in neural responses (Goh et al., 2010). The fusiform face area (FFA) is a known to be highly specialized for processing face stimuli (Kanwisher, McDermott, & Chun, 1997) such

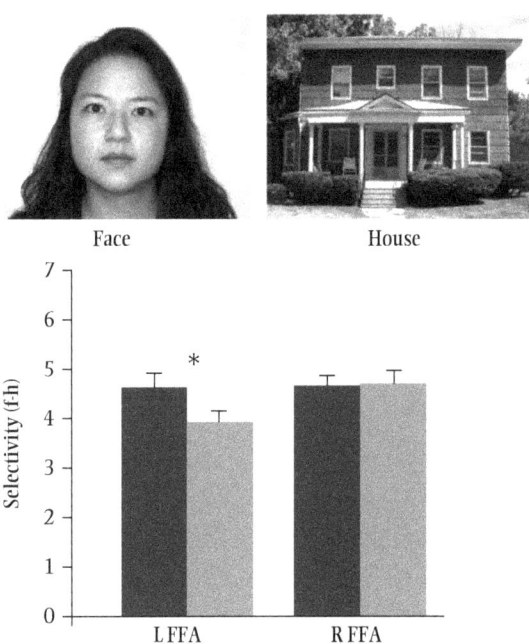

Figure 6.5. Face-house selectivity of neural activity (difference between responses to face and house stimuli) in bilateral fusiform face areas (FFA) of Westerners (blue) and East Asians during passive viewing. East Asians show significantly lower left FFA face-house selectivity associated with more right lateralized responses compared to Westerners who showed bilateral engagement. Adapted and reproduced with permission from Joshua O. S. Goh, A. C. Hebrank, B. P. Sutton, M. W. L. Chee, S. K. Y. Sim, & D. C. Park, 2013, Culture-related differences in default network activity during visuo-spatial judgments. *Social Cognitive and Affective Neuroscience, 8,* 134–142. © The Authors (2011). Published by Oxford University Press.

that lesions in this brain area results in prosopagnosia (Barton, Press, Keenan, & O'Connor, 2002; Bouvier & Engel, 2006)—the inability to recognize faces despite generally intact ability to process face and other visual features. Moreover, substantial evidence has shown that while the left FFA is more sensitive to specific facial features, the right FFA is specialized for processing face stimuli as a whole (Rhodes, 1985; Rossion et al., 2000; Rotshtein, Geng, Driver, & Dolan, 2007; Sergent, 1982). Goh et al. (2010) showed that under passive viewing conditions, Westerners engaged both left and right FFA whereas East Asians showed more right lateralized FFA sensitivity to faces relative to house stimuli as the baseline (Figure 6.5). Also, individuals with more reduced left FFA responses rated the value of security on the Schwartz Value Survey (Schwartz, 1992) with higher personal importance, indicating greater collectivism was associated with more holistic visual processing of faces. Such links between personal values, cognitive visual biases, neural processing, and behavioral and eye-movement responses add substantial credence to the notion that social expectations drive cultural differences in visual processing.

A domain where social expectations exert particularly strong culture-specific influences on visual facial processing is in emotional expressions. Using movie clips with transitioning emotional faces, Ishii, Miyamoto, Mayama, and Niedenthal (2011) found that East Asians were more sensitive to changes involving happy to neutral emotions or disappearance of happiness than Westerners were. Remarkably, recent findings of differences between Westerner and East Asian facial emotional mental representations also challenge the long-standing notion of the universality of emotional expressions and recognition (Jack, Caldara, & Schyns, 2012; Jack, Garrod, Yu, Caldara, & Schyns, 2012) (Figure 6.6). Western and East Asian participants categorized emotions depicted in a large range of computer-generated faces with varying degrees of expressions. Expressions were varied by virtually simulating select facial muscle movements associated with specific emotional expressions according to the Facial Action Coding System (FACS; Ekman & Friesen, 1977). The authors found that whereas categorizations in Westerners were relatively consistent with FACS guidelines and involved distinct facial muscle movement contributions, categorizations in East Asians adhered less to the FACS and also involved more overlap of muscle contributions across emotions. Moreover, whereas Westerners weighted the eyebrows and mouth with greater contribution to emotional expression, East Asians weighted more on the gaze direction (Jack, Caldara, et al., 2012).

Western and East Asian Differences in Number and Mathematical Processing

For many individuals, one of the strongest social influences arguably comes from formal education. Indeed, culture-related differences have also been observed in mathematical ability most likely related to differences in teaching environments. Several studies have documented better mathematical ability in East Asian compared to Westerners (Cantlon & Brannon, 2007; Hsin & Xie, 2014; Z. Luo, Jose,

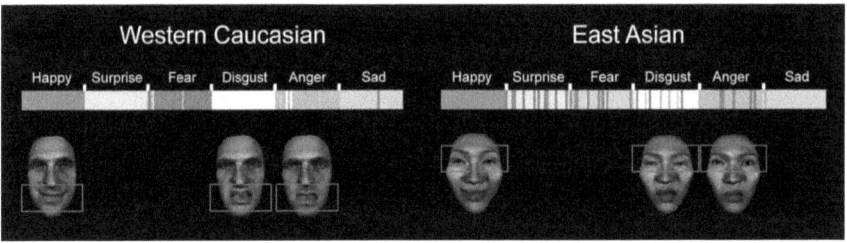

Figure 6.6. Clustering of computer generated emotional faces based on ratings from Western and East Asian participants in relation to the Facial Action Coding System (Ekman & Friesen, 1977; top colored bars). Western ratings yielded homogenous and distinctive clusters for each of the six facial emotion expressions as illustrated by relatively well separated colors for each emotion. By contrast, East Asian ratings yielded more heterogeneous and overlapping categorizations of the faces depicted by the scattered colors across emotions. Also, Westerners weighted the mouth as more important for emotional expression interpretation whereas East Asians weighted more the eyes (bottom faces). (See color plate) Adapted and reproduced with permission from R. E. Jack, O. G. B. Garrod, H. Yu, R. Caldara, & P. G. Schyns, 2012, Facial expressions of emotion are not culturally universal. *Proceedings of the National Academy of Sciences of the United States of America, 109,* 7241–7244.

Huntsinger, & Pigott, 2007; Miller, Smith, Zhu, & Zhang, 1995; Ng & Rao, 2010). This East Asian mathematical advantage is suggested to stem from more rigorous general academic expectations and habits (Hsin & Xie, 2014) or specific styles of upbringing (Luo et al., 2007). Of relevance to cultural effects on visual processing, different language systems across cultures also play an important role in the recognition and overall processing of numerical symbols (Miller et al., 1995; Ng & Rao, 2010; Rodic et al., 2015). In particular, among other property differences between Western languages and Chinese, Chinese language orthography contains more visuospatial complexity (Rodic et al., 2015). Also, written representations of numbers in Chinese are considered to be more straightforward compared to greater use of alternative morphological forms in English. For instance, to represent the number 36 (Arabic form), the Chinese characters are written as 三十六—literally three ten six—whereas the English word is written as "thirty-six", which involves inflecting "three" to "thir" and "ten" to "ty" (Ng & Rao, 2010). Although seemingly minor, the English transformations are thought to still require additional resources to process relative to Chinese. Thus, incipient habitual visual experience with the Chinese orthographical system might particularly facilitate numeral manipulation and some mathematical processes. It should be noted, however, that faster numerical and mathematical processing might come at a cost as although Chinese Canadians required less working memory resources and were faster for solving complex additions, they were less able to use adaptive strategies compared to English-speaking Canadians and Flemish-speaking Belgians (Imbo & LeFevre, 2009). Consistent with these behavioral findings, when performing the same mathematical task native Chinese speakers utilize a visuomotor brain network and English speakers rely more on the perisylvian area, suggesting different strategy use (Tang et al., 2006; Figure 6.7). Whereas the former brain regions might involve faster and more direct visual-to-motor processes, the latter regions are involved in more formal language processing and might afford greater strategic flexibility.

Overall, a wealth of evidence reliably shows proliferate differences in the way Westerners and East Asians process objects and scenes, facial features, and expressions—and even numbers and mathematics—that are related to their cultural values. Thus, as suggested in some studies, such findings are not consistent with the Darwinian notion of universality specifically for facial emotions (Darwin, 1999) and more generally for other cognitive domains so that even

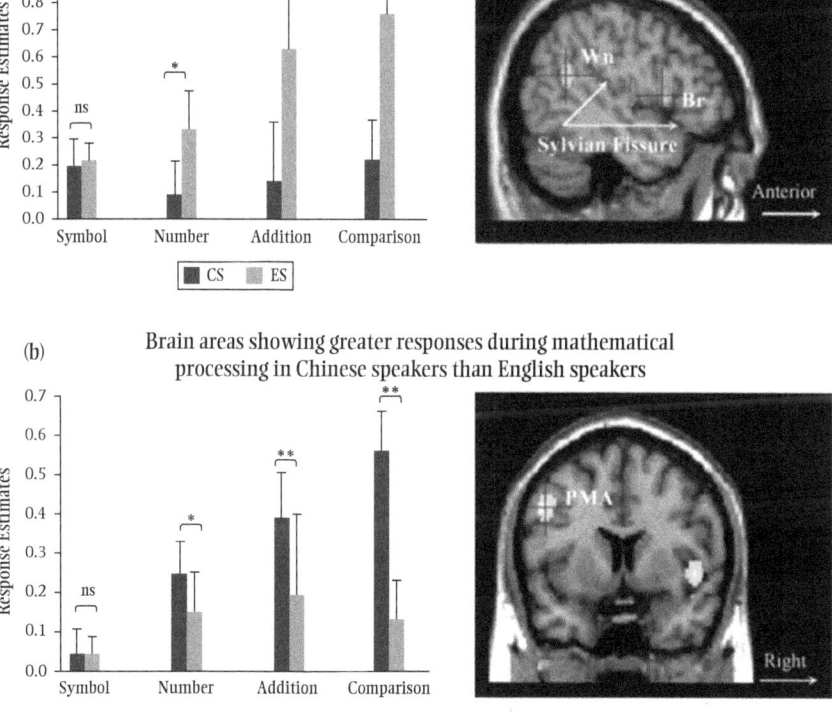

Figure 6.7. (a) The perisylvian area shows higher responses in English (ES) than Chinese (CS) speaking participants across different mathematical tasks. (b) The premotor and visual (not shown) areas show higher neural responses in CS than ES instead during mathematical processing. (See color plate) Adapted and reproduced with permission from Y. Tang, W. Zhang, K. Chen, S. Feng, Y. Ji, J. Shen, J., ... Y. Liu, 2006, Arithmetic processing in the brain shaped by cultures. *Proceedings of the National Academy of Sciences, 103,* 10775–10780. Copyright (2006) National Academy of Sciences, USA.

perceptions and representations of objective physical quantities are not immune to cultural differences. Such cultural influences in visual processing are not trivial as the real biases they produce results in different outcomes in various contexts. We now turn to consider reinforcement learning from outcomes as a mechanism for such profound cultural influences on visual processes, how they might be acquired, their neural correlates, and what might be the parameters with which they operate.

SOCIO-ENVIRONMENTAL REINFORCEMENT LEARNING AS A NEUROBIOLOGICAL MECHANISM FOR THE ACQUISITION OF CULTURAL DIFFERENCES IN VISUAL PERCEPTION

In reinforcement learning, the outcome reward or punishment of behaviors shape an individual's stimulus-response mappings for subsequent interactions with the environment and, consequently, cognitive representations about the environment. Specifically, cognitive representations associated with behaviors that result in expected outcomes are maintained or strengthened whereas those that result in unexpected outcomes are updated. If the outcome was an unexpected reward or more rewarding than predicted, a new or stronger representation is established. If the outcome was an unexpected punishment or less rewarding than predicted, the representation is dampened. For social interactions, the affirmation or disapproval of others reacting to our behaviors might constitute reward or punishment signals that modulate our subsequent behaviors.

How might the distal influence of social approval/disapproval signals ultimately impact on neurobiologically instantiated representations of the physical environment in the brain? We propose that as it implicates reinforcement learning, social feedback likely engages the dopaminergic reward system, which include the ventral tegmental area (VTA) and its projection targets in the frontal, subcortical, and limbic regions (Schultz, 2013; Schultz, Dayan, & Montague, 1997; Schultz & Dickinson, 2000; also see Mu et al., 2015, for similar discussion on the neurobiological acquisition of culture). Predicted social outcomes that yield unsurprising feedback should not modulate dopamine levels in this system, and thus the associated behavioral and cognitive processes are maintained. However, affirmative social feedback construed as reward for behavior should evoke increased synaptic dopamine in this system that provides a neural signal to strengthen synaptic connections that are associated with the behavior and its cognitive representations, which include connections involved in processing physical quantities. By contrast, discouraging social feedback reduces dopaminergic activity and weakens associated behavioral and cognitive processes. Note that while the role of dopamine in reinforcement learning of socially appropriate or inappropriate behaviors is highlighted here, other neurotransmitters are likely also involved. However, whereas dopamine is known to have more specific and defined actions with respect to prediction, reward, and punishment, much less is known about the contributions of other neurotransmitters to such Bayesian-like learning. Nevertheless, proposals and initial evidence have been offered supporting the role

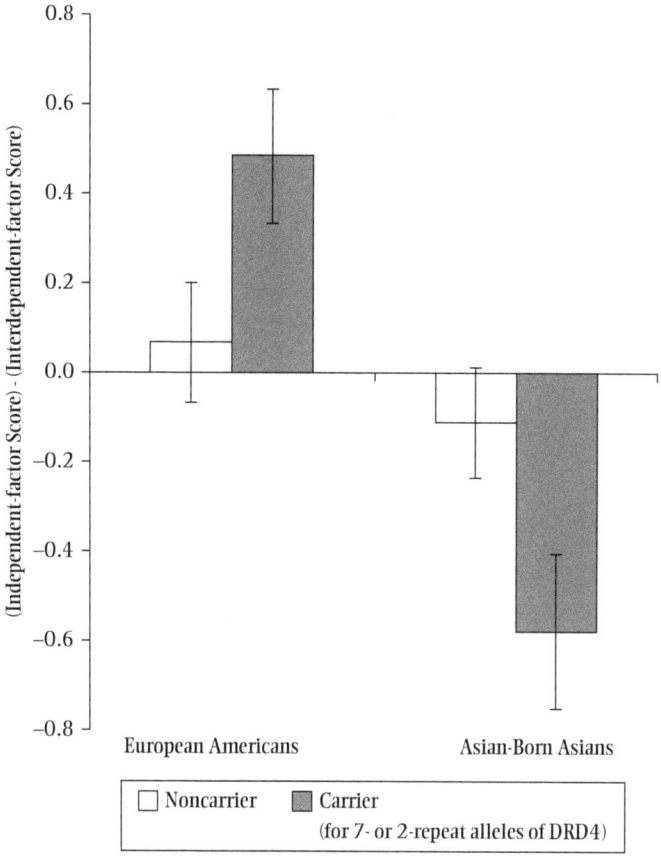

Figure 6.8. Carriers of the 7- or 2-repeat alleles of DRD4 gene for dopamine receptors show greater and culture-specific sensitivity for independence-interdependence values. American carriers showed greater independence scores whereas East Asians showed greater interdependence scores relative to their respective noncarrier counterparts. Adapted and reproduced with permission from S. Kitayama, A. King, C. Yoon, S. Tompson, S. Huff, & I. Liberzon, 2014, The dopamine D4 receptor gene (DRD4) moderates cultural difference in independent versus interdependent social orientation. *Psychological Science, 25,* 1169–1177.

of dopamine, serotonin, and oxytocin activity in cultural differences in cognition (Chiao & Blizinsky, 2010; Kim et al., 2010, 2011; Kim & Sasaki, 2014; Kitayama et al., 2014; Luo et al., 2015; Figure 6.8).

Critically, across cultures, specific sets of social behaviors are considered appropriate or inappropriate that constitutes the boundaries that define the different cultures. Evidence for this is seen in culture-related differences in the manner that Westerners and East Asians process stimuli depicting painful scenarios (Cheon et al., 2013), threat (Park & Kitayama, 2014), violations of social norms (Mu et al., 2015), or moral decisions (Han, Glover, & Jeong, 2014; Figure 6.9). In these studies, East Asians generally evince behaviors reflecting greater empathy

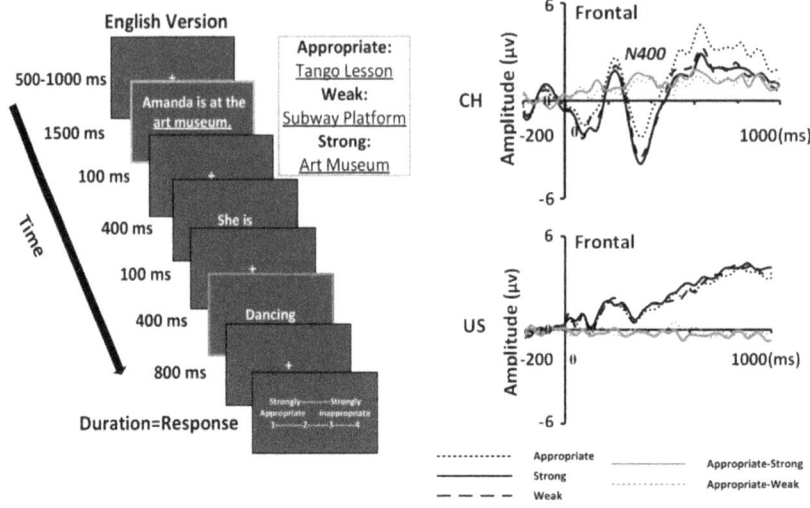

Figure 6.9. Sample stimuli describing behaviors that have different degrees of appropriateness under different contexts. East Asians showed greater N400 responses compared to Westerners to target words (red box) that describe norm violations relative to more appropriate behaviors. (See color plate) Adapted and reproduced with permission from Y. Mu, S. Kitayama, S. Han, & M. J. Gelfand, 2015, How culture gets embrained: Cultural differences in event-related potentials of social norm violations. *Proceedings of the National Academy of Sciences of the United States of America, 112,* 15348–15353.

and social sensitivity than Westerners. This is consistent with greater affirmation of behaviors reflecting interpersonal interdependence and a general discouragement of being overly independent from others in East Asian culture. By contrast, Western culture tends to affirm independent behavior in many contexts and discourages reliance on others.

Thus, as an individual navigates and interacts with their culture-specific physical and social terrain, specific behavior-outcome mappings or cognitive representations are differentially associated with dopaminergic activity (Figure 6.10). Subsequently, certain representations are reinforced over others because they may be associated with higher dopaminergic responses signaling greater congruence between the behavior and the intended outcome in the physical and social environment. Over extended time and exposure to the environment, the drive of the brain to automatize and maximize information processing under limited resources might then involve consolidating initially novel behaviors so that they become habitual or more reflexive (de Wit et al., 2012; Dolan & Dayan, 2013; Graybiel, 2008). Thus, culture-specific habits are formed from an individual's endogenous cognitive model of how the environment functions, which facilitates subsequent habitual behaviors when interacting with the environment so long as violations of the cognitive model do not occur. Indeed, habitual behaviors require minimal neural effort under expected circumstances but require additional contingent processing when norms are violated, consistent

Visual Cognition and Culture

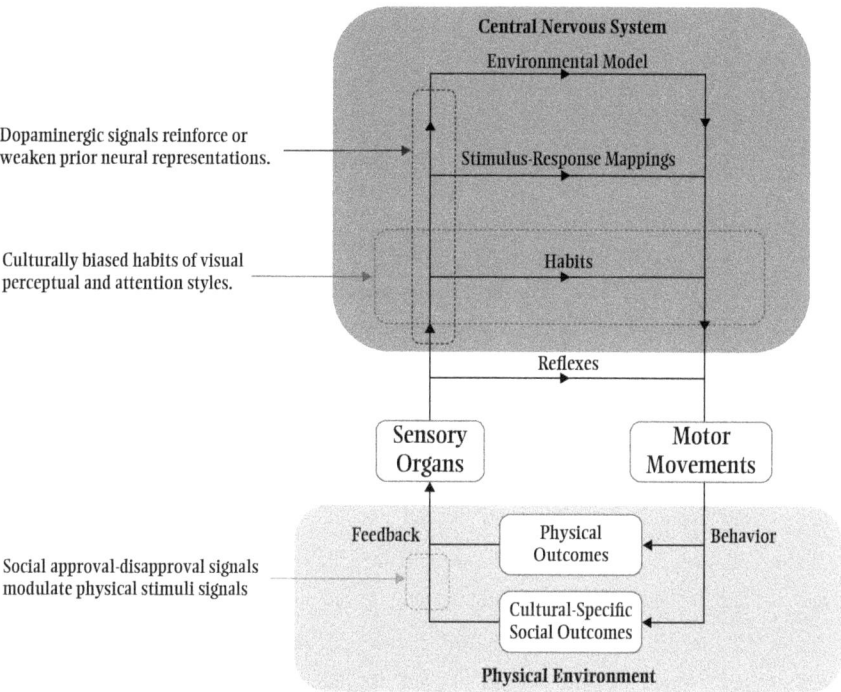

Figure 6.10. Schematic of how reinforcement learning that relies the dopaminergic reward system might be involved in the formation of culture-specific habits of visual perceptual and attention styles. A key contributor is the culture-specific social feedback that modulates information about physical stimuli in the environment.

with the neurophysiological studies reviewed above that find greater neural effort when engaging culturally non-preferred vs. preferred cognitive strategies. When norms are violated, the cognitive model is in error and the habitual behavior does not result in expected outcomes. As such, greater neural effort may be required to detect errors, update of the cognitive model, and overcome the established habit.

Given this present framework, certain concrete projections can be made about the operations of cultural influences in modulating visual processing. First, cultural differences in visual processing of physical information must be a function of the nature and degree of cultural differences in social feedback pertaining to the processing of that specific information. To the extent that social feedback experience about a physical element is absent, visual representations of that element must be more veridical and less culturally biased. Second, cultural differences in visual processing of physical information must also be a function of the sensitivity of the specific neural processes to dopaminergic (or other neurotransmitter) feedback signals. Neural processes that are unaffected by dopamine-related input should be more veridical with regard to the physical quantity. Third, the extent to which feedback is incongruent to a culture-specific bias on visual representation is measurable in terms of changes in dopaminergic activity relative to when feedback is congruent (or as expected) in relevant brain systems. Fourth, this latter measure also serves as an index of how established a culture-specific bias on visual representations is in the brain, which is then tied to the malleability of the representation or associated behavior with respect to the outcome. Strong cultural biases may be inaccessible to the awareness of the individual, or at least difficult to bypass. Weak cultural biases may be cognitively apprehended and circumvented with emphasis or strategies, albeit with the requirement of more neural resources. Finally, an individual's susceptibility to cultural influences on visual processing is a function of the sensitivity of his or her dopaminergic system to a specific social feedback that may be associated with genetic, epigenetic, and neurodevelopmental factors. Future investigations to evaluate cultural influences on visual cognition based on these considerations will be important to establish critical links between social, cognitive, and neurobiological mechanisms and the physical world. Indeed, some studies have already made initial links in other domains of cultural neuroscience including self-construal, emotional processing, social interaction, genetic processes, and clinical applications, and it is likely that such work will expand in this nascent field (Chiao, Li, & Seligman, 2016; Han et al., 2013; Kim & Sasaki, 2014).

APPLICATIONS IN NORMATIVE AND CLINICAL SETTINGS

The pervasiveness of cultural influences on visual perception and attention has implications on human self-construal and social interaction with others. Evidence shows that individuals tend to justify and patch their values rather than change their judgments when engaged in behaviors that are inconsistent with those personal values (Brehm, 1956; Festinger, 1962; Stone & Cooper, 2001). Importantly, there are cultural differences in the contexts under which individuals justify their behaviors as well as the strategies used at least between Westerners and East Asians (Kitayama, Snibbe, Markus, & Suzuki, 2004). Thus, it is plausible that certain cultural biases, including those for visual perceptions, might be so embedded in an individual's self-construal that it is difficult to process alternative models of the same physical phenomena adopted by other persons. Under such circumstances, there is potential for individual biases in physical perceptions such as in interpreting facial expressions to lead to miscommunications, misunderstandings, and disagreements in social interactions, however subtle. We suggest that better understanding of the presence and specific operations of cultural influences on visual cognition and associated neural mechanisms is critical in developing strategies that facilitate cross-cultural interpersonal and social interactions. Broadly, more effective communication strategies for instance can be adopted in formal settings such as in educational contexts, the marketplace, or political exchanges if there can be more clearly established common grounds despite mutual differences in physical perceptions.

Such clarifications might be most immediately relevant in clinical settings where cultural biases affect the expression of psychiatric symptoms, societal attitudes toward deviant behaviors, and the diagnosis and treatment of these behaviors (Alarcón et al., 2002; Chiao & Blizinsky, 2013; Kirmayer & Ban, 2013). For example, due to differences in beliefs about diseases and human behaviors, manifestations of schizophrenic symptoms are more accepted in Balinese culture than in Japanese culture (Kurihara, Kato, Sakamoto, Reverger, & Kitamura, 2000). Again, differences in the expression and interpretation of facial emotions in patients, caretakers, and physicians with different cultural backgrounds might be particularly relevant to this issue. It is also possible that cultural differences in visual perceptions about how objects are related to each other in these same agents are important in clinical scenarios. Thus, to aid diagnosis and treatment, there is a need for more objective determination of the contribution of cultural biases on the manifestations of mental illnesses. A culture-specific approach might be required in which the relevant beliefs of values associated with an observed psychiatric behavior must be characterized and its internal logic and neural underpinnings understood. Specifically, rather than conveniently applying widely used conventional screening tools including assessments that involve visual perception, validity across cultures should be closely examined and further adaptation should include adequate interpretation based on well-studied linguistically- and culturally-adapted normative data and cut-off scores. With sufficient data, clinical norms regarding these behaviors can be more accurately determined that account for cultural values. With such a profile of the culturally embedded individual's perceptions and representations, more effective care might then be provided.

In a similar vein, the consideration of cultural differences in visual cognition might also be critical in assessing normative versus pathological aging in different culture groups—an urgent issue given the aging world population. As suggested in the previously reviewed studies by Goh and colleagues (Chee et al., 2006; Goh et al., 2004, 2007), older East Asians compared to older Westerners appeared to be less able to engage effective object-focused processing of scenes during passive viewing although this was rescued under altered instructions. Importantly, the former result was associated with poorer medial temporal function—a region implicated for well-known age-related declines in episodic memory. Without accounting for the contribution of holistic (vs. analytic) visual style in normative older East Asians, a mistaken conclusion might have been made that older East Asians were more susceptible to age-related cognitive declines or dementia than their Western counterparts. Overall, biological aging and cultural influences both are longitudinal forces that operate over the lifespan, and by extension over generational cohorts (Graves et al., 1999; Hoffman, Hofer, & Sliwinski, 2011). As such, investigations involving such samples are unique opportunities to evaluate the relative contributions of neurobiological, experiential, and evolutionary mechanisms acting on human behavior or cognition (Gutchess & Goh, 2013).

Finally, computer and mobile technology is arguably the most important new cultural influence on visual perception in the human species as a whole in this information age. On a daily basis, billions of individuals are intensely engaging visuomotor processes to interact with visual information streaming into the brain from computer screens, and, recently, more immersive virtual reality environments and augmented reality applications. These visual stimuli are also coupled with social feedback signals, particular when using social media, multimedia, and programmed outcomes that are at times more addictive than real interactions (Andreassen et al., 2016; Hormes, Kearns, & Timko, 2014; Ryan, Chester, Reece, & Xenos, 2014). The same property that enables such media to actively engage individuals also affords a new avenue to explore its application in psychiatric or other clinical treatments and therapies over traditional methods that might be more time-consuming and less motivating (Bavelier et al., 2011; Shams et al., 2015). Importantly, many individuals can now readily receive social and physical information that is not limited by physical distance or even temporal delays, which vastly expands the variability of culture-specific social feedback signals encountered. Such availability and increasing reliance on this form of interaction with the physical and social environment has a profound impact on the formation of novel cultural boundaries between groups of individuals. Applying our framework to understand the neural consequences of this new cultural influence, it will thus be intriguing to see how different culture-specific cognitive styles for processing physical quantities might emerge from this phenomenon and, ultimately, how we are affected by our own devices. Indeed, this technologically driven evolution of the human species presents a unique challenge for us to apprehend and adapt to this new cultural environment, for which we do not yet have a map.

CONCLUSION

In this chapter, we have reviewed many findings showing that there are culture-related effects on cognitive processing even for objective physical visual information. We described how reinforcement learning and the reward processing system in the brain might utilize social congruence feedback signals to modulate visual representations thus giving rise to cultural differences in visual processing. Given this, cognitive research studies involving only select samples require re-evaluation, or at least new investigations that either validate or delineate the specific cognitive mechanisms that operate differentially across samples from different cultural backgrounds. With the rapid increase of interconnectedness worldwide, understanding culture-related contributions to behavior and cognition has also become a foremost agendum. Modern ease of access to the Internet and international travel has resulted in proliferation of cultural fronts, cultural interactions, and cultural mixing. Thus, such inquiry and considerations of the role of culture on visual cognition, among other cognitive processes, are critical to address societal and health issues that might be associated with the increasing frequency of individuals who encounter unfamiliar cultural interpretations of the physical environment in both normative and clinical contexts. Only by knowing what maps others are using, can we begin to reliably navigate the physical and social world together.

ACKNOWLEDGMENTS

The Taiwan Ministry of Science and Technology grants MOST103-2410-H-002-082-MY2 and MOST 105-2420-H-002-002-MY2 supported the authors in the duration of this work.

REFERENCES

Akechi, H., Senju, A., Uibo, H., Kikuchi, Y., Hasegawa, T., & Hietanen, J. K. (2013). Attention to eye contact in the West and East: Autonomic responses and evaluative ratings. *PLoS One, 8*(3). https://doi.org/10.1371/journal.pone.0059312

Alarcón, R. D., Bell, C. C., Kirmayer, L. J., Lin, K.-M., Üstün, B., & Wisner, K. L. (2002). Beyond the funhouse mirrors: Research agenda on culture and psychiatric diagnosis. In D. J. Kupfer, M. B. First, & D. A. Regier (Eds.), *A research agenda for DSM-V* (pp. 219–281). Arlington, VA: American Psychiatric Association.

Andreassen, C. S., Billieux, J., Griffiths, M. D., Kuss, D. J., Demetrovics, Z., Mazzoni, E., & Pallesen, S. (2016). The relationship between addictive use of social media and video games and symptoms of psychiatric disorders: A large-scale cross-sectional study. *Psychology of Addictive Behaviors, 30*, 252–262. https://doi.org/10.1037/adb0000160

Baciu, M., Koenig, O., Vernier, M. P., Bedoin, N., Rubin, C., & Segebarth, C. (1999). Categorical and coordinate spatial relations: fMRI evidence for hemispheric specialization. *Neuroreport, 10*, 1373–1378. https://doi.org/ 10.1097/00001756-199904260-00040

Barton, J. J. S., Press, D. Z., Keenan, J. P., & O'Connor, M. (2002). Lesions of the fusiform face area impair perception of facial configuration in prosopagnosia. *Neurology, 58*, 71–78. http://dx.doi.org/10.1212/WNL.58.1.71

Bavelier, D., Green, C. S., Han, D. H., Renshaw, P. F., Merzenich, M. M., & Gentile, D. A. (2011). Brains on video games. *Nature Reviews. Neuroscience, 12*, 763–768. https://doi.org/10.1038/nrn3135

Blais, C., Jack, R. E., Scheepers, C., Fiset, D., & Caldara, R. (2008). Culture shapes how we look at faces. *Public Library of Science One, 3*(8), e3022. https://doi.org/10.1371/journal.pone.0003022

Boduroglu, A., Shah, P., & Nisbett, R. E. (2009). Cultural differences in allocation of attention in visual information processing. *Journal of Cross-Cultural Psychology, 40*, 349–360. http://dx.doi.org/10.1177/0022022108331005

Bouvier, S. E., & Engel, S. A. (2006). Behavioral deficits and cortical damage loci in cerebral achromatopsia. *Cerebral Cortex, 16*, 183–191. https://doi.org/10.1093/cercor/bhi096

Brehm, J. W. (1956). Postdecision changes in the desirability of alternatives. *Journal of Abnormal and Social Psychology, 52*, 384–389. https://doi.org/10.1037/h0041006

Caldara, R., Zhou, X., & Miellet, S. (2010). Putting culture under the "spotlight" reveals universal information use for face recognition. *PloS One, 5*(3), e9708. https://doi.org/10.1371/journal.pone.0009708

Cantlon, J. F., & Brannon, E. M. (2007). Adding up the effects of cultural experience on the brain. *Trends in Cognitive Sciences, 11*, 1–4. http://dx.doi.org/10.1016/j.tics.2006.10.008

Chee, M. W., Goh, J. O. S., Venkatraman, V., Tan, J. C., Gutchess, A., Sutton, B., . . . Park, D. (2006). Age-related changes in object processing and contextual binding revealed using fMR adaptation. *Journal of Cognitive Neuroscience, 18,* 495–507. https://doi.org/10.1162/jocn.2006.18.4.495

Cheon, B. K., Im, D.-M., Harada, T., Kim, J.-S., Mathur, V. A., Scimeca, J. M., . . . Chiao, J. Y. (2013). Cultural modulation of the neural correlates of emotional pain perception: The role of other-focusedness. *Neuropsychologia, 51,* 1177–1186. https://doi.org/10.1016/j.neuropsychologia.2013.03.018

Chiao, J. Y., & Blizinsky, K. D. (2010). Culture-gene coevolution of individualism–collectivism and the serotonin transporter gene. *Proceedings. Biological Sciences/The Royal Society, 277,* 529–537. https://doi.org/10.1098/rspb.2009.1650

Chiao, J. Y., & Blizinsky, K. D. (2013). Population disparities in mental health: Insights from cultural neuroscience. *American Journal of Public Health, 103*(Suppl 1), S122–132. https://doi.org/10.2105/AJPH.2013.301440

Chiao, J. Y., Li, S.-C., & Seligman, R. (2016). *The Oxford handbook of cultural neuroscience.* New York, NY: Oxford University Press.

Chua, H. F., Boland, J. E., & Nisbett, R. E. (2005). Cultural variation in eye movements during scene perception. *Proceedings of the National Academy of Sciences USA, 102,* 12629–12633. https://doi.org/10.1073/pnas.0506162102

Darwin, C. E. (1999). *The expression of the emotions in man and animals* (3rd ed.). London, England: HarperCollins.

de Wit, S., Standing, H. R., Devito, E. E., Robinson, O. J., Ridderinkhof, K. R., Robbins, T. W., & Sahakian, B. J. (2012). Reliance on habits at the expense of goal-directed control following dopamine precursor depletion. *Psychopharmacology, 219,* 621–631. https://doi.org/10.1007/s00213-011-2563-2

Dolan, R. J., & Dayan, P. (2013). Goals and habits in the brain. *Neuron, 80,* 312–325. https://doi.org/10.1016/j.neuron.2013.09.007

Ekman, P., & Friesen, W. (1977). *Facial action coding system.* Palo Alto, CA: Consulting Psychologists Press.

Festinger, L. (1962). *A theory of cognitive dissonance.* Stanford, CA: Stanford University Press.

Goh, J. O. S., Chee, M. W., Tan, J. C., Venkatraman, V., Hebrank, A., Leshikar, E. D., . . . Park, D. C. (2007). Age and culture modulate object processing and object–scene binding in the ventral visual area. *Cognitive, Affective, & Behavioral Neuroscience, 7,* 44–52. https://doi.org/10.3758/CABN.7.1.44

Goh, J. O. S., Hebrank, A. C., Sutton, B. P., Chee, M. W. L., Sim, S. K. Y., & Park, D. C. (2013). Culture-related differences in default network activity during visuo-spatial judgments. *Social Cognitive and Affective Neuroscience, 8,* 134-42https://doi.org/10.1093/scan/nsr077

Goh, J. O. S., & Huang, C.-M. (2012). Images of the cognitive brain across age and culture. In P. Bright (Ed.), *Neuroimaging: Cognitive and clinical neuroscience.* London, England: InTech. Retrieved from http://www.intechopen.com/books/neuroimaging-cognitive-and-clinical-neuroscience/imaging-the-brain-across-culture-and-age

Goh, J. O. S., Leshikar, E. D., Sutton, B. P., Tan, J. C., Sim, S. K. Y., Hebrank, A. C., & Park, D. C. (2010). Culture differences in neural processing of faces and houses in the ventral visual cortex. *Social Cognitive and Affective Neuroscience, 5,* 227–235. https://doi.org/10.1093/scan/nsq060

Goh, J. O. S., & Park, D. C. (2009). Culture sculpts the perceptual brain. *Progress in Brain Research, 178,* 95–111. https://doi.org/10.1016/S0079-6123(09)17807-X

Goh, J. O. S., Siong, S. C., Park, D., Gutchess, A., Hebrank, A., & Chee, M. W. L. (2004). Cortical areas involved in object, background, and object–background processing revealed with functional magnetic resonance adaptation. *Journal of Neuroscience, 24,* 10223–10228. https://doi.org/10.1523/JNEUROSCI.3373-04.2004

Goh, J. O. S., Tan, J. C., & Park, D. C. (2009). Culture modulates eye-movements to visual novelty. *PloS One, 4,* e8238. https://doi.org/10.1371/journal.pone.0008238

Graves, A. B., Rajaram, L., Bowen, J. D., McCormick, W. C., McCurry, S. M., & Larson, E. B. (1999). Cognitive decline and Japanese culture in a cohort of older Japanese Americans in King County, WA: The Kame Project. *The Journals of Gerontology. Series B, Psychological Sciences and Social Sciences, 54,* S154–S161. http://dx.doi.org/10.1093/geronb/54B.3.S154

Graybiel, A. M. (2008). Habits, rituals, and the evaluative brain. *Annual Review of Neuroscience, 31,* 359–387. https://doi.org/10.1146/annurev.neuro.29.051605.112851

Gutchess, A. H., & Goh, J. O. S. (2013). Refining concepts and uncovering biological mechanisms for cultural neuroscience. *Psychological Inquiry, 24,* 31–36. https://doi.org/10.1080/1047840X.2013.765338

Han, H., Glover, G. H., & Jeong, C. (2014). Cultural influences on the neural correlate of moral decision making processes. *Behavioural Brain Research, 259,* 215–228. https://doi.org/10.1016/j.bbr.2013.11.012

Han, S., Northoff, G., Vogeley, K., Wexler, B. E., Kitayama, S., & Varnum, M. E. W. (2013). A cultural neuroscience approach to the biosocial nature of the human brain. *Annual Review of Psychology, 64,* 335–359. https://doi.org/10.1146/annurev-psych-071112-054629

Hedden, T., Ketay, S., Aron, A., Markus, H. R., & Gabrieli, J. D. E. (2008). Cultural influences on neural substrates of attentional control. *Psychological Science, 19,* 12–17. https://doi.org/10.1111/j.1467-9280.2008.02038.x

Hitokoto, H., Glazer, J., & Kitayama, S. (2016). Cultural shaping of neural responses: Feedback-related potentials vary with self-construal and face priming. *Psychophysiology, 53,* 52–63. https://doi.org/10.1111/psyp.12554

Hoffman, L., Hofer, S. M., & Sliwinski, M. J. (2011). On the confounds among retest gains and age-cohort differences in the estimation of within-person change in longitudinal studies: A simulation study. *Psychology and Aging, 26,* 778–791. https://doi.org/10.1037/a0023910

Hormes, J. M., Kearns, B., & Timko, C. A. (2014). Craving Facebook? Behavioral addiction to online social networking and its association with emotion regulation deficits. *Addiction, 109,* 2079–2088. https://doi.org/10.1111/add.12713

Hsin, A., & Xie, Y. (2014). Explaining Asian Americans' academic advantage over whites. *Proceedings of the National Academy of Sciences of the United States of America, 111,* 8416–8421. https://doi.org/10.1073/pnas.1406402111

Imbo, I., & LeFevre, J.-A. (2009). Cultural differences in complex addition: Efficient Chinese versus adaptive Belgians and Canadians. *Journal of Experimental Psychology. Learning, Memory, and Cognition, 35,* 1465–1476. https://doi.org/10.1037/a0017022

Ishii, K., Miyamoto, Y., Mayama, K., & Niedenthal, P. M. (2011). When your smile fades away: Cultural differences in sensitivity to the disappearance of smiles. *Social Psychological and Personality Science, 2,* 516–522. https://doi.org/10.1177/1948550611399153

Jack, R. E., Caldara, R., & Schyns, P. G. (2012). Internal representations reveal cultural diversity in expectations of facial expressions of emotion. *Journal of Experimental Psychology: General, 141,* 19–25. https://doi.org/10.1037/a0023463

Jack, R. E., Garrod, O. G. B., Yu, H., Caldara, R., & Schyns, P. G. (2012). Facial expressions of emotion are not culturally universal. *Proceedings of the National Academy of*

Sciences of the United States of America, 109, 7241–7244. https://doi.org/10.1073/pnas.1200155109

Jenkins, L. J., Yang, Y.-J., Goh, J. O. S., Hong, Y.-Y., & Park, D. C. (2010). Cultural differences in the lateral occipital complex while viewing incongruent scenes. *Social Cognitive and Affective Neuroscience, 5,* 236–241. https://doi.org/10.1093/scan/nsp056

Kanwisher, N., McDermott, J., & Chun, M. M. (1997). The fusiform face area: A module in human extrastriate cortex specialized for face perception. *Journal of Neuroscience, 17,* 4302–4311. https://doi.org/10.1523/JNEUROSCI.17-11-04302

Kim, H. S., & Sasaki, J. Y. (2014). Cultural neuroscience: Biology of the mind in cultural contexts. *Annual Review of Psychology, 65,* 487–514. https://doi.org/10.1146/annurev-psych-010213-115040

Kim, H. S., Sherman, D. K., Mojaverian, T., Sasaki, J. Y., Park, J., Suh, E. M., & Taylor, S. E. (2011). Gene–culture interaction oxytocin receptor polymorphism (OXTR) and emotion regulation. *Social Psychological and Personality Science, 2,* 665–672. https://doi.org/10.1177/1948550611405854

Kim, H. S., Sherman, D. K., Sasaki, J. Y., Xu, J., Chu, T. Q., Ryu, C., . . . Taylor, S. E. (2010). Culture, distress, and oxytocin receptor polymorphism (OXTR) interact to influence emotional support seeking. *Proceedings of the National Academy of Sciences of the United States of America, 107,* 15717–15721. https://doi.org/10.1073/pnas.1010830107

Kirmayer, L. J., & Ban, L. (2013). Cultural psychiatry: Research strategies and future directions. *Advances in Psychosomatic Medicine, 33,* 97–114. https://doi.org/10.1159/000348742

Kitayama, S., Duffy, S., Kawamura, T., & Larsen, J. T. (2003). Perceiving an object and its context in different cultures: A cultural look at new look. *Psychological Science, 14,* 201–206. https://doi.org/ 10.1111/1467-9280.02432

Kitayama, S., King, A., Yoon, C., Tompson, S., Huff, S., & Liberzon, I. (2014). The dopamine D4 receptor gene (DRD4) moderates cultural difference in independent versus interdependent social orientation. *Psychological Science, 25,* 1169–1177. https://doi.org/10.1177/0956797614528338

Kitayama, S., Snibbe, A. C., Markus, H. R., & Suzuki, T. (2004). Is there any "free" choice? Self and dissonance in two cultures. *Psychological Science, 15,* 527–533. https://doi.org/10.1111/j.0956-7976.2004.00714.x

Knox, P. C., & Wolohan, F. D. A. (2014). Cultural diversity and saccade similarities: Culture does not explain saccade latency differences between Chinese and Caucasian participants. *PloS One, 9*(4), e94424. https://doi.org/10.1371/journal.pone.0094424

Kurihara, T., Kato, M., Sakamoto, S., Reverger, R., & Kitamura, T. (2000). Public attitudes towards the mentally ill: A cross-cultural study between Bali and Tokyo. *Psychiatry and Clinical Neurosciences, 54,* 547–552. https://doi.org/10.1046/j.1440-1819.2000.00751.x

Luo, S., Ma, Y., Liu, Y., Li, B., Wang, C., Shi, Z., . . . Han, S. (2015). Interaction between oxytocin receptor polymorphism and interdependent culture values on human empathy. *Social Cognitive and Affective Neuroscience, 10,* 1273–1281. https://doi.org/10.1093/scan/nsv019

Luo, Z., Jose, P. E., Huntsinger, C. S., & Pigott, T. D. (2007). Fine motor skills and mathematics achievement in East Asian American and European American kindergartners and first graders. *British Journal of Developmental Psychology, 25,* 595–614. https://doi.org/10.1348/026151007X185329

Masuda, T., Ellsworth, P. C., Mesquita, B., Leu, J., Tanida, S., & Van de Veerdonk, E. (2008). Placing the face in context: Cultural differences in the perception of facial emotion. *Journal of Personality and Social Psychology, 94,* 365–381. https://doi.org/10.1037/0022-3514.94.3.365

Masuda, T., & Nisbett, R. E. (2001). Attending holistically versus analytically: Comparing the context sensitivity of Japanese and Americans. *Journal of Personality and Social Psychology, 81,* 922–934. http://dx.doi.org/10.1037/0022-3514.81.5.922

Masuda, T., & Nisbett, R. E. (2006). Culture and change blindness. *Cognitive Science, 30,* 381–399. https://doi.org/ 10.1207/s15516709cog0000_63

Miller, K. F., Smith, C. M., Zhu, J., & Zhang, H. (1995). Preschool origins of cross-national differences in mathematical competence: The role of number-naming systems. *Psychological Science, 6,* 56–60. https://doi.org/10.1111/j.1467-9280.1995.tb00305.x

Mu, Y., Kitayama, S., Han, S., & Gelfand, M. J. (2015). How culture gets embrained: Cultural differences in event-related potentials of social norm violations. *Proceedings of the National Academy of Sciences of the United States of America, 112,* 15348–15353. https://doi.org/10.1073/pnas.1509839112

Ng, S. S. N., & Rao, N. (2010). Chinese number words, culture, and mathematics learning. *Review of Educational Research, 80,* 180–206. https://doi.org/10.3102/0034654310364764

Nisbett, R. E. (2003). *The geography of thought: How Asians and Westerners think differently—And why.* New York, NY: Free Press.

Park, J., & Kitayama, S. (2014). Interdependent selves show face-induced facilitation of error processing: Cultural neuroscience of self-threat. *Social Cognitive and Affective Neuroscience, 9,* 201–208. https://doi.org/10.1093/scan/nss125

Rhodes, G. (1985). Lateralized processes in face recognition. *British Journal of Psychology,76*(Pt 2), 249–271. http://dx.doi.org/10.1111/j.2044-8295.1985.tb01949.x

Rodic, M., Zhou, X., Tikhomirova, T., Wei, W., Malykh, S., Ismatulina, V., . . . Kovas, Y. (2015). Cross-cultural investigation into cognitive underpinnings of individual differences in early arithmetic. *Developmental Science, 18,* 165–174. https://doi.org/10.1111/desc.12204

Rossion, B., Dricot, L., Devolder, A., Bodart, J. M., Crommelinck, M., De Gelder, B., & Zoontjes, R. (2000). Hemispheric asymmetries for whole-based and part-based face processing in the human fusiform gyrus. *Journal of Cognitive Neuroscience, 12,* 793–802. https://doi.org/10.1162/089892900562606

Rotshtein, P., Geng, J. J., Driver, J., & Dolan, R. J. (2007). Role of features and second-order spatial relations in face discrimination, face recognition, and individual face skills: Behavioral and functional magnetic resonance imaging data. *Journal of Cognitive Neuroscience, 19,* 1435–1452. https://doi.org/10.1162/jocn.2007.19.9.1435

Ryan, T., Chester, A., Reece, J., & Xenos, S. (2014). The uses and abuses of Facebook: A review of Facebook addiction. *Journal of Behavioral Addictions, 3,* 133–148. https://doi.org/10.1556/JBA.3.2014.016

Schultz, W. (2013). Updating dopamine reward signals. *Current Opinion in Neurobiology, 23,* 229–238. https://doi.org/10.1016/j.conb.2012.11.012

Schultz, W., Dayan, P., & Montague, P. R. (1997). A neural substrate of prediction and reward. *Science, 275,* 1593–1599. https://doi.org/10.1126/science.275.5306.1593

Schultz, W., & Dickinson, A. (2000). Neuronal coding of prediction errors. *Annual Review of Neuroscience, 23,* 473–500. https://doi.org/10.1146/annurev.neuro.23.1.473

Schwartz, S. H. (1992). Universals in the content and structure of values: Theoretical advances and empirical tests in 20 countries. *Advances in Experimental Social Psychology, 25,* 1–62. https://doi.org/10.1016/S0065-2601(08)60281-6

Sergent, J. (1982). About face: Left-hemisphere involvement in processing physiognomies. *Journal of Experimental Psychology: Human Perception and Performance, 8,* 1–14. http://dx.doi.org/10.1037/0096-1523.8.1.1

Shams, T. A., Foussias, G., Zawadzki, J. A., Marshe, V. S., Siddiqui, I., Müller, D. J., & Wong, A. H. C. (2015). The Effects of video games on cognition and brain structure: Potential implications for neuropsychiatric disorders. *Current Psychiatry Reports, 17,* 71. https://doi.org/10.1007/s11920-015-0609-6

Stone, J., & Cooper, J. (2001). A self-standards model of cognitive dissonance. *Journal of Experimental Social Psychology, 37,* 228–243. https://doi.org/10.1006/jesp.2000.1446

Tang, Y., Zhang, W., Chen, K., Feng, S., Ji, Y., Shen, J., . . . Liu, Y. (2006). Arithmetic processing in the brain shaped by cultures. *Proceedings of the National Academy of Sciences, 103,* 10775–10780. https://doi.org/10.1073/pnas.0604416103

Tulving, E. (2002). Episodic memory: From mind to brain. *Annual Review of Psychology, 53,* 1–25. https://doi.org/10.1146/annurev.psych.53.100901.135114

Wang, K., Umla-Runge, K., Hofmann, J., Ferdinand, N. K., & Chan, R. C. K. (2014). Cultural differences in sensitivity to the relationship between objects and contexts: Evidence from P3. *Neuroreport, 25,* 656–660. https://doi.org/10.1097/WNR.0000000000000152

7

Cognitive Reserve, Bilingualism, and the Aging Brain

BRIAN T. GOLD

INTRODUCTION

The world's population is aging at an unprecedented rate. For example, the proportion of individuals aged 60 or above increased from 9.2% in 1990 to 11.7% in 2013 and is projected to reach 21.1% of the world's population by 2050 (United Nations, 2013). Population aging in part reflects healthcare advances that constitute cause for rejoicing. However, older adults will only be able to reap the rewards of their unprecedented longevity if they have a healthy brain to go with it. As our population continues to age, increasing numbers of individuals will experience cognitive decline, which will place a large strain on individuals, families, and healthcare systems. The identification of lifestyle variables that may attenuate age-related cognitive declines has thus become a practical imperative.

Aging is associated with declines in multiple cognitive domains that appear to be most pronounced in the areas of general information processing speed (Salthouse, 1996), memory (Craik & Salthouse, 2008) and executive functions (Schaie, 1996). Aging is also the greatest risk factor for dementia, with Alzheimer's disease (AD) being the most common form. The clinical prototype for AD involves early impairment of memory functions followed by subsequent disruption of other cognitive domains such as executive function, visuospatial processes, and semantic memory (McKhann et al., 1984). Eventually AD results in the inability to communicate, recognize family members, or function independently.

Over the past few decades, much progress has been made in understanding the cellular-, molecular-, and systems-level bases of aging and AD. At the neural systems level, age-related cognitive declines have been linked to multiple forms of degenerative change, including atrophy of gray matter brain structures, disruption of white matter (WM) connections, reductions in vascular integrity, and depletion of neurotransmitter systems (Kemper, 1994; Raz & Kennedy, 2009). In

AD, early memory loss has been linked with damage to medial temporal lobe (MTL) regions such as the hippocampus (Convit et al., 1997; Jack et al., 1997), a structure that plays a critical role in the encoding of new memories (Smith & Squire, 2009).

At present, current drug treatments have only modest effects on the symptomatic course of AD (Birks, 2006; Farrimond, Roberts, & McShane, 2012). Although new interventions are under investigation, they are likely to be more successful at slowing rather than reversing neurodegeneration. On the positive side, it has been known for some time that the symptomatic course of AD can be modulated by environmental variables (Mayeux, 2003; Spires & Hannan, 2005). In particular, certain lifestyle variables appear to boost the brain's capacity to resist cognitive declines associated with age-related neurodegenerative diseases (Stern, 2002). Exciting recent evidence suggests that bilingualism may delay the onset of clinical AD symptoms by 4–5 years (Bialystok, Craik, & Freedman, 2007; Craik, Bialystok, & Freedman, 2010; Alladi et al., 2013; Bak, Nissan, Allerhand, & Deary, 2014).

The purpose of present review is to summarize the available evidence concerning bilingualism as a potential cognitive reserve (CR) variable against AD. I begin by summarizing the theory of CR and then describe key studies suggesting that bilingualism may delay the onset of AD symptoms. Next, the role of potential confounding factors in bilingual CR studies is discussed. I then describe findings from recent studies that appear to have addressed several of these potential confounds. Results from neuroimaging studies that provide support for bilingualism as a CR variable are then discussed. Based on the available data, it is hypothesized that bilingualism may delay AD symptom onset by boosting the functioning of frontostriatal and/or frontoparietal brain regions involved in executive control (EC) functions.

COGNITIVE RESERVE

The theory of CR arose as an explanation for the gap between brain health and cognitive functioning (Stern, 2002; Richards & Deary, 2005). For example, while the majority of individuals with who meet criteria for pathological AD also meet clinical AD criteria, a significant number remain cognitively normal (Mortimer, 1997; Valenzuela & Sachdev, 2006; Shaw et al., 2009). An early example of the gap between AD neuropathology and cognitive functioning came from the Kentucky Nun Study, which found that 32% of older adults with Braak Stage III and IV pathology (where Stage VI indicates the most severe pathology) had normal memory function before death (Riley, Snowdon, & Markesbery, 2002).

CR theory points to such findings as evidence that certain variables can moderate the relationship between brain integrity and clinical expression of disease (Stern, 2002; Richards & Deary, 2005). These cognitive factors are thought to improve the brain's ability to cope with damage, effectively mitigating its effects on cognition (Stern, 2002, 2009). Stern (2002) proposed two forms through which reserve may be instantiated in the brain. One passive form of reserve, brain reserve, was suggested to involve structural differences such as overall brain size and number of neurons/synapses. An active form of reserve, CR, was described as the ability for plastic functional brain reorganization of cognitive networks in response to injury, aging, or disease. While cognitive and brain reserve may ultimately boil down to similar underlying cellular/molecular mechanisms, they remain useful heuristics for describing findings from structural and functional neuroimaging experiments.

Putative CR variables include education, intelligence, socioeconomic status (SES) and aerobic fitness (Albert et al., 1995; Christensen, 2001; Hillman, Erickson, & Kramer, 2008; Steffener & Stern, 2012). Uncovering other CR variables represents an important step toward maximizing the ability of older adults to live independently. In addition, this line of research has implications for early detection of dementia. Because individuals with high CR present with greater brain burden, typical cognitive screening tests may be insufficient for their early detection. A more complete understanding of the range of effective CR variables may aid early detection of dementia.

EARLY EVIDENCE FOR BILINGUALISM AS A COGNITIVE RESERVE VARIABLE

Lifelong bilingualism (hereafter referred to simply as bilingualism) refers to speaking two languages on a regular basis since childhood. Bilingualism has garnered much interest as a potential CR variable because it is an environmental factor for which no special education or intelligence appears to be needed. The initial evidence for bilingualism as a CR came from two studies suggesting that lifelong bilinguals tend to develop clinical AD symptoms at an older age than monolinguals (Bialystok et al., 2007; Craik et al., 2010). In these studies, patient records were retrospectively classified into one of two groups: bilinguals (those who regularly used at least two languages since least early adulthood) and monolinguals (those who had regular exposure to only one language). The age of dementia onset was estimated as that at which the first clinical symptoms suggestive of dementia were reported to the neurologist by patients or their families.

In Bialystok et al.'s (2007) seminal study, the final sample consisted of 184 case records of patients meeting clinical criteria for probable AD or another form of dementia. The average age of symptom onset in the bilingual group of 75.5 was about 4 years later than the average age of symptom onset in the monolingual group of 71.4. In addition, the bilingual group was an average of 3 years older than the monolingual group at their first clinic appointment (75.4 vs. 78.6). The bilingual patients thus exhibited symptoms of dementia when they were 3–4 years older than the monolingual patients. In a follow-up study by Craik et al. (2010) with a different group of participants, bilingual patients were found to have been diagnosed with dementia when they were an average of 4.3 years older than monolingual patients.

CONTROL FOR POTENTIAL CONFOUNDS OF BILINGUALISM AS A COGNITIVE RESERVE VARIABLE

The initial studies described above on bilingualism as a CR variable controlled for the potentially confounding effects of education, IQ and SES levels. The influence of several other potentially confounding variables has been addressed more recently. One potential confound has concerned immigration status. In the initial studies, bilinguals tended to be much more likely to be immigrants than their monolingual counterparts. This trend reflects the practical reality that most studies have been conducted in geographical regions in which lifelong bilingualism tends to be rare/recent phenomenon in native-born individuals.

The potential influence of immigration on bilingual CR effects was first directly addressed in a study conducted in Montreal, Canada, a city in which bilinguals can be either immigrants or native-born Canadians who grew up speaking both English and French (Chertkow et al., 2010). Chertkow et al. (2010) found an overall delay in AD diagnosis of approximately 3 years for multilingual patients (including both immigrant and nonimmigrant groups) compared to monolingual patients. However, the results were weaker when only native-born Canadian groups were compared. While there was a trend for an older age of AD diagnosis in native-born Canadian bilinguals who spoke French as their first language, there was no comparable trend for native-born bilinguals who spoke English as their first language.

These findings raised the possibility that bilingualism may only confer CR through an interaction with immigration status. On first pass, the reason for such an interaction does not seem intuitively obvious. However, it is conceivable that immigrants may be more resilient than nonimmigrants for genetic and environmental reasons. For example, population-genetic studies have linked migratory patterns with frequency of novelty-seeking alleles of the DRD4 gene (Matthews & Butler, 2011), a trait that can predict positive outcomes when combined with persistence and curiosity (Josefsson et al., 2013). In addition, immigrants may also be more likely to live in extended families, which is relevant in that increased social stimulation may delay cognitive decline (Bennett, Schneider, Tang, Arnold, Wilson, & 2006).

However, results from a recent study conducted in India have provided strong evidence for bilingualism as a CR variable independent of immigration status (Alladi et al., 2013). Because multilingualism is extremely common in India, a large number of native-born, nonimmigrant participants who were bilingual or monolingual could be recruited for this study. In addition, the location of this study allowed for stronger control of over variables of literacy/education since bilingualism in India is typical even in illiterate individuals (Alladi et al., 2013). Participants were part of a large-sample of 648 individuals involved in a longitudinal dementia study, 391 of whom were multilingual. Results indicated that native-born bilingual patients developed clinical symptoms suggestive of AD or other dementia an average of 4.5 years later than native-born monolingual patients. In addition, and importantly, illiterate bilinguals were found to be an average of 6 years older than illiterate monolinguals at the first sign of dementia symptoms.

Finally, results from another recent large-scale study (N = 853) have provided evidence that the positive effects of bilingualism cannot be explained by childhood intelligence or verbal fluency (Bak et al., 2014). Participants were part of the Lothian Birth Cohort 1936 study and underwent baseline cognitive testing in 1947 (at 11 years of age) and were then retested in 2008–2010 (at a mean age of 72.5 years old). Results indicated that older adult bilinguals performed significantly better than their monolingual peers on measures of fluid intelligence, verbal fluency, and visual search speed (but not memory) even after controlling for childhood scores on these measures.

BILINGUALISM APPEARS TO BE A CR VARIABLE

A search of the PubMed databases (through October 2014, using the search terms *bilingual AND reserve, bilingual AND aging,* and *bilingual AND dementia*) was conducted as a way to summarize the current literature on bilingualism and dementia risk. This search indicated a total of 16 studies reporting protective effects of bilingualism against either age-related cognitive or brain declines or the clinical onset of dementia symptoms. The search also revealed two studies with large sample sizes reporting no protective effects of bilingualism (Yeung et al., 2014; Zahodne, Schofield, Farrell, Stern, & Manly, 2014). While these null findings should be noted, it should also be pointed out that null effects have also been reported for other putative CR variables. For example, longitudinal studies have reported conflicting results with respect to the relationship between educational level and age-related cognitive decline (cf. Christensen, 2001; Van Dijk, Van Gerven, Van Boxtel, Van der Elst, & Jolles, 2008).

This underscores the enormous complexity of human CR research. Findings in CR studies may be affected by age range of the sample, specific cognitive measures used, the length of the study and number of consecutive assessments, and confounding effects of health (e.g., preclinical AD), among many other variables. Notwithstanding these caveats, the available evidence suggests that bilingualism shows strong promise as a CR variable against AD. First, the finding that bilinguals develop clinical AD symptoms at an older age than monolinguals has now been reported a number of times by a number of research groups. Second, CR effects associated with bilingualism do not appear attributable to education, SES, adult IQ, childhood IQ, literacy, or immigration status. Together, these data suggest that bilingualism is a form of CR that appears to operate primarily through environmental rather than genetic influences.

Future research will be required to determine if bilingual protective effects are on par with more well-established CR variables. However, preliminary results are encouraging. For example, results from the Lothian Birth Cohort indicated effect sizes associated with bilingualism in delaying cognitive declines (Bak et al., 2014) that were comparable to those reported in the same participants for variation in the gene for apolipoprotein E, physical fitness, and (not) smoking (Deary,

Gow, Pattie, & Starr, 2012). In the following discussion, I consider several potential candidate neurocognitive systems which may underlie bilingual CR against AD.

EFFECTS OF BILINGUALISM ON MEMORY SYSTEMS IN AGING

Bilingualism could delay the onset of clinical AD symptoms by directly protecting memory circuits affected in early-stage AD or by enhancing other neural systems. Regions of the MTL form a critical portion of the neural circuitry for declarative memory (Squire, 1992), and are affected in the earliest stages of AD (Convit et al., 1997; Jack et al., 1997). The volume of MTL structures appear to be reduced in preclinical stages of amnestic mild cognitive impairment/AD (Martin, Smith, Collins, Schmitt, & Gold; Dickerson & Wolk, 2012). Other brain structures prominently affected in preclinical stages of AD include midline parietal and basal frontal regions (Dickerson et al., 2009; Vlassenko, Benzinger, & Morris, 2012). In addition, microstructural properties of WM tracts that connect MTL structures with midline parietal and basal frontal regions such as the cingulum and the fornix, respectively, are also altered in early-stage AD (Chua, Wen, Slavin, & Sachdev, 2008; Gold et al., 2012; Stebbins & Murphy, 2009).

Does bilingualism delay AD by directly protecting MTL-based declarative memory systems affected in early-stage AD? At present, there appears to be little evidence to support this possibility. For example, bilingualism does not attenuate age-related declines in the ability to overcome memory interference (Fernandes, Craik, Bialystok, & Kreuger, 2007). Similarly, in the large-sample study by Bak et al. (2014), older adult bilinguals performed significantly better than monolinguals on several EC measures (discussed in the following text) but showed no differences on memory measures. However, older adult bilinguals have been found to show larger gray matter volume in left temporal pole than their monolingual peers (Abutalebi et al., 2014). The temporal pole contributes to semantic memory, which is negatively affected in AD, and protection of this region could thus contribute to reserve against AD.

Nevertheless, the earliest AD-related declines involve reductions in episodic memory and MTL structures (Convit et al., 1997; Jack et al., 1997). At present, there exist two neuroimaging studies with findings directly related to the relative structural integrity of MTL regions in older adult bilinguals and monolingual groups (Schweizer, Ware, Fischer, Craik, & Bialystok, 2012; Gold, Johnson, & Powell, 2013). Results from these studies have suggested that bilingual CR effects against AD do not appear to require maintenance of high MTL structural integrity in aging.

Prior to summarizing results from these two studies, it is important to describe their recruitment methods, which bear directly on their results. Both of these studies followed the original approach to the study of CR, which involves rigorous matching of sub-groups on a large number of demographic variables and neuropsychological test scores to equate cognitive functioning as closely as possible (Stern, Alexander, Prohovnik, & Mayeux, 1992). CR theory predicts that individuals with higher reserve should require greater structural brain decline than those with lower reserve before they manifest cognitive declines associated with aging or dementia. When subgroups are rigorously matched for cognitive functioning, the high CR subgroup is therefore predicted to show *poorer brain structure*, particularly in neural regions that are less relevant contributors to active CR mechanisms.

The study by Schweizer et al. (2012) focused on groups of bilingual and monolingual older adults who were diagnosed with mild AD and showed similar levels of cognitive impairment. The two AD patient groups were compared on several estimates of brain volume derived from computed tomography (CT) scans. Results showed that bilinguals had a larger width of the temporal horn ratio, suggesting more atrophy of MTL structures in the bilingual AD patients compared to the monolingual AD patients. Bilingual CR effects in this study were therefore not based upon neuroprotection of MTL-based memory circuits because the bilingual group had more damage to MTL structures than monolinguals.

The study by Gold, Johnson et al. (2013) compared cognitively normal older adult bilingual and monolingual groups on several measures of WM microstructure. Bilinguals showed poorer WM microstructure in several tracts with MTL connections. For instance, the bilingual group showed lower WM in the fornix, a tract containing major cholinergic projections from basal frontal structures in the septal region to the hippocampus. In addition, bilinguals showed poorer WM microstructure in portions of the inferior longitudinal fasciculus that contain connections between MTL structures and visual association cortex. These findings suggested that bilingual older adults can retain similar cognitive functioning as their monolingual peers despite significantly more damage to MTL memory circuits.

It is worth reiterating that neither the findings from the volumetric of Schweizer et al. (2012) nor those from our diffusion tensor imaging (DTI) study (Gold, Johnson et al., 2013) suggest that bilingualism causes MTL atrophy. Instead, findings from these studies suggest that bilinguals appear to be able to tolerate significant MTL damage without showing the expected cognitive impairments. These findings, along with the available behavioral evidence, suggest that the principal brain circuits underlying bilingual CR effects appear likely to be located outside of classic MTL-based episodic memory systems.

EFFECTS OF BILINGUALISM ON EXECUTIVE CONTROL SYSTEMS IN AGING

EC functions decline markedly with aging (Schaie, 1996). Considerable evidence suggests that age-related declines in EC functions are less steep in older adult bilinguals compared to their monolingual peers (reviewed in Bialystok et al., 2012). Older adult bilinguals have been found to outperform their monolingual peers even on those EC tasks using nonlinguistic perceptual stimuli, suggesting that lifelong bilingualism may strengthen general-purpose EC systems (Costa, Hernandez, & Sebastian-Galles, 2008; Bialystok & Craik, 2010). For example, Bialystok, Craik, and Ryan (2006) compared older adult monolingual and bilinguals using a modified antisaccade task. Older bilinguals showed proportionally smaller reaction time suppression costs compared to their monolingual peers, suggesting better maintained inhibitory control functions (Bialystok et al., 2006).

Tasks tapping EC functions such as inhibitory control and switching are subserved by frontostriatal and frontoparietal networks (Miller & Cohen, 2001; Robbins & Arnsten, 2009; Kim, Cilles, Johnson, & Gold, 2012). These neural systems undergo significant age-related neurodegenerative changes. For example, aging is associated with marked atrophy in frontostriatal systems including prefrontal cortex, the caudate, putamen (Good et al., 2001; Resnick, Pham, Kraut, Zonderman, & Davatzikos, 2003; Tisserand et al., 2004; Raz et al., 2005; Smith, Chebrolu, Wekstein, Schmitt, & Markesbery, 2007). Aging also affects the integrity of the WM connections within the frontostriatal and frontoparietal EC circuitries. For example, DTI studies have documented age-related declines in tracts connecting frontal and striatal structures, such as the anterior limb of the internal capsule, and tracts connecting frontal and parietal regions, such as the superior longitudinal fasciculus (reviewed in Madden et al., 2012).

Several neuroimaging studies relevant to EC circuitry have now been conducted comparing older adult bilingual and monolingual groups (Abutalebi et al., 2015; Gold, Kim, Johnson, Kryscio, & Smith, 2013; Grady, Luk, Craik, & Bialystok, 2015; Luk, Bialystok, Craik, & Grady, 2011). The first neuroimaging study to suggest that bilingualism may attenuate age-related declines in EC systems was conducted by Luk et al. (2011), who compared older adult bilingual and monolingual groups on measures of DTI and resting-state functional connectivity. The bilingual group showed stronger resting-state connectivity between an inferior frontal region and multiple posterior structures in temporal, parietal, and occipital cortices (Figure 7.1). In addition, bilinguals showed better structural WM microstructure than the monolinguals in several tracts, including the superior longitudinal fasciculus, which connects frontal and parietal portions of the EC network.

Further evidence of resting state connectivity differences between bilingual and monolingual older adult groups has recently been reported (Grady et al., 2015). These authors explored potential differences in resting state connectivity in both EC and memory networks between older adult bilingual and monolingual groups. Results indicated that the bilingual group showed stronger resting-state connectivity between an inferior frontal region and the overall frontoparietal control network. The frontoparietal control includes several lateral frontal structures and the posterior parietal lobes and is thought to act as a switch to flexibly control the specific EC processes needed to meet task demands (Spreng, Sepulcre, Turner, Stevens, & Schacter, 2013). In contrast, the bilingual group did not show higher FC at rest between MTL or lingual gyrus and their respective mean overall networks. These findings are consistent with a view that bilingualism may affect EC networks more than memory networks.

There is also evidence suggesting different patterns of frontal cortex activation during EC task performance in older adult bilinguals compared to their

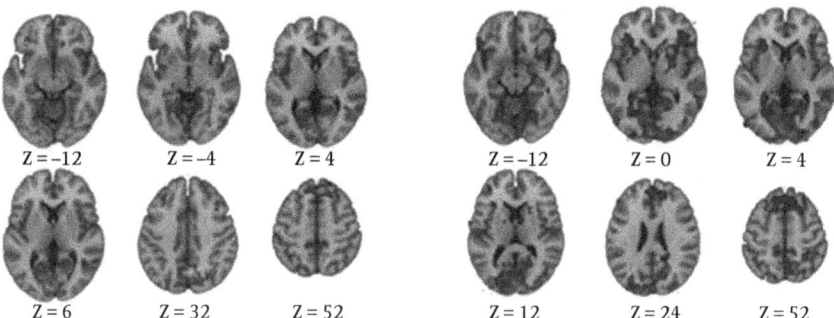

Figure 7.1. Resting-state functional connectivity for monolinguals and bilinguals. Regions showing correlated activity with the inferior frontal seed voxels. Regions of positively correlated activity with the seed were observed in the frontal–posterior network, which was strongly represented in bilinguals (warm-colored voxels). Regions of negatively correlated activity with the seed were observed in the anterior left–right network, which was similar for both monolinguals and bilinguals (cool-colored voxels). (See color plate) Adapted from G. Luk, E. Bialystok, F. I. M. Craik, & C. L. Grady, 2011, Lifelong bilingualism maintains white matter integrity in older adults. *Journal of Neuroscience, 31,* 16808–16813. Used with permission.

monolingual peers (Gold, Kim et al., 2013). In this functional magnetic resonance imaging (fMRI) study, groups performed a task switching paradigm. Results showed that the older adult bilingual group was faster than their monolingual peers at switching between tasks. Interestingly, the fMRI results showed that older bilinguals had lower blood oxygen level dependent response in several task-relevant regions of frontal cortex (left DLPFC, left VLPFC, and ACC) than their monolingual peers (Figure 7.2). The observation that bilinguals required a lower metabolic response to outperform their monolingual peers suggests that bilingualism may boost functional properties of some frontal regions in aging. As discussed in the following text, such a frontal boost could emerge from the continuous need for bilinguals to inhibit the language not under current use, which may serve as a form of implicit practice effects on frontal brain systems involved in inhibitory control processes (Green, 1998).

Finally, there is recent evidence suggesting that bilingualism may enhance and/or protect the gross structural integrity parietal cortices in aging (Abutalebi et al., 2015). This study found larger gray matter volume in the inferior parietal lobules in older adult Chinese bilinguals than their monolingual peers. Interestingly, within bilinguals, more prominent volumetric increases were observed in speakers of languages that were linguistically similar (Cantonese–Mandarin) than linguistically distinct (Cantonese–English). This finding suggests that increased need for inhibitory control associated with suppression of similar/conflicting linguistic codes may contribute to the maintained structural integrity of parietal cortex in aging (Abutalebi et al., 2015).

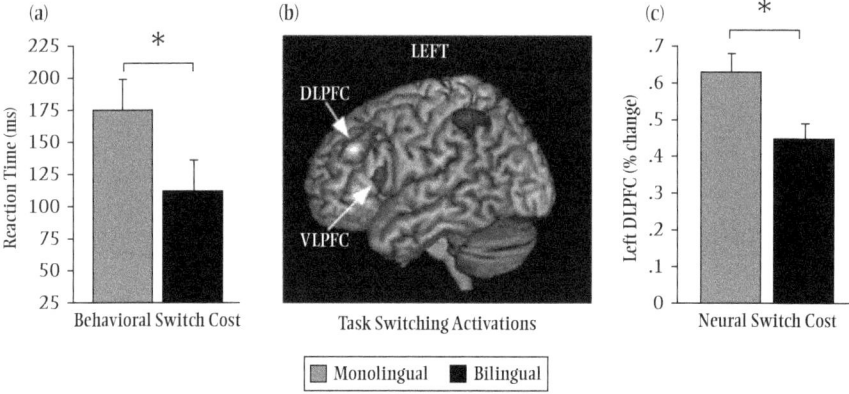

Figure 7.2. (a) Task switching behavioral and brain activation differences for monolinguals and bilinguals. Older bilinguals showed smaller behavioral switch costs than their monolingual peers. (b) Task switching (switch–nonswitch contrast) was associated with the activation of frontostriatal and frontoparietal regions across participants, including left frontal activations seen here in the DLPFC and VLPFC. (c) Older adult bilinguals showed lower task switching fMRI activation (i.e., lower neural switch costs) in several frontal regions, including the left DLPFC. (See color plate) Adapted from B.T. Gold, C. Kim, N. F. Johnson, R. J. Kryscio, & C. D. Smith, 2013, Lifelong bilingualism maintains neural efficiency for cognitive control in aging. *Journal of Neuroscience, 33,* 387–396. Used with permission. *p < 0.05.

PROTECTING EC PATHWAYS THROUGH EXPERIENCE

The combined evidence from behavioral and neuroimaging studies suggest that bilingualism may delay AD symptoms by strengthening EC pathways. In considering how the experience of bilingualism may strengthen EC circuits, it is important to point out that bilingual control of two languages involves frontostriatal and frontoparietal structures sharing close correspondence with structures involved in EC processes more generally such as the DLPFC, VLPFC, insula, ACC, basal ganglia, thalamus, and posterior parietal cortex (Green & Abutalebi, 2013). The continuous need for bilinguals to inhibit the language not currently under use may thus have implicit benefits on neurocognitive circuits involved in a broad range of EC functions (Green, 1998).

The precise neural mechanisms underlying CR variables such as bilingualism are unknown due to the coarseness of most available neuroimaging methods. I have elsewhere discussed a number of candidate mechanisms that could contribute to bilingual CR effects (Gold, 2015). Briefly, as shown in Figure 7.3, increased metabolic activity within EC circuits, and related increases in delivery of oxygen and glucose within those circuits, may result in a synergistic cascade of beneficial effects in the bilingual brain. This may include increased myelination of axons within EC circuits, as increased neuronal activity has been linked with

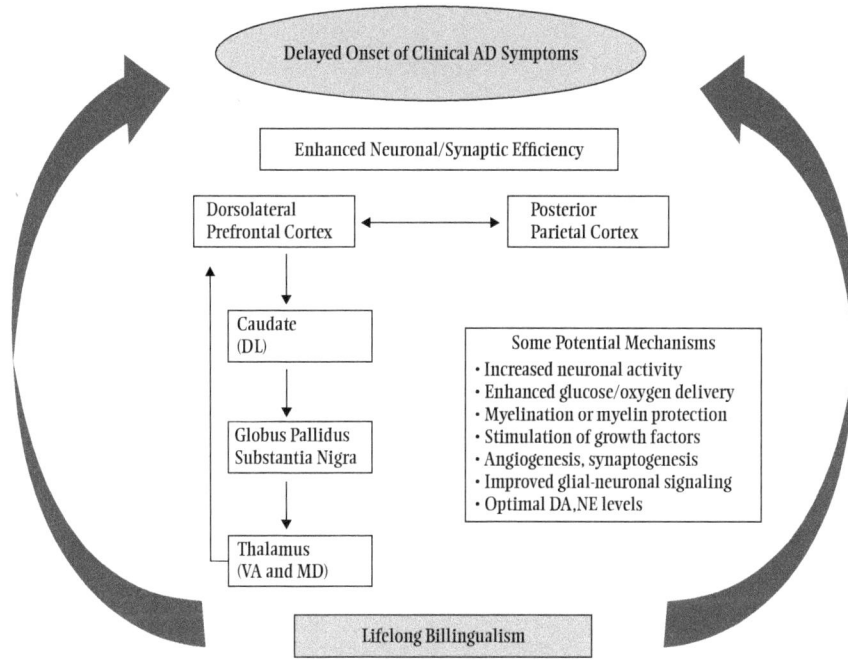

Figure 7.3. A schematic representation of potential bilingual cognitive reserve (CR) mechanisms. Bilingualism may delay the onset of clinical Alzheimer's disease symptoms through positive effects on the functioning of frontostriatal and frontoparietal brain systems involved in executive control functions. The DLPFC represents a key hub in both frontoparietal and frontostriatal networks relevant to executive functions. DA = dopamine; NE = norepinephrine. See text for discussion of potential mechanisms that may contribute to bilingual CR effects.

greater myelination (Gyllensten & Malmfors, 1963; Bradl & Lassmann, 2010). Increased neuronal metabolism could also promote angiogenesis, which would appear to be a relevant mechanistic contributor to bilingual CR effects in that age-related small vessel disease appears to predominantly disrupt frontostriatal circuits (Schmidtke & Hull, 2005) that appear to be protected by bilingualism.

The regular activation of specific brain pathways over the lifetime could also maintain synaptic efficiency through protection of terminal arbours, presynaptic boutons, or the density/affinity of neurotransmitter receptors within these circuits (Petrosini et al., 2009). Ultimately, protection of synapses associated with CR would be expected to offset age-related declines in neurochemical signaling within relevant brain circuits. Multiple neurotransmitter systems are likely to be implicated in CR and dopamine in particular may be especially relevant to bilingual CR effects given its ubiquity in frontostriatal systems and established role of DA projections from the midbrain in modulating frontally based EC functions (Robbins & Arnsten, 2009).

FUTURE DIRECTIONS

The potential individual and joint contributions of various linguistic factors to bilingual differences in EC remain to be elucidated. A number of these linguistic factors, such as degree of usage, proficiency, and competition between languages, are likely to be highly correlated and difficult to tease apart. Nonetheless, future studies should attempt to explore relative contributions of these and other variables to bilingual CR effects. In addition, it will be of high practical relevance to work toward understanding the minimum usage/proficiency required for observable bilingual CR effects.

Longitudinal studies would be helpful on many levels in testing/understanding bilingual CR effects. For example, it will be important to determine if older adult monolinguals show neurocognitive benefits resulting from second language learning. In Bak et al.'s (2014) study, cognitive advantages were observed even in those older adults who acquired their second language in young adulthood. However, it remains to be determined if similar cognitive benefits can be attained by older adults. Similarly, the potential influence of second language acquisition on brain structure and function in older adults is unknown. However, there is a strong rationale to test for such possible effects in older adults given recent results showing positive effects of second language acquisition on brain structure in younger adults (Martensson et al., 2012; Stein et al., 2012).

Proof-of-concept studies of bilingualism as a CR variable are also required. For example, it would be informative to determine if bilinguals decline more rapidly on neuropsychological tests within a few years of developing AD than monolinguals. Faster progression of cognitive decline in the prodromal phase of AD has been demonstrated in individuals with higher compared to lower levels of education, and is thought to indicate the "exhaustion of compensation" in the group with high reserve (Amieva et al., 2005). Similarly, proof-of-concept studies could also attempt to determine if bilinguals who stop speaking their second language lose any previous cognitive advantage compared to monolinguals.

Longitudinal positron emission tomography amyloid binding studies would help inform the mechanistic basis of bilingual CR protection against AD. For example, these studies could assess whether bilingualism affects the progression rates of amyloid deposition or the clinical response to AD pathology. Finally, it will be important to establish whether bilingualism delays symptom onset in non-AD dementias. Preliminary support for such protective effects comes from the study by Alladi et al. (2013), in which bilinguals were found to be significantly older than monolinguals when diagnosed with frontotemporal dementia or vascular dementia (in addition to AD). These findings are particularly interesting in that both frontotemporal dementia and vascular dementia prominently affect frontally based EC functions. Further studies of this kind that integrate neuroimaging methods will provide an additional avenue for testing the hypothesis that bilingual CR effects relate to strengthening of frontally based EC neurocognitive systems.

OVERALL CONCLUSION

Bilingualism appears to delay the onset of clinical symptoms associated with AD. Neuroimaging studies are in their infancy but converge on the notion that bilingual CR effects appear to relate to strengthening of EC circuits rather than through the direct protection of memory circuits affected in early-stage AD. The regular use of two languages and its protective effects on fronto-subcortical and/or frontoparietal EC circuits may allow bilinguals to compensate for MTL neurodegenerative changes typically associated with clinical AD. The next decade of studies incorporating emerging neuroimaging techniques should lead to exciting advancements in understanding the potential neural and molecular bases of bilingual CR effects.

ACKNOWLEDGMENTS

I thank Dr. Fergus Craik for helpful comments on a previous version of this manuscript. I gratefully acknowledge my valued colleagues at the Sanders-Brown Center on Aging and the Magnetic Resonance Imaging and Spectroscopy Center who contributed to some of the studies described here. I also wish to thank Dr. Otto Pedraza and his staff for devoting their time and effort in serving as editors of this volume. This work was supported in part by a grant from the National Institutes of Health Institute of Aging (R01-AG033036). The content is solely the responsibility of the author and does not necessarily represent the official views of the NIH.

REFERENCES

Abutalebi, J., Canini, M., Della Rosa, P. A., Sheung, L. P., Green, D.W., & Weekes, B. S. (2014). Bilingualism protects anterior temporal lobe integrity in aging. *Neurobiology of Aging, 35,* 2126–2133. https://doi.org/10.1016/j.neurobiolaging.2014.03.010

Abutalebi, J., Canini, M., Della Rosa, P. A., Sheung, L. P., Green, D. W., & Weekes, B. S. (2015). The neuroprotective effects of bilingualism upon the inferior parietal lobule: A structural neuroimaging study in aging Chinese bilinguals. *Journal of Neurolinguistics, 33,* 3–13. https://doi.org/10.1016/j.jneuroling.2014.09.008

Albert, M. S., Jones, K., Savage, C. R., Berkman, L., Seeman, T., Blazer, D., & Rowe, J. W. (1995). Predictors of cognitive change in older persons: MacArthur studies of successful aging. *Psychology and Aging, 10,* 578–589.

Alladi, S., Bak, T. H., Duggirala, V., Surampudi, B., Shailaja, M., Shukla, A. K., . . . Kaul, S. (2013). Bilingualism delays age at onset of dementia, independent of education and immigration status. *Neurology, 81,* 1938–1944. https://doi.org/10.1212/01.wnl.0000436620.33155.a4

Amieva, H., Jacqmin-Gadda, H., Orgogozo, J. M., Le Carret, N., Helmer, C., Letenneur, L., . . . Dartigues, J. F. (2005). The 9 year cognitive decline before dementia of the Alzheimer type: A prospective population-based study. *Brain, 128,* 1093–1101. https://doi.org/10.1093/brain/awh451

Bak, T. H., Nissan, J. J., Allerhand, M. M., & Deary, I. J. (2014). Does bilingualism influence cognitive aging? *Annals of Neurology, 75,* 959–963. https;//doi.org/ 10.1002/ ana.24158

Bennett, D. A., Schneider, J. A., Tang, Y., Arnold, S. E., & Wilson, R. S. (2006). The effect of social networks on the relation between Alzheimer's disease pathology and level of cognitive function in old people: A longitudinal cohort study. *The Lancet Neurology, 5,* 406–412. https://doi.org/10.1016/S1474-4422(06)70417-3

Bialystok, E., & Craik, F. I. M. (2010). Cognitive and linguistic processing in the bilingual mind. *Current Directions in Psychological Science, 19,* 19–23. https://doi.org/ 10.1177/0963721409358571

Bialystok, E., Craik, F. I. M., & Freedman, M. (2007). Bilingualism as a protection against the onset of symptoms of dementia. *Neuropsychologia, 45,* 459–464. https://doi.org/ 10.1016/j.neuropsychologia.2006.10.009

Bialystok, E., Craik, F. I. M., & Luk, G. (2012). Bilingualism: Consequences for mind and brain. *Trends in Cognitive Sciences, 16,* 240–250. https://doi.org/ 10.1016/ j.tics.2012.03.001

Bialystok, E., Craik, F. I. M., & Ryan, J. (2006). Executive control in a modified antisaccade task: Effects of aging and bilingualism. *Journal of Experimental Psychology: Learning, Memory, and Cognition, 32,* 1341–1354. https://doi.org/10.1037/0278-7393.32.6.1341

Birks, J. (2006). Cholinesterase inhibitors for Alzheimer's disease. *Cochrane Database of Systematic Reviews, 1,* CD005593. https://doi.org/10.1002/14651858.CD005593

Bradl, M., & Lassmann, H. (2010). Oligodendrocytes: Biology and pathology. *Acta Neuropathologica, 119,* 37–53. https://doi.org/10.1007/s00401-009-0601-5

Chertkow, H., Whitehead, V., Phillips, N., Wolfson, C., Atherton, J., & Bergman, H. (2010). Multilingualism (but not always bilingualism) delays the onset of Alzheimer disease: Evidence from a bilingual community. *Alzheimer Disease and Associated Disorders, 24,* 118–125. https://doi.org/ 10.1097/WAD.0b013e3181ca1221

Christensen, H. (2001). What cognitive changes can be expected with normal ageing? *Australian and New Zealand Journal of Psychiatry, 35,* 768–775. https://doi.org/ 10.1046/j.1440-1614.2001.00966.x

Chua, T. C., Wen, W., Slavin, M. J., & Sachdev, P. S. (2008). Diffusion tensor imaging in mild cognitive impairment and Alzheimer's disease: A review. *Current Opinion in Neurology, 21,* 83–92. https://doi.org/10.1097/WCO.0b013e3282f4594b

Convit, A., De Leon, M. J., Tarshish, C., De Santi, S., Tsui, W., Rusinek, H., & George, A. (1997). Specific hippocampal volume reductions in individuals at risk for Alzheimer's disease. *Neurobiology of Aging, 18,* 131–138.

Costa, A., Hernandez, M., & Sebastian-Galles, N. (2008). Bilingualism aids conflict resolution: Evidence from the ANT task. *Cognition, 106,* 59–86. https://doi.org/ 10.1016/ j.cognition.2006.12.013

Craik, F. I. M., Bialystok, E., & Freedman, M. (2010). Delaying the onset of Alzheimer disease: Bilingualism as a form of cognitive reserve. *Neurology, 75,* 1726–1729. https:// doi.org/ 10.1212/WNL.0b013e3181fc2a1c

Craik, F. I. M., & Salthouse, T.A. (Eds.). (2008). *The handbook of aging and cognition* (3rd ed.). New York, NY: Psychology Press.

Deary, I. J., Gow, A. J., Pattie, A., & Starr, J. M. (2012). Cohort profile: The Lothian birth cohorts of 1921 and 1936. *International Journal of Epidemiology, 41,* 1576–1584. https://doi.org/10.1093/ije/dyy022

Dickerson, B. C., Bakkour, A., Salat, D. H., Feczko, E., Pacheco, J., Greve, D. N., ... Buckner, R. L. (2009). The cortical signature of Alzheimer's disease: Regionally specific cortical thinning relates to symptom severity in very mild to mild AD dementia and is detectable in asymptomatic amyloid-positive individuals. *Cerebral Cortex, 19,* 497–510. https://doi.org/10.1093/cercor/bhn113

Dickerson, B. C., & Wolk, D. A. (2012). MRI cortical thickness biomarker predicts AD-like CSF and cognitive decline in normal adults. *Neurology, 78,* 84–90. https://doi.org/10.1212/WNL.0b013e31823efc6c

Farrimond, L. E., Roberts, E., & McShane, R. (2012). Memantine and cholinesterase inhibitor combination therapy for Alzheimer's disease: A systematic review. *BMJ Open, 2*(3), e000917. https://doi.org/ 10.1136/bmjopen-2012-000917

Fernandes, M. A., Craik, F. I. M., Bialystok, E., & Kreuger, S. (2007). Effects of bilingualism, aging, and semantic relatedness on memory under divided attention. *Canadian Journal of Experimental Psychology, 61,* 128–141. https://doi.org/10.1007/978-3-531-91596-8_4

Gold, B. T. (2015). Lifelong Bilingualism and Neural Reserve against Alzheimer's disease: A review of findings and potential mechanisms. *Behavioural Brain Research, 281C,* 9–15. https://doi.org/10.1016/j.bbr.2014.12.006

Gold, B. T., Johnson, N. F., & Powell, D. K. (2013). Lifelong bilingualism contributes to cognitive reserve against white matter integrity declines in aging. *Neuropsychologia, 51,* 2841–2846. https://doi.org/10.1016/j.neuropsychologia.2013.09.037

Gold, B. T., Johnson, N. F., Powell, D. K., & Smith, C. D. (2012). White matter integrity and vulnerability to Alzheimer's disease: Preliminary findings and future directions. *Biochimica et Biophysica Acta, 1822,* 416–422. https://doi.org/10.1016/j.bbadis.2011.07.009

Gold, B. T., Kim, C., Johnson, N. F., Kryscio, R. J., & Smith, C. D. (2013). Lifelong bilingualism maintains neural efficiency for cognitive control in aging. *Journal of Neuroscience, 33,* 387–396. https://doi.org/10.1523/JNEUROSCI.3837-12.2013

Good, C. D., Johnsrude, I. S., Ashburner, J., Henson, R. N., Friston, K. J., & Frackowiak, R. S. (2001). A voxel-based morphometric study of ageing in 465 normal adult human brains. *NeuroImage, 14,* 21–36. https://doi.org/10.1006/nimg.2001.0786

Grady, C. L., Luk, G., Craik, F. I. M., & Bialystok, E. (2015). Brain network activity in monolingual and bilingual older adults. *Neuropsychologia, 66,* 170–181. https://doi.org/ 10.1016/j.neuropsychologia.2014.10.042

Green, D. W. (1998). Mental control of the bilingual lexico-semantic system. *Bilingualism: Language and Cognition, 1,* 67–81. https://doi.org/10.1017/S1366728998000133

Green, D.W., & Abutalebi, J. (2013). Language control in bilinguals: The adaptive control hypothesis. *Journal of Cognitive Psychology 25,* 515–530. https://doi.org/10.1080/20445911.2013.796377

Gyllensten, L., & Malmfors, T. (1963). Myelinization of the optic nerve and its dependence on visual function: A quantitative investigation in mice. *Journal of Embryology and Experimental Morphology, 11,* 255–266.

Hillman, C. H., Erickson, K. I., & Kramer, A. F. (2008). Be smart, exercise your heart: Exercise effects on brain and cognition. *Nature Reviews Neuroscience, 9,* 58–65. https://doi.org/10.1038/nrn2298

Jack, C. R., Jr., Petersen, R. C., Xu, Y. C., Waring, S. C., O'Brien, P. C., Tangalos, E. G., . . . Kokmen, E. (1997). Medial temporal atrophy on MRI in normal aging and very mild Alzheimer's disease. *Neurology, 49,* 786–794. https://doi.org/ 10.1212/wnl.49.3.786

Josefsson, K., Jokela, M., Cloninger, C. R., Hintsanen, M., Salo, J., Hintsa, T., ... Keltikangas-Jarvinen, L. (2013). Maturity and change in personality: Developmental trends of temperament and character in adulthood. *Development and Psychopathology, 25,* 713–727. https://doi.org/10.1017/S0954579413000126

Kemper, T. L. (1994). Neuroanatomical and neuropathological changes during aging and in dementia. In M. L. Albert & E. J. E. Knoepfel (Eds.), *Clinical neurology of aging* (pp. 3–67). New York, NY: Oxford University Press.

Kim, C., Cilles, S. E., Johnson, N. F., & Gold, B. T. (2012). Domain general and domain preferential brain regions associated with different types of task switching: A meta-analysis. *Human Brain Mapping, 33,* 130–142. https;//doi.org/10.1002/hbm.21199

Luk, G., Bialystok, E., Craik, F. I. M., & Grady, C. L. (2011). Lifelong bilingualism maintains white matter integrity in older adults. *Journal of Neuroscience, 31,* 16808–16813. https://doi.org/10.1523/JNEUROSCI.4563-11.2011

Madden, D. J., Bennett, I. J., Burzynska, A., Potter, G. G., Chen, N. K., & Song, A. W. (2012). Diffusion tensor imaging of cerebral white matter integrity in cognitive aging. *Biochimica et Biophysica Acta, 1822,* 386–400. https://doi.org/10.1016/j.bbadis.2011.08.003

Martensson, J., Eriksson, J., Bodammer, N. C., Lindgren, M., Johansson, M., Nyberg, L., & Lovden, M. (2012). Growth of language-related brain areas after foreign language learning. *NeuroImage, 63,* 240–244. https://doi.org/10.1016/j.neuroimage.2012.06.043

Martin, S. B., Smith, C. D., Collins, H. R., Schmitt, F. A., & Gold, B. T. (2010). Evidence that volume of anterior medial temporal lobe is reduced in seniors destined for mild cognitive impairment. *Neurobiology of Aging, 31,* 1099–1106. https://doi.org/10.1016/j.neurobiolaging.2008.08.010

Matthews, L. J., & Butler, P. M. (2011). Novelty-seeking DRD4 polymorphisms are associated with human migration distance out-of-Africa after controlling for neutral population gene structure. *American Journal of Physical Anthropology, 145,* 382–389. https://doi.org/10.1002/ajpa.21507

Mayeux, R. (2003). Epidemiology of neurodegeneration. *Annual Review of Neuroscience, 26,* 81–104. https://doi.org/10.1146/annurev.neuro.26.043002.094919

McKhann, G., Drachman, D., Folstein, M., Katzman, R., Price, D., & Stadlan, E. M. (1984). Clinical diagnosis of Alzheimer's disease: Report of the NINCDS-ADRDA Work Group under the auspices of Department of Health and Human Services Task Force on Alzheimer's Disease. *Neurology, 34,* 939–944.

Miller, E. K., & Cohen, J. D. (2001). An integrative theory of prefrontal cortex function. *Annual Review of Neuroscience, 24,* 167–202. https://doi.org/10.1146/annurev.neuro.24.1.167

Mortimer, J. A. (1997). Brain reserve and the clinical expression of Alzheimer's disease. *Geriatrics, 52*(Suppl 2), S50–S53.

Petrosini, L., De Bartolo, P., Foti, F., Gelfo, F., Cutuli, D., Leggio, M. G., & Mandolesi, L. (2009). On whether the environmental enrichment may provide cognitive and brain reserves. *Brain Research Reviews, 61,* 221–239. https://doi.org/10.1016/j.brainresrev.2009.07.002

Raz, N., & Kennedy, K. M. (2009). A systems approach to age-related change: Neuroanatomical changes, their modifiers, and cognitive correlates. In W. Jagust & M. Desposito (Eds.), *Imaging the aging brain* (pp. 43–70). New York, NY: Oxford University Press.

Raz, N., Lindenberger, U., Rodrigue, K. M., Kennedy, K. M., Head, D., Williamson, A., . . . Acker, J. D. (2005). Regional brain changes in aging healthy adults: General trends, individual differences and modifiers. *Cerebral Cortex, 15,* 1676–1689. https://doi.org/10.1093/cercor/bhi044

Resnick, S. M., Pham, D. L., Kraut, M. A., Zonderman, A. B., & Davatzikos, C. (2003). Longitudinal magnetic resonance imaging studies of older adults: A shrinking brain. *Journal of Neuroscience, 23,* 3295–3301. https://doi.org/10.1523/JNEUROSCI.23-08-03295.2003

Richards, M., & Deary, I. J. (2005). A life course approach to cognitive reserve: A model for cognitive aging and development? *Annals of Neurology, 58,* 617–622. https://doi.org/10.1002/ana.20637

Riley, K. P., Snowdon, D. A., & Markesbery, W. R. (2002). Alzheimer's neurofibrillary pathology and the spectrum of cognitive function: Findings from the Nun Study. *Annals of Neurology, 51,* 567–577. https://doi.org/10.1002/ana.10161

Robbins, T. W., & Arnsten A. F. (2009). The neuropsychopharmacology of fronto-executive function: Monoaminergic modulation. *Annual Review of Neuroscience, 32,* 267–287. https://doi.org/ 10.1146/annurev.neuro.051508.135535

Salthouse, T. A. (1996). The processing-speed theory of adult age differences in cognition. *Psychological Review, 103,* 403–428.

Schaie, K. W. (1996). *Intellectual development in adulthood: The Seattle Longitudinal Study.* Cambridge, England: Cambridge University Press.

Schmidtke, K., & Hull, M. (2005). Cerebral small vessel disease: How does it progress? *Journal of Neurological Sciences, 229–230,* 13–20. https://doi.org/ 10.2147/CIA.S90871

Schweizer, T. A., Ware, J., Fischer, C. E., Craik, F. I. M., & Bialystok, E. (2012). Bilingualism as a contributor to cognitive reserve: Evidence from brain atrophy in Alzheimer's disease. *Cortex, 48,* 991–996. https://doi.org/10.1016/j.cortex.2011.04.009

Shaw, L. M., Vanderstichele, H., Knapik-Czajka, M., Clark, C. M., Aisen, P. S., Petersen, R. C., . . . Trojanowski, J. Q. (2009). Cerebrospinal fluid biomarker signature in Alzheimer's disease neuroimaging initiative subjects. *Annals of Neurology, 65,* 403–413. https://doi.org/0.1002/ana.21610

Smith, C. D., Chebrolu, H., Wekstein, D. R., Schmitt, F. A., & Markesbery, W. R. (2007). Age and gender effects on human brain anatomy: A voxel-based morphometric study in healthy elderly. *Neurobiology of Aging, 28,* 1075–1087. 10.1016/j.neurobiolaging.2006.05.018

Smith, C. N., & Squire, L. R. (2009). Medial temporal lobe activity during retrieval of semantic memory is related to the age of the memory. *Journal of Neuroscience, 29,* 930–938. https://doi.org/10.1523/JNEUROSCI.4545-08.2009

Spires, T. L., & Hannan, A. J. (2005). Nature, nurture and neurology: Gene-environment interactions in neurodegenerative disease. FEBS Anniversary Prize Lecture delivered on 27 June 2004 at the 29th FEBS Congress in Warsaw. *FEBS Journal, 272,* 2347–2361. https://doi.org/10.1111/j.1742-4658.2005.04677.x

Spreng, R. N., Sepulcre, J., Turner, G. R., Stevens, W. D., & Schacter, D. L. (2013). Intrinsic architecture underlying the relations among the default, dorsal attention, and frontoparietal control networks of the human brain. *Journal of Cognitive Neuroscience, 25,* 74–86. https://doi.org/10.1162/jocn_a_00281

Squire, L. R. (1992). Memory and the hippocampus: A synthesis from findings with rats, monkeys, and humans. *Psychological Review, 99,* 195–231.

Stebbins, G. T, & Murphy, C. M. (2009). Diffusion tensor imaging in Alzheimer's disease and mild cognitive impairment. *Behavioural Neurology, 21,* 39–49. https://doi.org/10.3233/BEN-2009-0234

Steffener, J., & Stern, Y. (2012). Exploring the neural basis of cognitive reserve in aging. *Biochimica et Biophysica Acta, 1822,* 467–473. 10.1016/j.bbadis.2011.09.012

Stein, M, Federspiel, A, Koenig, T, Wirth, M, Strik, W, Wiest, R, . . . Dierks, T. (2012). Structural plasticity in the language system related to increased second language proficiency. *Cortex, 48,* 458–465. https://doi.org/10.3389/fpsyg.2014.01116

Stern, Y. (2002). What is cognitive reserve? Theory and research application of the reserve concept. *Journal of the International Neuropsychological Society, 8,* 448–460. https://doi.org/10.1017/S1355617702813248

Stern, Y. (2009). Cognitive reserve. *Neuropsychologia, 47,* 2015–2028. https://doi.org/10.1016/j.neuropsychologia.2009.03.004

Stern, Y., Alexander, G. E., Prohovnik, I., & Mayeux, R. (1992). Inverse relationship between education and parietotemporal perfusion deficit in Alzheimer's disease. *Annals of Neurology, 32,* 371–375. https://doi.org/10.1002/ana.410320311

Tisserand, D. J., van Boxtel, M. P., Pruessner, J. C., Hofman, P., Evans, A. C., & Jolles, J. (2004). A voxel-based morphometric study to determine individual differences in gray matter density associated with age and cognitive change over time. *Cerebral Cortex, 14,* 966–973. https://doi.org/10.1093/cercor/bhh057

United Nations, Department of Economic and Social Affairs, Population Division (2013). World Population Ageing 2013. ST/ESA/SER.A/348.

Valenzuela, M. J., & Sachdev, P. (2006). Brain reserve and dementia: A systematic review. *Psychological Medicine, 36,* 441–454. https://doi.org/10.1017/S0033291706007744

Van Dijk, K. R. A., Van Gerven, P. W. M., Van Boxtel, M. P. J., Van der Elst, W., & Jolles, J. (2008). No protective effects of education during normal cognitive aging: Results from the 6-year follow-up of the Maastricht Aging Study. *Psychology and Aging, 23,* 119–130. https://doi.org/10.1037/0882-7974.23.1.119

Vlassenko, A. G., Benzinger, T. L., & Morris, J. C. (2012). PET amyloid-beta imaging in preclinical Alzheimer's disease. *Biochimica et Biophysica Acta, 1822,* 370–379. https://doi.org/ 10.1016/j.bbadis.2011.11.005

Yeung, C. M., St John, P. D., Menec, V., & Tyas, S. L. (2014). Is bilingualism associated with a lower risk of dementia in community-living older adults? Cross-sectional and prospective analyses. *Alzheimer Disease and Associated Disorders, 28,* 326–332. 10.1097/WAD.0000000000000019

Zahodne, L. B., Schofield, P. W., Farrell, M. T., Stern, Y., & Manly, J. J. (2014). Bilingualism does not alter cognitive decline or dementia risk among Spanish-speaking immigrants. *Neuropsychology, 28,* 238–246. https://doi.org/10.1097/WAD.0000000000000019

8

Neuropsychological Assessment of Non-English Speakers

OCTAVIO A. SANTOS, DARYL E. M. FUJII, AND OTTO PEDRAZA

INTRODUCTION

Sixty million individuals in the United States speak a language other than English at home. The majority (62%) speak Spanish, while 18% speak other Indo-European languages and 16% speak Asian and Pacific Islander languages (Ryan, 2013). Of those who are foreign-born, the majority speak a language other than English at home, although this figure varies considerably by state (from approximately 90% in Texas and California to 49% in Montana) and region of birth (from 89% of those from Latin America to 28% of those from Canada and Australia). This language diversity will continue to expand as the foreign-born U.S. population grows by a projected 85% between 2014 and 2060, a fourfold increase compared with the growth of the native U.S. population (Colby & Ortman, 2015). Consequently, psychologists will face an ever-growing demand to deliver professional services to non-English monolingual or bilingual speakers (Casas et al., 2012; Strutt, Burton, Resendiz, & Peery, 2016).

U.S. national data on the language skills of mental health providers are lacking, but available practitioner surveys show that the proportion of psychologists who identify themselves as Hispanic is approximately 3%, while those who identify themselves as Asian is about 2% (American Psychological Association, 2010; Elbulok-Charcape, Rabin, Spadaccini, & Barr, 2014; Fujii, 2011; U.S. Department of Health and Human Services, 2001). Because ethnicity is an imperfect proxy for language proficiency, these figures likely overestimate the number of psychologists who are fluent in respective non-English languages. The discrepancy between projected immigration and demographic trends and the availability of providers fluent in non-English languages suggests that psychologists will be increasingly likely to evaluate non-English speakers with the use of translated tests, bilingual psychometrists, or interpreters (Casas et al., 2012; Searight & Searight, 2009).

The present chapter considers these demographic shifts and surveys the neuropsychological assessment of non-English monolingual and bilingual adults in the United States, with a particular focus on the two predominant immigrant groups: Latin Americans and Asian Pacific Islanders. We will first review the diversity of languages in the United States and briefly consider the neural and clinical aspects of bilingualism. Next, we will review several important points regarding test translation and the use of interpreters. This will lead us to consider unique aspects in the assessment of speakers of Spanish as well as Asian and Pacific Islander languages, including a description of selected neuropsychological tests. Finally, we will propose an algorithm to help clinicians choose the language of examination based on guidelines put forth by professional organizations.

Before proceeding, we acknowledge that a single chapter is insufficient in doing justice to the topics here discussed. Comprehensive treatments are readily available (e.g., Ardila, Rosselli, & Puente, 1994; Benuto, 2013; Davis & D'Amato, 2014; Ferraro, 2016; Fletcher-Janzen, Strickland, & Reynolds, 2000; Fujii, 2011, 2016; Llorente, 2008; Pontón & León-Carrión, 2001; Uzzell, Pontón, & Ardila, 2007), and the reader is referred to those primary sources for a detailed elaboration of the concepts.

LANGUAGE DIVERSITY IN THE UNITED STATES

The United States is a melting pot of linguistic features and modes of bilingualism. The 2011 American Community Survey (Ryan, 2013) highlights this diversity, tabulating at least 14 non-English languages each spoken at home by more than half a million people: Spanish, French, French Creole, Italian, Portuguese, German, Russian, Polish, Hindi, Chinese, Korean, Vietnamese, Tagalog, and Arabic. Each survey respondent classifies how well does he or she speak English by choosing "very well," "well," "not well," or "not at all." Of the 60 million people in the United States who speak a language other than English at home, almost 38 million are Spanish speakers, including those who are bilingual (74% who consider themselves to speak English "very well" or "well") or monolingual (26% who consider themselves to speak English "not well" or "not at all"). These 38 million Spanish speakers constitute a 232% increase over a 30-year period (Ryan, 2013). Not surprisingly, in 2014 the United States surpassed Spain as the second-largest Spanish-speaking country after Mexico, and by 2050 it is projected that the United States will become the largest Spanish-speaking nation on Earth (Instituto Cervantes, 2014).

Spanish will remain the predominant non-English language for the foreseeable future. However, Asians recently surpassed Hispanics as the largest group of U.S. immigrants (Pew Research Center, 2012). As a result, the percentage change in non-English languages spoken at home between 2000 and 2011 was greater for Hindi, Arabic, Gujarati, Chinese, Urdu, Vietnamese, and Thai than for Spanish (Ryan, 2013). Sixteen percent of Asian households in the United States speak a language other than English, predominantly Chinese (4.8%), Tagalog (2.6%), and Vietnamese (2.3%). The majority of those speakers (70% of Chinese, 93% of

Tagalog, and 67% of Vietnamese) also consider themselves able to speak English "well" or "very well." Those figures are comparable to the proportion of Spanish speakers who speak English "well" or "very well" (74%) and suggest that the majority of non-English speakers in the United States consider themselves to be conversationally bilingual.

Bilingualism

Individuals learn multiple languages either through simultaneous acquisition or sequential learning. A child raised in a dual-language home and who is exposed regularly to both languages will acquire each language simultaneously, with negligible detriment to their overall linguistic development (Bialystok, Craik, & Luk, 2012). The extent to which children remain proficient in both languages as they develop into adulthood depends partly on their continued exposure to those languages. In contrast, a monolingual child or adult may choose to learn a second language once they have mastered their primary language. In this sequential approach, the ease or difficulty of learning the second language varies as a function of the person's learning style and capacity, parental attitudes toward learning languages, formal education in the native language, degree of literacy in the home, and degree of acculturation (Cascallar & Arnold, 2001; Centeno & Obler, 2001; French & Llorente, 2008). Accordingly, bilingual adults attain varying degrees of proficiency, with relatively few becoming fully proficient in both languages (Hoffman, 1991). Most bilingual adults demonstrate unbalanced skills (i.e., better reading, writing, speaking, and listening in one language over the other) or mixed skills (i.e., better reading and writing in one language and better speaking and listening in the other language, or vice versa). Even when a person professes full bilingualism and upon initial contact with the clinician appears to have equivalence in both languages, closer inspection may reveal domain-specific proficiency (Puente, Zink, Hernandez, Jackman-Venanzi, & Ardila, 2013). For instance, most bilingual adults who converse equally well in either language nevertheless rely on their primary language to perform mathematical calculations (Salillas & Wicha, 2012). In the context of neuropsychological assessment, this distinction suggests that a bilingual patient may perform most clinical tasks in one language, yet revert to the other language for one or more specific cognitive operations.

Numerous premigration and postmigration variables are associated with higher rates of bilingualism, including high educational attainment in the country of origin, younger age at the moment of migration, employment in highly skilled or professional careers, living outside of predominantly immigrant neighborhoods, marriage to an English speaker, longer period residing in the United States, and becoming a naturalized U.S. citizen (Espenshade & Fu, 1997; Espinoza & Massey, 1997). Across all groups, 72% of foreign-born individuals who obtained at least a bachelor's degree speak English at home "well" or "very well," compared with 31% of those with less than a high school diploma (Gambino, Acosta, & Grieco, 2014). The association between high educational attainment and bilingualism is particularly notable among recent Asian immigrants, who hail mostly from China, the

Philippines, and India and are considered the most highly educated foreign-born cohort in U.S. history (Pew Research Center, 2012).

It is worth pointing out that for most U.S. immigrant groups, an inexorable shift toward English monolingualism occurs within three generations (Hurtado & Vega, 2004; Portes & Rumbaut, 2006; Veltman, 1983). For instance, six percent of children born in Spanish-speaking countries (first-generation immigrants) are English monolinguals, compared with 15% of U.S.-born children with immigrant parents (second generation) and 72% of U.S.-born children with immigrant grandparents (third generation). Most second-generation immigrant children continue to speak Spanish at home, yet 92% also speak English "well" or "very well." These percentages vary slightly as a function of parental country of origin, with 97% of second-generation Cuban, Colombian, Ecuadorian, and Peruvian children able to speak English "well" or "very well" compared with 91% of second-generation Mexican children.[1] This is consistent with the geographical trend toward English monolingualism for families who reside away from the U.S.–Mexico border (Portes & Rumbaut, 2006). By the third generation, 97% of children speak English "well" or "very well," and only a minority speak Spanish at home (Alba, 2004). Data for Hispanic adults demonstrate a similar effect: 42% of second-generation and 76% of third-generation adults communicate predominantly in English (Krogstad & Gonzalez-Barrera, 2015).

NEURAL BASES OF BILINGUALISM

Bilingualism may shape the structure and function of the brain across the lifespan and, in doing so, promote cognitive resources that confer protection as the individual ages (Bialystok & Craik, 2010; Bialystok, Craik, & Luk, 2012; Gold, 2015; Gold, Kim, Johnson, Kryscio, & Smith, 2013; Kroll, Bobb, & Hoshino, 2014; Li, et al., 2017; Perani et al., 2017). The neural bases of bilingualism are the focus of burgeoning investigational research and debate (e.g., Bialystok, 2016; de Bruin & Della Salla, 2016; García-Pentón, Fernandez Garcia, Costello, Duñabeitia, & Carreiras, 2016a, 2016b; Green & Abutalebi, 2016; Kroll & Chiarello, 2016; Luk & Pliatsikas, 2016; Paap, 2016). Thus far, evidence from multiple sources suggests that comparable neural networks exist between monolinguals and bilinguals, these networks start to become established during the second half of the first year of life, and they remain analogous throughout the lifespan (Costa & Sebastián-Gallés, 2014).

Morphology studies demonstrate increased gray matter density along the inferior parietal cortex in bilinguals compared with monolinguals, a result that correlates positively with the degree of second language proficiency and is more pronounced for bilinguals who acquired their second language before the age of five (Mechelli et al., 2004). Additional studies have shown greater cortical gray matter volume for bilinguals along the anterior temporal pole and cingulate cortex (Abutalebi, et al., 2012, 2014), inferior frontal gyrus (Klein, Mok, Chen,

---

1. Calculations performed on data from Alba (2004, Table 1).

& Watkins, 2014), and primary auditory cortex (Ressel, et al., 2012). Studies using diffusion tensor imaging additionally suggest that bilingual adults demonstrate higher fractional anisotropy (an index of white matter integrity) along the corpus callosum and extending to bilateral superior longitudinal fasciculi, uncinate fasciculi, and inferior frontal-occipital fasciculi when compared with age-matched monolingual adults (Luk, Bialystok, Craik, & Grady, 2011; Pliatsikas, Moschopoulou, & Saddy, 2015). Recently, Burgaleta, Sanjuán, Ventura-Campos, Sebastián-Gallés, and Ávila (2016) extended these anatomical investigations into subcortical structures and found morphological expansion of the bilateral putamen and thalamus, left pallidum, and right caudate nucleus in bilingual compared with monolingual young adults. Taken together, these findings suggest that bilinguals recruit neural structures traditionally associated with language processing in monolinguals, but those structures exhibit increased gray matter density and white matter integrity.

CLINICAL ASPECTS OF BILINGUALISM

Bilingualism appears to delay the onset of dementia by up to 5 years, as bilingual adults require a greater burden of neuropathology before the symptoms of dementia become manifest (Alladi et al., 2015; Craik, Bialystok, & Freedman, 2010; Schweizer, Ware, Fischer, Craik, & Bialystok, 2012). The putative cognitive benefits of bilingualism stem from increased efficiency in mechanisms of cognitive control (Bialystok, Craik, & Luk, 2012). Bilinguals recruit executive control functions to resolve conflicts that arise from two parallel language systems that are not independent of each other but instead are jointly active and in frequent interaction. Qualitative and quantitative meta-analyses of bilinguals across various language pairs suggest that a distributed network predominantly involving the left prefrontal cortex and caudate nucleus bilaterally is essential for bilingual language control (Abutalebi & Green, 2008; Luk, Green, Abutalebi, & Grady, 2011). This network overlaps with known frontal-subcortical circuits involved in error detection, response control, and goal-directed behavior. The regular use of such mechanisms throughout the lifespan is thought to enhance attentional control for language and nonlanguage tasks. This benefit becomes more salient as the person ages, presumably because lifelong bilingualism attenuates age-related decline in neural efficiency for task selection and control (Bialystok, Craik, & Luk, 2012; Gold, Kim, Johnson, Kryscio, & Smith, 2013).

Although bilingualism may confer a protective effect against dementia, bilingual adults demonstrate reduced lexical knowledge within each language compared with monolingual speakers of either language. Bilinguals generate fewer words to semantic categories and obtain lower scores on visual naming tasks, a disadvantage relative to monolinguals that persists even if test responses are permitted in either language (Gollan, Fennema-Notestine, Montoya, & Jernigan, 2007; Roberts, Garcia, Desrochers, & Hernandez, 2002; Rosselli, Ardila, Jurado, & Salvatierra, 2012). The bilingual disadvantage in breadth of lexical knowledge has been attributed to the reduced frequency of using each language in everyday discourse compared with monolingual speakers. It remains to be established whether the protective effect of bilingualism and the associated disadvantage in lexical skills extends to bilingual speakers of English and Asian languages.

TRANSLATION, BACK TRANSLATION, AND USE OF INTERPRETERS

Translation is the process of converting one written language into another, whereas interpreting is the process of converting one spoken or signed language into another. A translator must write well in the target language and usually has sufficient time to explore the nuances of word choice and the relevance of cultural factors. Because most bilinguals are not fully balanced bilinguals, professional translators typically convert text from their second language into their native language. Interpreters, however, have to convert oral information bidirectionally, often at a very fast pace and without the benefit of an accessible dictionary. An interpreter functions as a near-instantaneous conduit between two parties and, additionally, must possess a wide range of communication, interpersonal, and decision-making skills (National Council on Interpreting in Health Care and American Translators Association, 2010).

Translation and Back Translation

Twenty-five years ago, few neuropsychological tests were available in non-English languages, and only a handful of those were developed using extensive translation-back translation methods (Ardila, Rosselli, & Puente, 1994, Fujii, 2011). Interest in cross-cultural research within the fields of psychology and education led to a surge in peer-reviewed articles addressing the translation or adaptation of tests (Hambleton & Zenisky, 2011). Today, neuropsychologists choose from hundreds of translated tests that span the developmental age spectrum (Benuto & Leany, 2013; Leany, Benuto, & Thaler, 2013; Salazar, Perez Garcia, & Puente, 2007). And while the increased availability of translated tests is an exciting development for the field, significant caution remains warranted. Direct translations are problematic because they achieve linguistic equivalence (i.e., each word matches its counterpart in the other language) but may lack equivalent meaning across the languages. Moreover, tests are often translated without the appropriate copyright permissions or in an idiosyncratic manner that can render the results difficult to interpret or generalize (Echemendia & Harris, 2004; Puente & Ardila, 2000).

Two procedures exist to develop adequate test translations: a translation-back translation approach and a committee approach (Brislin, 1970; Hambleton, 2005; Van de Vijver & Tanzer, 2004). In the former, a test is translated to the target language, followed by a translation back to its original language and a detailed reconciliation of the original and the back-translated forms. At that point, errors can be identified and corrected. Back translation works well if there are few idiomatic expressions or other linguistic features difficult to translate (Saklofske, Van de Vijver, Oakland, Mpofu, & Susuki, 2015). Alternatively, in a committee approach a group of experts cooperate to prepare a consensus version of the translated test.

These two methods are particularly useful during the initial test development phase, when tests can be generated simultaneously in each of the target languages and cross-referenced for back translation. However, in practice, the most common approach is to develop successive translations of an already-established test (Tanzer, 2005). In a successive translation, the clinician or researcher has three options: application, adaptation, or assembly of the test (Van de Vijver & Poortinga, 2005). *Application* consists of the literal translation of the test to the target language, "applying" it to the new cultural group. This is the most common practice in the social sciences due to simplicity, cost effectiveness, and face validity (Van de Vijver & Tanzer, 2004). *Adaptation* refers to the modification of a few items to accommodate idioms or cultural aspects that might not be captured in a direct translation. A core portion of the test is translated literally while a subset of the items is adapted to reflect the construct in a culturally appropriate manner. Finally, if an adaptation is insufficient, a relatively new test may need to be *assembled*. An assembled test will be suitable for the target culture and language but may preclude cross-cultural or cross-language test score comparisons.

A successful test translation does not imply that the different language versions are equivalent (Sireci, 2005). For instance, the back translator may have performed an excellent job, but if the original translation was done poorly the result is a nonequivalent version in the target language. Kolen and Brennan (2014) provide a real-world example of this problem, in which the idiom "out of sight, out of mind" was translated as "invisible, insane." Ideally, psychometric studies should be performed after a test has been translated to establish the equivalence of the different versions (Hambleton, 2005; International Test Commission, 2005). Unfortunately, most neuropsychological tests developed in English and translated into other languages lack demonstration of equivalence.

Interpreters

Neuropsychologists are unlikely to have regular access to a professional interpreter. Commercial health insurance carriers seldom cover language services, and few U.S. states pay for healthcare interpreters (Searight & Searight, 2009). Consequently, neuropsychologists often rely on bilingual yet untrained office staff or the patient's relatives in the evaluation of a non-English speaker. Although well-meaning, interpreters are not necessarily well versed in psychological or medical terminology, which can impede the clarity of communication and lead to misunderstandings (Echemendia, Harris, Congett, Diaz, & Puente, 1997; Dugbartey, 2014; Puente & Ardila, 2000). Furthermore, an interpreter fluent in the target language may not have sufficient knowledge of the cultural factors specific to the patient's country of origin. Interpreters constitute a third-party observer and actor in the assessment process, which raises ethical and practical concerns (American Academy of Clinical Neuropsychology, 2001; Committee on Psychological Tests and Assessment, American Psychological Association, 2007; National Academy of Neuropsychology, 2000).

If working with interpreters is absolutely necessary, Searight and Searight (2009) offer several general principles that can mitigate any negative impact on test results. First, a presession conversation with the interpreter is advised to discuss the interpreter's qualifications and expectations from the neuropsychologist, as well as discuss the specific test administration procedures and scoring rules. Next, during the interview process, the neuropsychologist should interact directly with the patient while discouraging any cross-communication between patient and interpreter. Questions should be presented using brief statements that allow for a cadence or rhythm in the exchanges between psychologist–interpreter and patient–interpreter. Finally, in a postsession, the neuropsychologist can use this opportunity with the interpreter to seek clarification on any issues and provide suggestions that may help the interpreter improve their skills.

CULTURAL AND HISTORICAL CONSIDERATIONS WITH SPANISH SPEAKERS

Numerous cultural and historical factors can exert an overt or covert influence upon the clinical evaluation of Spanish speakers. The impact of cultural factors may become manifest during the clinical interview; in the responses provided to medical and family history questionnaires; in the responses provided to mood, anxiety, and personality inventories; in the extent of collaboration with a psychometrist and/or interpreter; and in the responsiveness to recommendations. Ardila (2005) identifies eight cultural values implicit in neuropsychological testing that may not represent universal attitudes or behaviors and thus need to be considered in the interpretation of assessment results. These cultural values include (a) the one-to-one relationship between examiner and examinee, (b) the subordinate relationship of examinee to examiner, (c) the competitive, optimal performance expected of an examinee, (d) the isolated physical environment of the testing room, (e) the formality in relationship and communication style between examiner and examinee, (f) the time and speed component in test performance, (g) the private and intimate nature of clinical assessment, and (h) the physical elements (e.g., blocks, figures) that constitute the test materials. In addition to those values, several aspects conspicuous in the Hispanic culture, including machismo, fatalism, and religion, can interfere with the evaluation's *process* (i.e., the interpersonal dynamic between patient or family member and the psychologist, psychometrist, and/or interpreter) and *content* (i.e., the quantity and type of information disclosed during interview and provided as test performance).

Machismo, Fatalism, and Religious Beliefs

Machismo refers to the masculine ideal that invokes a prosocial role to protect and provide for family members, with historical roots in the chivalric code of medieval Spain, and distorts it to a level of domination, sexual aggression, and misogyny (Meana, Oliver, & Jones, 2013). It represents a dimensional construct that juxtaposes family-centered, nurturing values against sexist, chauvinistic values

(Arciniega, Anderson, Tovar-Blank, & Tracey, 2008). Traditional machismo is associated with elevated rates of alcohol abuse; history of aggression, arrest, and traumatic brain injury; and low frequency of healthcare utilization. It can hinder the assessment process when the gender-power roles contrast with machismo expectations (e.g., a male patient evaluated by a female psychologist or psychometrist).

Fatalism refers to the belief that one must resign oneself to destiny or God's will and thus accept circumstances as abiding and ever-present. In some respects, it reflects an adaptive psychological mechanism to cope with life stressors, whereby a locus of control external to the individual allows for the attribution of life's difficulties toward fate instead of personal choices (Alonzo, 2013; Champagne, Fox, Mills, Sadler, & Malcarne, 2015). But a fatalist mindset can dampen the motivation to seek clinical assessment, limit the depth of detail provided during clinical interview, and curtail the willingness to complete the examination once started or welcome recommendations for rehabilitation or treatment.

Fatalism and *religious beliefs* may be understood as complementary world views, and, indeed, both are linked along the locus of control dimension (Champagne, Fox, Mills, Sadler, & Malcarne, 2015). Whereas fatalism reflects a scarcity of motivation for decision-making, beliefs in supernatural forces presuppose a motivation to affect change through prayer or rituals. These rituals may take the form of scripted interchanges with the clergy, for instance in organized religions such as Catholicism, or informal worship of saints and supernatural beings as found in *Santería, brujería* (witchcraft), and other syncretic spiritual practices (Pontón, 2001). Belief in the supernatural is an important consideration in neuropsychological assessment insofar as those beliefs may be misinterpreted as clinical manifestations of disease (e.g., hearing voices or seeing persons in the context of *Santería* is mistaken as psychosis or signs of dementia). Alternatively, integration of the person's religious beliefs into a neuropsychological framework for rehabilitation can enhance the therapeutic process (Pontón, 2001).

Historical Influences on Contemporary Spanish

In addition to the aforementioned factors, it is important to note that the modern Spanish language comprises a rich amalgamation of historical influences. Contemporary Spanish retains the lexical and syntactical architecture of its Castilian-Andalusian ancestry, yet has adopted into its corpus the contributions from centuries of synergistic contact with other cultures (Penny, 2002; Pharies, 2007). Words such as *chofer* (from the French *chauffeur*); *aceituna* (olive, from the Arabic *azzaytuna*); and *sándwich* (from the English *sandwich*) are in common usage. In the American continents, contemporary Spanish has been shaped by the partial adoption of the indigenous lexicon and the assimilation of African linguistic features, the latter a particular influence in the Caribbean islands.

As the conquistadors displaced or mixed with the indigenous populations, native American lexical forms were incorporated into Spanish with varying degrees of geographical penetration. For instance, Nahuatl was the dominant language of the Aztecs and words such as *xokolatl* (chocolate) are now recognized in Spanish and other languages throughout the world, whereas words such as *papalotl* (butterfly) have lost their original meaning and instead are recognized as Spanish only in select geographic regions (e.g., *papalote* is *kite* in Mexico, Honduras, and Cuba).[2] Similarly, the dominant language of the Inca civilization, Quechua, provided words such as *llama* and *puma* that now are universally recognized, and words such as *chacra* (farm), recognized as Spanish mainly in the southern half of South America. Quechua remains the most widely spoken indigenous language in South America, with approximately 9 million speakers concentrated in Bolivia, Peru, and Ecuador (Lewis, Simons, & Fennig, 2015). The vast majority, however, are Spanish–Quechua bilinguals, and monolingual Quechua speakers are only found in remote rural regions (A. G. Muntendam, personal communication, December 2015).

In the Caribbean, contemporary Spanish arguably represents the most complex and integrated expression of Castilian-Andalusian syntax, vernacular from the indigenous Taíno and Ciboney Arawaks, and lexical forms brought by African slaves. Hundreds of words currently recognized as Spanish were adopted from the vernacular Arawak, and, as the case with Central and South American indigenous languages, those words remain in contemporary use either within a focal geographic area (e.g., *jimagua* [twins] in Cuba) or widely throughout the Greater Antilles (e.g., *cacique* [tribal chief]). African slaves were drawn from a vast geographic area stretching from west Africa southward into present-day Angola and deep into the Congo, leading to substantial variation in their spoken languages (e.g., Atlantic, Bantu, Congo-Benue, Kru, Kwa, Mande). This heterogeneity in African languages and the uneven geographic dispersion of those slaves throughout the Caribbean basin contribute to the localized distribution of many African lexical forms despite insular proximity. For instance, popular words such as the Afro-Cuban *asere* (buddy or friend) are seldom recognized outside of the Spanish-speaking Caribbean island in which they came into use.

Code Switching and Spanglish

Code switching or code mixing refers to the common phenomenon among bilinguals to alternate from one language to the other, particularly within a sentence or phrase unit. Psycholinguistic studies have demonstrated that the switch seldom violates the morphological and grammatical constraints of either language (Lipski, 2008). As such, a bilingual person would not express half of a word in one language and the remaining half in another, nor use a second-language word in the location where it would violate grammatical structure within the sentence. Code switching is not a simple combination of two grammars, but the integration of one language into another, requiring proficiency in both languages before the

---

2. The Spanish word for *kite* is variably *cometa* in Spain and Colombia, *chiringa* in Puerto Rico, *chichigua* in the Dominican Republic, and *volantín* in Chile.

person can switch (Lipski, 2008; Wei, 2007). It is a condition of bilingualism and, for most individuals, serves a variety of beneficial functions, including a pragmatic role in facilitating expression or comprehension, conveying meaning or emphasis, and allowing flexibility during oral narration.

Spanglish refers to a related occurrence among Spanish–English bilinguals, particularly those who reside in the United States. While there is no established definition for Spanglish, most linguists consider it an extension of the base, primary Spanish language, with the addition of lexical forms from English (Johnson, 2000; Lipski, 2008; Pountain, 1999). One pragmatic definition suggests that Spanglish is a form of code switching in which the Hispanic culture dominates the mix of Spanish and English languages but additionally incorporates frequent "borrowings" and "calques" (Johnson, 2000; Lipski, 2008). Borrowing is the common practice of adopting and adapting words from another language, and the modern borrowings from English represent a natural contemporary expansion of the Spanish lexicon (Hoffman, 1991; Lipski, 2008). Borrowings are incorporated into Spanish in their exact morphology (e.g., zipper, record, popcorn) or in slightly modified morphological forms that retain their meaning (e.g., *cloche* [clutch], *troca* [truck], *estrés* [stress]). In verbs, the most common form of lexical borrowing consists of adding the *-ear* termination to the English root word (e.g., *chequear* [to check], *cliquear* [to click], *parquear* [to park]). There is no clear dividing line between borrowing and code switching, although borrowing usually involves morphological adaptation whereas switching does not (Hoffman, 1991). In contrast, calques represent a strategy for accommodating semantic information without actual code switching or borrowings (Johnson, 2000). It constitutes the semantic extension of a word with specific meaning in monolingual Spanish, which now takes on a new or dual meaning (Potowski & Carreira, 2010). A common example of a calque is the word *librería*, which in Spanish traditionally refers to a bookstore or a place where books are sold. However, in the United States, the word has taken on the same meaning as a library, a place where books are kept for reading or reference, for which the corresponding word in Spanish should be *biblioteca*. Calques, borrowings, and other Anglicisms are common among Spanish–English bilinguals and are important to consider in the scoring and interpretation of language-based neuropsychological tests.

NEUROPSYCHOLOGICAL ASSESSMENT OF SPANISH SPEAKERS

The assessment of Spanish speakers is hindered not only by the limited availability of translated instruments and concerns about construct equivalence but also by the relative scarcity of Spanish-language norms and their unequal distribution compared with those in English (Gasquoine, Croyle, Cavazos-Gonzalez, & Sandoval, 2007). By one estimate, barely 15% of all psychological and neuropsychological tests are available in Spanish, and, of those, only a handful provide a test manual in Spanish and corresponding normative data (Salinas, Bordes Edgar, & Puente, 2016). As noted by Nell (2000), writing about all of cross-cultural neuropsychology but particularly applicable to the assessment of non-English speakers, a dialectical tension exists between the need to demonstrate construct equivalence and the need to collect and provide normative data for our tests. For every neuropsychological test available in Spanish, the bulk of the relevant peer-reviewed literature consists of a few validation studies or a few normative studies, but seldom both or in large scale. The reasons are understandable. Many neuropsychological tests are translated into Spanish by individual clinicians or small groups of investigators, often in an idiosyncratic manner to satisfy a specific clinical need (Echemendia & Harris, 2004; Puente & Ardila, 2000). Without access to grant funding or additional resources, those clinical investigators contribute to the field by introducing a new instrument or a translated version of an existing instrument. But a detailed study of test variables, the role of language or ethnicity on the construct, and clinical validation across neurologic conditions requires large data sets for which convenience samples are not ideal (Romero et al., 2009). Fortunately, this bottom–up approach by individual clinicians is increasingly being complemented by top–down contributions from institutional researchers and test publishing companies. Large-scale projects, including the SENAS, NEURONORMA, Latin American normative project, and others provide extensive normative data that can aid the clinician in choosing the most appropriate test for their Spanish-speaking patient. Programmatic research projects afford the opportunity to translate or develop instruments in Spanish, collect a large set of representative norms, and perform construct validation studies.

In the paragraphs that follow, an attempt is made to summarize a few of the tests and batteries that provide normative data for Spanish-speaking adults. This summary is not exhaustive, and the reader should consult the original sources for more detailed information. The Hispanic Neuropsychological Society additionally maintains on their website a compendium of nearly 200 tests, with annotations on those with and without representative norms. Furthermore, the Buros Center for Testing now provides an index of commercially published tests that are available in part or wholly in Spanish (Schlueter, Carlson, Geisinger, & Murphy, 2013).

The NEUROPSI consists of two separate batteries: a screener version that typically takes 30 minutes to administer and a comprehensive battery of attention, executive functions, and memory that takes approximately 60 minutes to administer (Ostrosky-Solís, Ardila, & Rosselli, 1999; Ostrosky-Solís et al., 2007). The screener battery was normed on 800 monolingual speakers from Mexico aged 16 to 85 years old. The comprehensive battery, called NEUROPSI: Atención y Memoria (NEUROPSI: Attention and Memory), is currently in its second edition and was normed on 950 monolinguals aged 6 to 85 years old. The educational level of the normative sample ranged from illiteracy to over 10 years, and education was more strongly associated with test scores than age (Ardila, Ostrosky-Solis, Rosselli, & Gómez, 2000). The NEUROPSI screener test manual indicates that the battery has 92% accuracy in differentiating mild-to-moderate dementia from normal cognition, but additional details are not provided. A version of the NEUROPSI screener translated into Portuguese differentiated cognitively healthy older adults from patients with mild or moderate Alzheimer's disease (AD) (Abrisqueta-Gomez, Ostrosky-Solis, Bertolucci, & Bueno, 2008). Additionally, the NEUROPSI screener performed comparably to the Mini-Mental State Examination (MMSE) in a small sample of Brazilian adults with AD or vascular dementia (Matioli & Caramelli, 2012). Despite the NEUROPSI's tremendous potential, unfortunately few clinical validation studies have been reported.

The Neuropsychological Screening Battery for Hispanics (NeSBHIS) was normed on 300 Spanish speakers aged 16 to 75 predominantly from Mexico, Central America, Cuba, and Puerto Rico (Pontón et al., 1996). Seventy percent of the normative sample consisted of Spanish monolinguals and 30% of bilinguals. The NeSBHIS includes tests of attention, language, visuospatial skills, episodic memory, and psychomotor skills. Latent structure analyses support the proposed 5-factor model, albeit with a few minor modifications (Bender et al., 2009; Pontón, Gonzalez, Hernandez, Herrera, & Higareda, 2000).

The Spanish and English Neuropsychological Assessment Scales (SENAS) is a comprehensive neuropsychological battery for adults age 60 and older. It was developed using methods from item response theory to minimize the impact of language- and education-based differential item functioning between Spanish and English speakers (Mungas, Reed, Marshall, & González, 2000; Mungas, Reed, Haan, & González, 2005; Mungas, Reed, Crane, Haan, & González, 2004). The current version consists of 16 subtests that take approximately 4 hours to administer. The SENAS demonstrates criterion validity in differentiating normal functioning from cognitive impairment and dementia. The combination of two subtests in particular, list-learning and object-naming, demonstrates better than 80% sensitivity and 80% specificity (Mungas, Reed, Farias, & Decarli, 2005). The SENAS battery, instructions, and normative data can be obtained upon request from the first author, D. Mungas.

The neuropsychological battery from the Consortium to Establish a Registry for Alzheimer's Disease (CERAD) consists of 7 subtests: MMSE; Boston Naming Test, 15-item version; animal fluency; constructional praxis; and word list learning, recall, and recognition. A version normed on 88 cognitively normal Spanish-speaking adults was translated and back-translated by a seasoned group of dementia investigators with experience using the original English version (Fillenbaum, Kuchibhatla, Henderson, Clark, & Taussig, 2007). The Spanish version of the CERAD appears to differentiate cognitively normal adults from those with mild AD, as well as mild AD from moderate AD. However, the Boston Naming subtest may be too easy for dementia patients (Marquez de la Plata et al., 2009). Another Spanish version of the CERAD was translated and normed in 848 adults age 50 and older from Colombia (Aguirre-Acevedo et al., 2007). A subsequent validation study in a cohort of presenilin-1 mutation carriers and 1,500 asymptomatic adults showed that the CERAD total score had high classification accuracy for dementia but poor specificity in the classification of low-education adults with mild cognitive impairment (Aguirre-Acevedo et al., 2016).

The Batería Neuropsicológica de Funciones Ejecutivas y Lóbulos Frontales (Neuropsychological Battery of Executive Functions and Frontal Lobes; BANFE; Flores Lázaro, Ostrosky-Solís, & Lozano, 2012) consists of 15 frontal-executive tasks such as the Wisconsin Card Sorting Test, Iowa Gambling Task, Stroop

Color-Word, and the Tower of Hanoi. The tests were selected with the purpose of sampling the functions of the prefrontal cortex. The initial normative sample consisted of 300 individuals from Mexico aged 6 to 80. Unfortunately, peer-reviewed validation studies are not yet available.

The Uniform Data Set (UDS) from the National Institute on Aging includes a comprehensive battery developed as a collaborative effort to characterize participants in the Alzheimer's Disease Research Centers across the United States (Weintraub et al., 2009). The UDS battery consists of tests of attention, executive functions, processing speed, episodic memory, and language. In 2005, a workgroup was established to develop a Spanish translation and facilitate uniform data collection of Spanish-speaking participants throughout the country. The Spanish translation and adaptation process is well described (Acevedo et al., 2009). Although "normal cognition" was defined separately by each Alzheimer's Disease Research Center, all cognitively normal adults were free of major psychiatric and neurologic conditions. The normative sample consisted of 276 Spanish-speaking adults, more than half of whom were age 70 and over. Summary statistics for each of the Spanish UDS subtests are provided, but clinical validation studies are yet to be reported (Benson, de Felipe, Xiaodong, & Sano, 2014).

The Repeatable Battery for the Assessment of Neuropsychological Status (RBANS; Randolph, 1998) consists of 12 subtests that evaluate attention, language, visuospatial skills, and episodic memory. As noted by Bender (2015), the Spanish version provided by the test publisher underwent translation and back translation, with an equivalence study conducted in a sample of Spanish speakers. However, comprehensive normative data for that version are not yet available. De la Torre et al. (2014) created a version of the RBANS translated and back-translated in southern Spain and normed on 336 adults ranging in age from 20 to 90 years. The Spanish RBANS demonstrated high internal consistency (Cronbach's alpha = 0.92) and test-retest stability (0.90) over a 6-month period, although this very high level of alpha also may reflect item redundancy. The RBANS showed very good specificity (0.96) but poor sensitivity (0.30) to detect mild AD. In a separate study, the same version of the RBANS had good sensitivity (0.88) and specificity (0.86) in classifying healthy adults from patients with schizophrenia (De la Torre, Perez, Ramallo, Randolph, & González-Villegas, 2015). Another version of the RBANS underwent translation/back translation in Barcelona, Spain, and was normed in a preliminary sample of 73 adults aged 40 to 80 years (Muntal Encinas, Gramunt-Fombuena, Badenes Guia, Casas Hernanz, & Aguilar Barbera, 2012). Internal consistency was adequate (Cronbach's alpha = 0.73), and the authors noted that additional normative data were being collected.

The NEURONORMA project is a multicenter study in Spain created to provide neuropsychological norms and clinical validation for the longitudinal follow-up of cognitive decline in aging and dementia (Peña-Casanova et al., 2009). The NEURONORMA battery includes tests of attention, language, executive functions, visuospatial skills, and episodic memory. Tests were translated and back-translated and the normative data for 356 adults aged 50 to 90 years old were published as a manuscript series in the *Archives of Clinical Neuropsychology*

(Volume 24, Issue 4). A recent study using the NEURONORMA battery differentiated cognitively normal adults from those with mild cognitive impairment and AD (Sánchez-Benavides et al., 2014).

In a special issue of *Neurorehabilitation,* Arango-Lasprilla (2015), the guest editor, reported the results of a multicenter normative study from 20 cities in 12 Spanish-speaking Latin American countries: Argentina, Bolivia, Chile, Colombia, Cuba, El Salvador, Guatemala, Honduras, Mexico, Paraguay, Peru, and Puerto Rico. The investigators provided norms from 5,402 adults age 18 to 90 for the Boston Naming Test, Verbal Fluency (F-A-S and animal category), Modified Wisconsin Card Sorting Test, Stroop Color-Word Test, Symbol Digit Modalities Test, Trail Making Test, Brief Test of Attention, Rey–Osterrieth Complex Figure, Hopkins Verbal Learning Test–Revised, and the Test of Memory Malingering. This represents the largest normative study in Central and South America, with over 2,500 participants combined from Colombia and Mexico and between 200 to 300 participants from each of the remaining countries. The authors noted that clinical validation studies using these norms in adults with various neurological conditions are forthcoming.

The Mattis Dementia Rating Scale-2 (DRS; Jurica, Leitten, & Mattis, 2001) consists of five subscales measuring attention, initiation/perseveration, visuoconstruction, conceptualization, and memory. An early study using a Spanish translation showed that it differentiated cognitively normal adults from those with AD, but the sample consisted only of 18 participants (Taussig, Henderson, & Mack, 1992). Subsequently, the DRS underwent detailed translation/back translation and minor adaptation, with normative data obtained from Spanish-speaking Mexican American older adults residing in southern Texas (Arnold, Cuellar, & Guzman, 1998). Using a cut-off value 2 standard deviations (SD) below the mean to define cognitive impairment, the Spanish version of the DRS had poor sensitivity (0.40) and very high specificity (0.97). Adjusting the cut-off to −1.3 SD yielded sensitivity of 55% and specificity of 92%. Additional normative data for the Arnold translation were provided by Lyness, Hernandez, Chui, and Teng (2006) in a sample of adults aged 55 to 89 years predominantly from Mexico, Cuba, Peru, and El Salvador. Fifty-nine percent of the sample considered themselves to be Spanish monolinguals. In a comparison with age- and education-matched English speakers, the Spanish-speaking sample obtained lower scores in the total DRS and in the attention, conceptualization, and memory subscales. The authors suggested that norms for the DRS based on English-speakers are not suitable for Spanish-speaking adults and re-emphasized the need for representative norms in Spanish. Lastly, in an almost direct translation, Hohl, Salmon, Thomas, and Thal (1999) matched Spanish-speaking Hispanic and English-speaking, non-Hispanic white AD patients by their MMSE scores. Both groups of patients in the mild-to-moderate range of dementia performed comparably on the DRS, but the more severely impaired Spanish speakers performed significantly worse than the English speakers. The authors attributed this discrepancy to possible cultural bias, differences in education between the groups, or greater sensitivity of the DRS to detect severe cognitive impairment in Hispanic adults.

CULTURAL AND HISTORICAL CONSIDERATIONS WITH SPEAKERS OF ASIAN AND PACIFIC ISLANDER LANGUAGES

Similar to Hispanics, Asian Americans are ethnically heterogeneous with 24 separate ethnicities listed in the census; the ten largest ethnic groups are Chinese (4,010,114; 23.2%), Filipino (3,416,840; 19.7%), Asian Indian (3,183,063; 18.4%), Vietnamese (1,737,433; 10.0%), Korean (1,706,822; 9.9%), Japanese (1,304,286; 7.5%), Pakistani (409,163; 2.4%), Cambodian (276,677; 1.6%), Hmong (260,073; 1.5%), and Thai (232,130; 1.3%; Hoeffel, Rastogi, Kim, & Shahid, 2012). Each country is unique in sociopolitical histories, economies, and educational systems, as well as reason for and timing of persons immigrating to the United States. Thus, when evaluating an Asian American patient, it is important to research their specific culture for contextual understanding to facilitate collection of accurate data and provide a context to interpret data and make useful recommendations (Fujii, 2016).

Fujii (2016) describes three categories of cultural information that neuropsychologists should research to develop a cultural understanding of an Asian-American patient. Each has specific implications for conducting a culturally informed neuropsychological evaluation. First, macrosocietal structures such a geography of country, population, economics, government, general history, languages spoken, and educational system can provide a context for understanding environmental factors and opportunities that shape a patient's general functional abilities. This conceptual context of the patient is crucial for both test selection and interpretation of test results. Second, values, beliefs, and social structures including world view, family structures, social roles, expectation for interactions, boundary issues, and religion and beliefs (Judd & Beggs, 2005), can provide clues for optimal approaches for developing rapport as well as generating culturally relevant recommendations. Learning styles (Van Hamme, 1996) and concepts of intelligence (Sternberg, 2004) are useful for guiding test selection and interpretation of test results. Third, common neurological and psychological disorders and recent history of exposure to events such as war or illnesses, such as Zika, that could cause a brain injury are useful for generating hypotheses about diagnoses. Attitudes and beliefs regarding health and illness are useful for making culturally relevant recommendations (Judd & Beggs, 2005).

Despite the numerous cultural differences between Asian countries, there is a pan-Asian culture that can guide general understanding when working with Asian American patients. Most Asian countries, with the exception of Pakistan and Malaysia, which are predominantly Muslim, and the Philippines, which is predominantly Catholic, have historically been influenced by Eastern philosophical traditions such as Hinduism, Buddhism, Taoism, and Confucianism that emphasize harmony and oneness with others, nature, and the universe (Fujii, 2015; Guo & Uhm, 2014). Thus, it follows that Asian societies are collective where the group is stressed over the individual and harmonious interpersonal relationships are valued. Communication is indirect to avoid conflict or embarrassment, emotions are moderated, social roles are well defined and respected, and high expectations dictate proper conduct (for a review, see Guo & Uhm, 2014).

Pan-Asian values and behaviors have important implications for communicating with and diagnosing Asian American patients. For example, due to a strong deference to authority, in many Asian cultures nodding to an authority figure means "I am listening to you" versus "yes," as it is impolite to disagree with a person of higher stature (Fujii, 2011). Similarly, clinicians not familiar with the indirect communication style of an Asian patient may miss nonverbal cues important for communication and, as a result, come across as rude during the interviewing process, which could impact rapport (Fujii, 2016). In addition, neuropsychologists who are not aware of Asian Americans' tendency to somaticize emotional problems (Maffini & Wong, 2014) or downplay problems during the interview due to the stigma associated with mental illness can easily misdiagnose a patient's psychiatric condition (Chu & Sue, 2011).

When developing a cultural context, it is important to consider the impact of an Asian American patient's generation in the United States and ethnic identity, which are moderating factors for cultural influence. For a more thorough discussion of specific Asian cultures, see Fujii (2011).

NEUROPSYCHOLOGICAL ASSESSMENT OF SPEAKERS OF ASIAN AND PACIFIC ISLANDER LANGUAGES

A significant consideration when testing Asian Americans is that, unlike all Hispanic ethnicities who speak Spanish, for the most part Asian ethnicities each have their own distinct language and writing system. In some countries, multiple languages and dialects are spoken. For example, 54 languages are spoken in China (Chan, Leung, & Cheung, 2011), while India has 22 official languages and 1,652 dialects (Kumar, 2011). This increased heterogeneity within a significantly smaller population carries several ramifications. First, procuring translated, validated, and normed tests is likely to be much more difficult, as the challenges identified in test development for Hispanics is multiplied. A scarcity of available tests is particularly apparent for lower resource countries such as Laos or Pakistan. Indeed, most available commercial tests or those validated and normed by researchers exist primarily for the countries with the largest economies or gross domestic product per capita. For example, commercial versions of the Wechsler Intelligence tests are only available in Chinese, Japanese, Hindi, Korean, and Vietnamese, while versions of the Wechsler Memory tests are only available in Chinese, Japanese, and Korean. The following references provides a list of tests translated, normed, and validated in Chinese (Chan, Leung, & Cheung, 2011), Korean (Chey & Park, 2011), Japanese (Isomura-Motoki & Mimura, 2011), and Hindi (India; Kumar, 2011), while tests for other Asian ethnicities are reported in chapters from *The Neuropsychology of Asian Americans* (Fujii, 2011) and *Conducting a Culturally Informed Neuropsychological Evaluation* (Fujii, 2016). Second, a related issue is less accessibility even if translated tests were available. Unlike Spanish, which is related to English and written using the same alphabet, Asian languages have their own orthographies and are much more different in sound and structure. Third, the availability of a neuropsychologist who speaks a specific Asian language is substantially less common than for those who speak Spanish; thus, neuropsychologists may be more ethically obligated to provide services due to a paucity of colleagues who speak the same language. Fourth, in many areas there may be few interpreters who speak specific Asian languages.

For countries without available norms or test translations, clinicians will have to do much more preparation to select, administer, translate, and interpret appropriate neuropsychological tests with cross-cultural validity and obtain an estimate of premorbid abilities. If there are no tests validated in a patient's country of origin, clinicians should look for tests or subtests of larger batteries validated with similar cultures or across a broad range of cultures. For example, the RBANS (Randolph, 1998) has been validated in 16 countries including China, Japan, Singapore, and Sri Lanka. Thus, it could be argued that subtests or similarly structured tests—for example, Rey Auditory Verbal Learning Test as an analogue to List Learning—would be appropriate for use in other Asian countries. Of course, the neuropsychologist would have to ultimately decide the appropriateness of tests based upon the specific characteristics, experiences, and educational history of the patient (for specific strategies in test selection, see Fujii, 2016).

When interpreting test scores, Fujii (2016) recommends using the individual comparison method, which uses an estimate of premorbid intellectual abilities as the benchmark to which current performances on neuropsychological tests are compared (Gasquoine, 2009). A two-step process is proposed for estimating premorbid intelligence when no measures are available (Fujii, 2016). The first step is to procure an estimate of how persons from the patient's country of origin score on Western intelligence tests. Western is emphasized as many cultural factors including differences in conception of intelligence (Sternberg, 2004), comfort level with the testing situation, test biases, biases in accessibility or ability to respond to test items, biases in test validity, and differences in opportunities for learning can impact the fairness of a test for a culturally different client (American Education Research Association, American Psychological Association, & National Council on Measurement in Education, 2014). Intelligence estimates can be procured from the literature or calculated from national data that is correlated with intelligence, such as national economy or student scores on international standardized measures (Lynn & Vanhanen, 2012).

The second step is to determine a client's functional level within their country of origin and then adjust the IQ score accordingly. Characteristics that can factor into an estimate of function level would include occupational status, level of education, urban versus rural living, and reason for immigration, which typically results in selection bias in the representativeness of persons coming from a given country (Fujii, 2016). For example, Thailand has an estimated IQ of 91 (Lynn & Vanhanen, 2012) and 12.3 expected years of education ("Countries Compared by Education," n.d.). Thus, a medical doctor immigrating to the United States for residency may be roughly estimated to be one to two standard deviations above

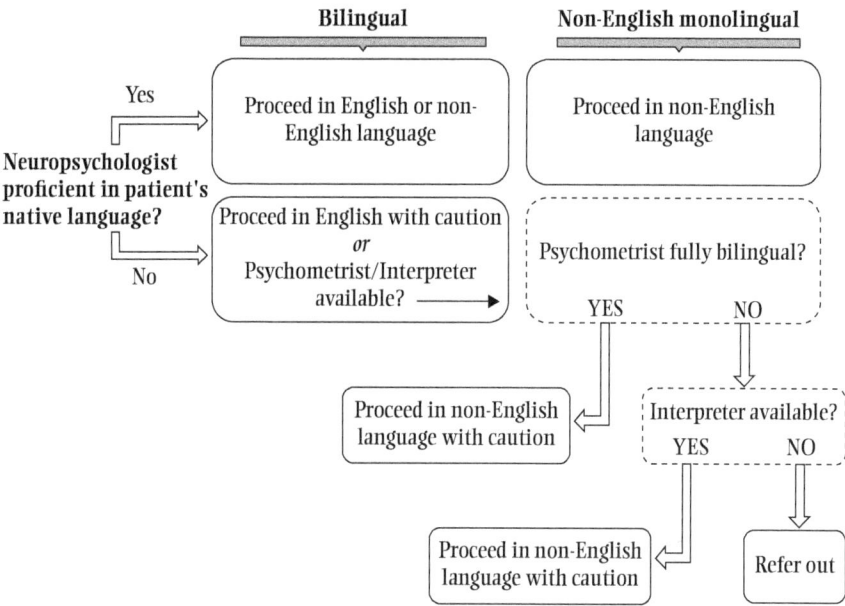

Figure 8.1. Decision algorithm for the assessment of non-English speakers.

the mean for an IQ range of about 106 to 121. As illustrated, this method is not an exact science; however, it is argued that estimations based on actual cultural data will provide a culturally grounded context for interpreting neuropsychological test data.

EXAMINATION LANGUAGE

Clinical neuropsychologists are often faced with the difficult task of determining the language in which to evaluate a non-English speaker. To help clinicians make this determination, we provide a decision algorithm drawn from converging professional opinion (Figure 8.1). While not exhaustive of all possible scenarios, it serves as a visual synthesis of the core considerations outlined in position papers by the National Academy of Neuropsychology and American Academy of Clinical Neuropsychology (Board of Directors, 2007; Judd et al., 2009). Based on those recommendations, clinical neuropsychologists should:

1. Determine their *own* level of proficiency in the patient's non-English language.
    a. Neuropsychologists who are proficient in the patient's non-English language make efforts to understand any unique linguistic and cultural variables pertaining to the patient's heritage and country of origin.

b. If not proficient in the patient's language, neuropsychologists may consider working with a qualified psychometrist or interpreter. When working with a psychometrist or interpreter, neuropsychologists still endeavor to obtain sufficient training and familiarize themselves with the relevant cultural factors.
   c. If a qualified psychometrist or interpreter is unavailable, neuropsychologists who are not proficient in the patient's language should refer the patient to a colleague who is proficient in that language.
   d. In those circumstances in which a qualified psychometrist or interpreter is unavailable, and there is no opportunity to refer the patient to a colleague proficient in the patient's language, neuropsychologists who proceed with an evaluation using nonverbal tasks or putative culture-free or culture-fair tasks must determine whether any potential harm to the patient from proceeding with the assessment outweighs any potential harm from declining the referral.
2. Determine the *patient's* level of proficiency in their non-English language.
   a. Neuropsychologists need to establish the patient's ability to read, write, speak, and understand their non-English language, and the potential impact of education, acculturation, and regional dialect on their test responses and scores.
   b. For a balanced bilingual patient, the neuropsychological assessment may be obtained in either language whenever the language of the evaluation is irrelevant to the purpose of the referral.
   c. Neuropsychologists are aware that older immigrant adults are more likely to be non-English monolingual or nonbalanced bilinguals.
3. Select appropriate tests and normative reference data.
   a. Neuropsychologists understand that despite an expanding selection of tests available in many languages, most lack rigorous translation, evidence of construct equivalence, or representative norms.
   b. Neuropsychologists should review the available information on the development and psychometric validation of each test to determine its usefulness for that particular patient's clinical needs.
   c. Neuropsychologists recognize the linguistic and cultural heterogeneity among non-English speakers and, accordingly, the heterogeneity of the available normative data. Every effort is made to seek and apply the most appropriate norm-referenced tests for that patient's linguistic, cultural, and educational background.
4. Provide the relevant documentation in the clinical report.
   a. Neuropsychologists should identify in their clinical report the version of any translated test and the normative data used.
   b. If the services of an interpreter are used, neuropsychologists document the name of the interpreter and any behavioral observations during clinical interview and testing that may bear on score interpretation and diagnostic disposition.

c. Neuropsychologists acknowledge the limitations inherent in evaluating patients using interpreters or administering tests with insufficient validation or norms. If applicable, neuropsychologists document in the clinical report whether caution is warranted in the interpretation of test results.

SUMMARY AND CONCLUSIONS

The neuropsychological assessment of non-English speakers is anything but uniform. Demographic, migrational, cultural, and socioeconomic variables interact with language to delineate each patient's unique mode of oral and written expression. Code switching, borrowings, and calques are omnipresent in the discourse of Spanish monolingual and bilingual adults and can be difficult to decode, even for a native speaker. Asian American countries each have their own unique spoken and written languages, which can significantly differ phonemically and grammatically in comparison to English. Moreover, indirect communication styles can add complexity to the assessment process. Neuropsychologists who evaluate non-English speakers need to consider these and many other factors, including the details of test development, clinical validation, and normative sampling; the patient's unique cultural heritage and regional dialect; and the impact of acculturation and other variables, always mindful not to presume universality of those cultural factors. Culture filters our perceptions of self, society, and health, and acknowledgment of the role of those factors can lead to a more humanistic and thorough appreciation of the patient's presenting clinical concerns.

There is much to be optimistic about in the neuropsychological assessment of non-English speakers. Consensus papers from our professional organizations provide needed guidelines and the rising tide of test development and normative projects herald a brighter future for the assessment of non-English speakers compared with just two decades ago. Still, there is room for improvement. The relative proportion of neuropsychologists who speak Spanish and Asian languages does not appear to be keeping pace with the demographic trends for the U.S. population at large, which by necessity implies a larger role for bilingual psychometrists or interpreters. Whether this shift leads to a proliferation of less-than-competent evaluations of non-English speakers remains to be seen. Finally, normative studies are necessary but insufficient; the need for construct validation studies is tremendous and hopefully will constitute the next wave of cross-cultural neuropsychological investigation.

REFERENCES

Abrisqueta-Gomez, J., Ostrosky-Solis, F., Bertolucci, P. H., & Bueno, O. (2008). Applicability of the abbreviated neuropsychological battery (NEUROPSI) in Alzheimer disease patients. *Alzheimer Disease and Associated Disorders, 22,* 72–78. https://doi.org/10.1097/WAD.0b013e3181665397

Abutalebi, J., Canini, M., Della Rossa, P. A., Sheung, L. P., Green, D. W., & Weekes, B. S. (2014). Bilingualism protects anterior temporal lobe integrity in aging. *Neurobiology of Aging, 35,* 2126–2133. https://doi.org/10.1016/j.neurobiolaging.2014.03.010

Abutalebi, J., Della Rosa, P. A., Green, D. W., Hernandez, M., Scifo, P., Keim, R., ... Costa, A. (2012). Bilingualism tunes the anterior cingulate cortex for conflic monitoring. *Cerebral Cortex, 22,* 2076–2086. https://doi.org/10.1093/cercor/bhr287

Abutalebi, J., & Green, D. W. (2008). Control mechanisms in bilingual language production: Neural evidence from language switching studies. *Language and Cognitive Processes, 23,* 557–582. https://doi.org/10.1080/01690960801920602

Acevedo, A., Krueger, K. R., Navarro, E., Ortiz, F., Manly, J. J., Padilla-Vélez, M. M., . . . Mungas, D. (2009). The Spanish translation and adaptation of the Uniform Data Set of the National Institute on Aging Alzheimer's Disease Centers. *Alzheimer's Disease and Associated Disorders, 23,* 102–109. https://doi.org/10.1097/WAD.0b013e318193e376

Aguirre-Acevedo, D. C., Gómez, R. D., S, M., Henao-Arboleda, E., Motta, M. C., Arana, A., ... Lopera, F. (2007). Validez y fiabilidad de la batería neuropsicológica CERAD-Col [Validity and reliability of the CERAD-Col neuropsychological battery]. *Revista de Neurología, 45,* 655–660.

Aguirre-Acevedo, D. C., Jaimes-Barragán, F., Henao, E., Tirado, V., Muñoz, C., Reiman, E. M., ... Lopera, F. (2016). Diagnostic accuracy of CERAD total score in a Colombian cohort with mild cognitive impairment and Alzheimer's disease affected by E280A mutation on presenilin-1 gene. *International Psychogeriatrics 28,* 503–510. https://doi.org/10.1017/S1041610215001660

Alba, R. (2004). *Language assimilation today: Bilingualism persists more than in the past, but English still dominates.* Lewis Mumford Center for Comparative Urban and Regional Research at the University of Albany Working Paper No. 11. Center for Comparative Immigration Studies, University of California–San Diego, La Jolla, CA. Retrieved from http://escholarship.org/uc/item/0j5865nk

Alladi, S., Bak, T. H., Duggirala, V., Surampudi, B., Shailaja, M., Shukla, A. K., ... Kaul, S. (2015). Bilingualism delays age at onset of dementia, independent of education and immigration status. *Neurology, 81,* 1938–1944. https://doi.org/10.1212/01.wnl.0000436620.33155.a4

Alonzo, D. M. (2013). Assessing suicidality with the Hispanic adult. In L. T. Benuto (Ed.), *Guide to psychological assessment with Hispanics* (pp. 141–151). New York, NY: Springer.

American Academy of Clinical Neuropsychology. (2001). Policy statement on the presence of third party observers in neuropsychological assessments. *The Clinical Neuropsychologist, 15,* 433–439. https://doi.org/10.1076/clin.15.4.433.1888

American Education Research Association, American Psychological Association, and the National Council on Measurement in Education. (2014). *Standards for educational and psychological testing* (2nd ed.). Washington, DC: American Education Research Association.

Michalski, D., Mulvey, T., & Kohout, J. (2010). 2008 APA survey of psychology health service providers. *American Psychological Association Center for Workforce Studies.* Retrieved from http://www.apa.org/workforce/publications/08-hsp/report.pdf

Arango-Lasprilla, J. C. (2015). Commonly used neuropsychological tests for Spanish speakers: Normative data from Latin America. *Neurorehabilitation, 37,* 489–491. https://doi.org/10.3233/NRE-151276

Arciniega, G. M., Anderson, T. C., Tovar-Blank, Z. G., & Tracey, T. J. (2008). Toward a fuller conception of machismo: Development of a traditional machismo and caballerismo scale. *Journal of Counseling Psychology, 55,* 19–33. https://doi.org/10.1037/0022-0167.55.1.19

Ardila, A. (2005). Cultural values underlying psychometric cognitive testing. *Neuropsychology Review, 15,* 185–195. https://doi.org/10.1007/s11065-005-9180-y

Ardila, A., Ostrosky-Solis, F., Rosselli, M., & Gómez, C. (2000). Age-related cognitive decline during normal aging: The complex effect of education. *Archives of Clinical Neuropsychology, 15,* 495–513. https://doi.org/10.1016/S0887-6177(99)00040-2

Ardila, A., Rosselli, M., & Puente, A. E. (1994). *Neuropsychological evaluation of the Spanish speaker.* New York, NY: Plenum.

Arnold, B. R., Cuellar, I., & Guzman, N. (1998). Statistical and clinical evaluation of the Mattis Dementia Rating Scale–Spanish adaptation: An initial investigation. *Journal of Gerontology: Psychological Sciences, 53B,* P364–P369. https://doi.org/10.1093/geronb/53B.6.P364

Bender, H. A. (2015). The neuropsychological assessment of culturally and linguistically diverse epilepsy patients. In W. B. Barr & C. Morrison (Eds.), *Handbook on the neuropsychology of epilepsy* (pp. 317–344). New York, NY: Springer.

Bender, H. A., Cole, J. R., Aponte-Samalot, M., Cruz-Laureano, D., Myers, L., Vasquez, B. R., & Barr, W. B. (2009). Construct validity of the Neuropsychological Screening Battery for Hispanics (NeSBHIS) in a neurological sample. *Journal of the International Neuropsychological Society, 15,* 217–224. https://doi.org/10.1017/S1355617709090250

Benson, G., de Felipe, J., Xiaodong, & Sano, M. (2014). Performance of Spanish-speaking community-dwelling elders in the United States on the Uniform Data Set. *Alzheimer's & Dementia, 10,* S338–S343. https://doi.org/10.1016/j.jalz.2013.09.002

Benuto, L. T. (Ed.). (2013). *Guide to psychological assessment with Hispanics.* New York, NY: Springer.

Benuto, L. T., & Leany, B. D. (2013). Assessment of dementia in the Hispanic client: A neuropsychological perspective. In L. T. Benuto (Ed.), *Guide to psychological assessment with Hispanics* (pp. 243–262). New York, NY: Springer.

Bialystok, E. (2016). How hazy views become full pictures. *Language, Cognition and Neuroscience, 31,* 328–330. https://doi.org/10.1080/23273798.2015.1074255

Bialystok, E., & Craik, F. I. (2010). Cognitive and linguistic processing in the bilingual mind. *Current Directions in Psychological Science, 19,* 19–23. https://doi.org/10.1177/0963721409358571

Bialystok, E., Craik, F. I., & Luk, G. (2012). Bilingualism: Consequences for mind and brain. *Trends in Cognitive Science, 16,* 240–250. https://doi.org/10.1016/j.tics.2012.03.001

Board of Directors. (2007). American Academy of Clinical Neuropsychology (AACN) practice guidelines for neuropsychological assessment and consultation. *The Clinical Neuropsychologist, 21,* 209–231. https://doi.org/10.1080/13825580601025932

Brislin, R. W. (1970). Back-translation for cross-cultural research. *Journal of Cross-Cultural Psychology, 1,* 185–216.

Burgaleta, M., Sanjuán, A., Ventura-Campos, N., Sebastián-Gallés, N., & Ávila, C. (2016). Bilingualism at the core of the brain: Structural differences between bilinguals and monolinguals revealed by subcortical shape analysis. *NeuroImage, 125,* 437–445. https://doi.org/10.1016/j.neuroimage.2015.09.073

Casas, R., Guzmán-Vélez, E., Cardona-Rodrigues, J., Rodriguez, N., Quiñones, G., Izaguirre, B., & Tranel, D. (2012). Interpreter-mediated neuropsychological testing

of monolingual Spanish speakers. *The Clinical Neuropsychologist, 26,* 88–101. https://doi.org/10.1080/13854046.2011.640641
Cascallar, E. C., & Arnold, J. (2001). Second language acquisition. In M. O. Pontón, & J. León-Carrión (Eds.), *Neuropsychology and the Hispanic patient: A clinical handbook* (pp. 59–74). Mahwah, NJ: Erlbaum.
Centeno, J. G., & Obler, L. K. (2001). Principles of bilingualism. In M. O. Pontón, & J. León-Carrión (Eds.), *Neuropsychology and the Hispanic patient: A clinical handbook* (pp. 75–86). Mahwah, NJ: Erlbaum.
Champagne, B. R., Fox, R. S., Mills, S. D., Sadler, G. R., & Malcarne, V. L. (2015). Multidimensional profiles of health locus of control in Hispanic Americans. *Journal of Health Psychology, 21,* 2376–2385. https://doi.org/10.1177/1359105315577117
Chan, A. S., Leung, W., & Cheung, M. C. (2011). Neuropsychology in China. In D. Fujii (Ed.). *The neuropsychology of Asian Americans* (pp. 201–218). New York, NY: Taylor & Francis.
Chey, J., & Park, H. (2011). Neuropsychology in Korea. In D. Fujii (Ed.). *The neuropsychology of Asian Americans* (pp. 247–268). New York, NY: Taylor & Francis.
Chu, J. P., & Sue, S. (2011). Asian American Mental Health: What We Know and What We Don't Know. *Online Readings in Psychology and Culture, 3*(1). https://doi.org/10.9707/2307-0919.1026
Colby, S. L., & Ortman, J. M. (2015). Projections of the size and composition of the U.S. population: 2014 to 2060. *U.S. Census Bureau.* Current Population Reports No. P25-1143. Retrieved from https://www.census.gov/content/dam/Census/library/publications/2015/demo/p25-1143.pdf
Committee on Psychological Tests and Assessment. (2007). Statement on third-party observers in psychological testing and assessment: A framework for decision making. *American Psychological Association.* Retrieved from https://www.apa.org/science/programs/testing/third-party-observers.pdf
Costa, A., & Sebastián-Gallés, N. (2014). How does the bilingual experience sculpt the brain? *Nature Reviews Neuroscience, 15,* 336–345. https://doi.org/10.1038/nrn3709
Craik, F. I., Bialystok, E., & Freedman, M. (2010). Delaying the onset of Alzheimer disease: Bilingualism as a form of cognitive reserve. *Neurology, 75,* 1726–1729. https://doi.org/10.1212/WNL.0b013e3181fc2a1c
Countries compared by education: Average years of schooling of adults by country: UNESCO. (n.d.). *NationMaster.* http://www.NationMaster.com/graph/edu_ave_yea_of_sch_of_adu-education-average-years-schooling-adults
Davis, J. M., & D'Amato, R. C. (Eds.). (2014). *Neuropsychology of Asians and Asian-Americans: Practical and theoretical considerations.* New York, NY: Springer.
de Bruin, A., & Della Salla, S. (2016). The importance of language use when studying the neuroanatomical basis of bilingualism. *Language, Cognition and Neuroscience, 31,* 335–339. https://doi.org/10.1080/23273798.2015.1082608
De la Torre, G. G., Perez, M. J., Ramallo, M. A., Randolph, C., & González-Villegas, M. B. (2015). Screening of cognitive impairment in schizophrenia: Reliability, sensitivity, and specificity of the Repeatable Battery for the Assessment of Neuropsychological Status in a Spanish sample. *Assessment, 23,* 221–231. https://doi.org/10.1177/1073191115583715
De la Torre, G. G., Suárez-Llorens, A., Caballero, F. J., Ramallo, M. A., Randolph, C., Lleó, A., . . . Sánchez, B. (2014). Norms and reliability for the Spanish version of the Repeatable Battery for the Assessment of Neuropsychological Status (RBANS) Form

A. *Journal of Clinical and Experimental Neuropsychology, 36,* 1023–1030. https://doi.org/10.1080/13803395.2014.965664

Dugbartey, A. T. (2014). Ethical considerations in neuropsychological assessment of Asian heritage clients. In J. M. Davis, & R. C. D'Amato (Eds.), *Neuropsychology of Asians and Asian-Americans: Practical and theoretical considerations* (pp. 17–26). New York, NY: Springer.

Echemendia, R. J., & Harris, J. G. (2004). Neuropsychological test use with Hispanic/Latino populations in the United States: Part II of a national survey. *Applied Neuropsychology, 11,* 4–12. https://doi.org/10.1207/s15324826an1101_2

Echemendia, R. J., Harris, J. G., Congett, S. M., Diaz, M. L., & Puente, A. E. (1997). Neuropsychological training and practices with Hispanics: A national survey. *The Clinical Neuropsychologist, 11,* 229–243. https://doi.org/10.1080/13854049708400451

Elbulok-Charcape, M. M., Rabin, L. A., Spadaccini, A. T., & Barr, W. B. (2014). Trends in the neuropsychological assessment of ethnic/racial minorities: A survey of clinical neuropsychologists in the United States and Canada. *Cultural Diversity and Ethnic Minority Psychology, 20,* 353–361. https://doi.org/10.1037/a0035023

Espenshade, T. J., & Fu, H. (1997). An analysis of English-language proficiency among U.S. immigrants. *American Sociological Review, 62,* 288–305. https://doi.org/10.2307/2657305

Espinoza, K. E., & Massey, D. S. (1997). Determinants of English proficiency among Mexican migrants to the United States. *International Migration Review, 31,* 28–50. https://doi.org/10.2307/2547256

Ferraro, F. R. (2016). *Minority and cross-cultural aspects of neuropsychological assessment: Enduring and emerging trends* (2nd ed.). New York, NY: Taylor & Francis.

Fillenbaum, G. G., Kuchibhatla, M., Henderson, V. W., Clark, C. M., & Taussig, I. M. (2007). Comparison of performance on the CERAD neuropsychological battery of Hispanic patients and cognitively normal controls at two sites. *Clinical Gerontologist, 30,* 1–22. https://doi.org/10.1097/JGP.0b013e3181f7d881

Fletcher-Janzen, E., Strickland, T. L., & Reynolds, C. R. (Eds.). (2000). *Handbook of cross-cultural neuropsychology.* New York, NY: Springer.

Flores Lázaro, J. C., Ostrosky-Solís, F., & Lozano, A. (2012). *Batería neuropsicológica de funciones ejecutivas y lóbulos frontales.* Mexico City, Mexico: Manual Moderno.

French, C., & Llorente, A. M. (2008). Language: Development, bilingualism, and abnormal states. In A. M. Llorente (Ed.), *Principles of neuropsychological assessment with Hispanics* (pp. 78–120). New York, NY: Springer.

Fujii, D. E. (Ed.). (2011). *The Neuropsychology of Asian Americans.* New York, NY: Taylor & Francis.

Fujii, D. (2015, December 9). How to conduct a culturally-informed neuropsychological evaluation with Asian-Americans [webinar]. *National Academy of Neuropsychology.* Retrieved from https://www.nanonline.org/NAN/_Education/Recorded_Webinars.aspx

Fujii, D. (2016). *Conducting a culturally informed neuropsychological evaluation.* Washington, DC: APA.

Gambino, C. P., Acosta, Y. D, & Grieco, E. M. (2014, June). English-speaking ability of the foreign-born population in the United States: 2012. *U.S. Census Bureau.* American Community Survey Reports No. 26. https://www2.census.gov/library/publications/2014/acs/acs-26.pdf

García-Pentón, L., Fernandez Garcia, Y., Costello, B., Duñabeitia, J. A., & Carreiras, M. (2016a). "Hazy" or "jumbled"? Putting together the pieces of the bilingual puzzle. *Language, Cognition and Neuroscience, 31,* 353–360. https://doi.org/10.1080/23273798.2015.1135247

García-Pentón, L., Fernandez Garcia, Y., Costello, B., Duñabeitia, J. A., & Carreiras, M. (2016b). The neuroanatomy of bilingualism: How to turn a hazy view into the full picture. *Language, Cognition and Neuroscience, 31,* 303–327. https://doi.org/10.1080/23273798.2015.1068944

Gasquoine, P. G. (2009). Race-norming of neuropsychological tests. *Neuropsychological Review, 19,* 250–262. https://doi.org/10.1007/s11065-009-9090-5

Gasquoine, P. G., Croyle, K. L., Cavazos-Gonzalez, C., & Sandoval, O. (2007). Language of administration and neuropsychological test performance in neurologically intact Hispanic American bilingual adults. *Archives of Clinical Neuropsychology, 22,* 991–1001. https://doi.org/10.1016/j.acn.2007.08.003

Gold, B. T. (2015). Lifelong bilingualism and neural reserve against Alzheimer's disease: A review of findings and potential mechanisms. *Behavioural Brain Research, 281,* 9–15. https://doi.org/10.1016/j.bbr.2014.12.006

Gold, B. T., Kim, C., Johnson, N. F., Kryscio, R. J., & Smith, C. D. (2013). Lifelong bilingualism maintains neural efficiency for cognitive control in aging. *Journal of Neuroscience, 33,* 387–396. https://doi.org/10.1523/JNEUROSCI.3837-12.2013

Gollan, T. H., Fennema-Notestine, C., Montoya, R. I., & Jernigan, T. L. (2007). The bilingual effect on Boston Naming Test performance. *Journal of the International Neuropsychological Society, 13,* 197–208. https://doi.org/10.1017/S1355617707070038

Green, D. W., & Abutalebi, J. (2016). Language control and the neuroanatomy of bilingualism: In praise of variety. *Language, cognition and neuroscience, 31,* 340–344. https://doi.org/10.1080/23273798.2015.1084428

Guo, T., & Uhm, S. Y. (2014). Society and acculturation in Asian American communities. In J. M. Davis & R. C. D'Amato (Eds.), *Neuropsychology of Asians and Asian Americans* (pp. 55–76). New York, NY: Springer Science.

Hambleton, R. K. (2005). Issues, designs, and technical guidelines for adapting tests into multiple languages and cultures. In R. K. Hambleton, P. F. Merenda, & C. D. Spielberger (Eds.), *Adapting educational and psychological tests for cross-cultural assessment* (pp. 3–38). Mahwah, NJ: Erlbaum.

Hambleton, R. K., & Zenisky, A. L. (2011). Translating and adapting tests for cross-cultural assessments. In D. Matsumoto, & F. J. Van de Vijver (Eds.), *Cross-cultural research methods in psychology* (pp. 46–70). Cambridge, England: Cambridge University Press.

Hoeffel, E. M., Rastogi, S., Kim, M. O., & Shahid, H. (2012, March). The Asian population 2010. *2010 Census Briefs.* Retrieved from https://www.census.gov/prod/cen2010/briefs/c2010br-11.pdf

Hoffman, C. (1991). *An introduction to bilingualism.* London, England: Routledge.

Hohl, U., Grundman, M., Salmon, D. P., Thomas, R., & Thal, L. J. (1999). Mini-Mental State Examination and Mattis Dementia Rating Scale performance differs in Hispanic and non-Hispanic Alzheimer's disease patients. *Journal of the International Neuropsychological Society, 5,* 301–307. https://doi.org/10.1017/S1355617799544019

Hurtado, A., & Vega, L. A. (2004). Shift happens: Spanish and English transmission between parents and their children. *Journal of Social Issues, 60,* 137–155. https://doi.org/10.1111/j.0022-4537.2004.00103.x

Instituto Cervantes. (2014). *El español: Una lengua viva.* Retrieved from http://www.cervantes.es/

International Test Commission. (2005). ITC guidelines for translating and adapting tests (Version 1.0). ITC-G-TA-20140617. Retrieved from https://www.intestcom.org/files/guideline_test_adaptation.pdf

Isomura-Motoki, A., & Mimura, M. (2011). Neuropsychology in Japan. In D. Fujii (Ed.). *The neuropsychology of Asian Americans* (pp. 237–246). New York, NY: Taylor & Francis.

Johnson, F. L. (2000). *Speaking culturally: Language diversity in the United States*. Thousand Oaks, CA: SAGE.

Judd, T., & Beggs, B. (2005). Cross cultural forensic neuropsychological assessment. In K. Barrett & W. George (Eds.), *Race, culture, psychology & law* (pp. 193–205). Thousand Oaks, CA: SAGE.

Judd, T., Capetillo, D., Carrión-Baralt, J., Mármol, L M., Miguel-Montes, L. S., Navarrete, M. G., . . . NAN Policy and Planning Committee. (2009). Professional considerations for improving the neuropsychological evaluation of Hispanics: A National Academy of Neuropsychology education paper. *Archives of Clinical Neuropsychology, 24,* 127–135. https://doi.org/10.1093/arclin/acp016

Jurica, P. J., Leitten, C. L., & Mattis, S. (2001). *Dementia Rating Scale-2: Professional manual*. Lutz, FL: Psychological Assessment Resources.

Klein, D., Mok, K., Chen, J.-K., & Watkins, K. E. (2014). Age of language learning shapes brain structure: A cortical thickness study of bilingual and monolingual individuals. *Brain and Language, 131,* 20–24. https://doi.org/10.1016/j.bandl.2013.05.014

Kolen, M. J., & Brennan, R. L. (2014). *Test equating, scaling, and linking: Methods and practices* (3rd ed.). New York, NY: Springer.

Krogstad, J. M., & Gonzalez-Barrera, A. (2015, March 24). 2013 national survey of Latinos. *Pew Research Center*. Retrieved from http://www.pewresearch.org/fact-tank/2015/03/24/a-majority-of-english-speaking-hispanics-in-the-u-s-are-bilingual/

Kroll, J. F., Bobb, S. C., & Hoshino, N. (2014). Two languages in mind: Bilingualism as a tool to investigate language, cognition, and the brain. *Current Directions in Psychological Science, 23,* 159–163. https://doi.org/10.1177/0963721414528511

Kroll, J. F., & Chiarello, C. (2016). Language experience and the brain: Variability, neuroplasticity, and bilingualism. *Language, Cognition and Neuroscience, 31,* 345–348. https://doi.org/10.1080/23273798.2015.1086009

Kumar, J. K. (2011). Neuropsychology in India. In D. Fujii (Ed.). *The neuropsychology of Asian Americans* (pp. 219–238). New York, NY: Taylor & Francis.

Leany, B. D., Benuto, L. T., & Thaler, N. S. (2013). Neuropsychological assessment with Hispanic clients. In L. T. Benuto (Ed.), *Guide to psychological assessment with Hispanics* (pp. 351–376). New York, NY: Springer.

Lewis, M. P., Simons, G. F., & Fennig, C. D. (Eds.). (2015). *Ethnologue: Languages of the world* (18th ed.). Dallas, TX: SIL International.

Li, L., Abutalebi, J., Emmorey, K., Gong, G., Yan, X., Feng, X., . . . Ding, G. (2017). How bilingualism protects the brain from aging: Insightws from bimodal bilinguals. *Human Brain Mapping, 38,* 4109–4124. https://doi.org/10.1002/hbm.23652

Lipski, J. M. (2008). *Varieties of Spanish in the United States*. Washington DC: Georgetown University Press.

Llorente, A. M. (Ed.). (2008). *Principles of neuropsychological assessment with Hispanics: Theoretical foundations and clinical practice*. New York, NY: Springer.

Luk, G., & Pliatsikas, C. (2016). Converging diversity to unity: Commentary on "The neuroanatomy of bilingualism." *Language, Cognition and Neuroscience,* 349–352. https://doi.org/10.1080/23273798.2015.1119289

Luk, G., Bialystok, E., Craik, F. I., & Grady, C. L. (2011). Lifelong bilingualism maintains white matter integrity in older adults. *The Journal of Neuroscience, 31,* 16808–16813. https://doi.org/10.1523/JNEUROSCI.4563-11.2011

Luk, G., Green, D. W., Abutalebi, J., & Grady, C. (2011). Cognitive control for language switching in bilinguals: A quantitative meta-analysis of functional neuroimaging studies. *Language and Cognitive Processes, 27,* 1479–1488. https://doi.org/10.1080/01690965.2011.613209

Lyness, S. A., Hernandez, I., Chui, H. C., & Teng, E. L. (2006). Performance of Spanish speakers on the Mattis Dementia Rating Scale (MDRS). *Archives of Clinical Neuropsychology, 21,* 827–836. https://doi.org/10.1016/j.acn.2006.09.003

Lynn, R., & Vanhanen, T. (2012). National IQs: A review of their educational, cognitive, economic, political, demographic, sociological, epidemiological, geographic and climatic correlates. *Intelligence, 40,* 226–234. https://doi.org/10.1016/j.intell.2011.11.004

Maffini, C. S., & Wong, Y. J. (2014). Assessing somatization with Asian American clients. In L. T. Benuto, N. S. Thaler, & B. D. Leany (Eds.), *Guide to psychological assessment with Asians* (pp. 347–360). New York, NY: Springer.

Marquez de la Plata, C., Arango-Lasprilla, J. C., Alegret, M., Moreno, A., Tárraga, L., Lara, M., . . . Cullum, C. M. (2009). Item analysis of three Spanish naming tests: A cross-cultural investigation. *Neurorehabilitation, 24,* 75–85. https://doi.org/10.3233/NRE-2009-0456

Matioli, M. N., & Caramelli, P. (2012). NEUROPSI battery subtest profile in subcortical vascular dementia and Alzheimer's disease. *Dementia e Neuropsychologia, 6,* 170–174.

Meana, M., Oliver, T. L., & Jones, S. C. (2013). Assessing sexual dysfunction in Hispanic clients. In L. T. Benuto (Ed.), *Guide to psychological assessment with Hispanics* (pp. 183–199). New York, NY: Springer.

Mechelli, A., Crinion, J. T., Noppeney, U., O'Doherty, J., Ashburner, J., Frackowiak, R. S., & Price, C. J. (2004). Neurolinguistics: Structural plasticity in the bilingual brain. *Nature, 431,* 757. https://doi.org/10.1038/431757a

Mungas, D., Reed, B. R., Crane, P. K., Haan, M. N., & González, H. M. (2004). Spanish and English Neuropsychological Assessment Scales (SENAS): Further development and psychometric characteristics. *Psychological Assessment, 16,* 347–359. https://doi.org/10.1037/1040-3590.16.4.347

Mungas, D., Reed, B. R., Farias, S. T., & Decarli, C. (2005). Criterion-referenced validity of a neuropsychological test battery: Equivalent performance in elderly Hispanics and non-Hispanic Whites. *Journal of the International Neuropsychological Society, 11,* 620–630. https://doi.org/10.1017/S1355617705050745

Mungas, D., Reed, B. R., Haan, M. N., & González, H. M. (2005). Spanish and English Neuropsychological Assessment Scales: Relationship to demographics, language, cognition, and independent function. *Neuropsychology, 19,* 466–475. https://doi.org/10.1037/0894-4105.19.4.466

Mungas, D., Reed, B. R., Marshall, S. C., & González, H. M. (2000). Development of psychometrically matched English and Spanish language neuropsychological tests for older persons. *Neuropsychology, 14,* 209–223. https://doi.org/10.1037/0894-4105.14.2.209

Muntal Encinas, S., Gramunt-Fombuena, N., Badenes Guia, D., Casas Hernanz, L., & Aguilar Barbera, M. (2012). Spanish translation and adaptation of the Repeatable Battery for the Assessment of Neuropsychological Status (RBANS) Form A in a pilot sample. *Neurología, 27,* 531–546.

National Academy of Neuropsychology. (2000). Presence of third party observers during neuropsychological testing. *Archives of Clinical Neuropsychology, 15,* 379–380. https://doi.org/10.1093/arclin/15.5.379

National Council on Interpreting in Health Care and American Translators Association. (2010). What's in a word? A guide to understanding interpreting and translation in health care. *National Health Law Program.* Retrieved from https://ncihc.memberclicks.net/assets/documents/publications/Whats_in_a_Word_Guide.pdf

Nell, V. (2000). *Cross-cultural neuropsychological assessment: Theory and practice.* Mahwah, NJ: Erlbaum.

Ostrosky-Solís, F., Ardila, A., & Rosselli, M. (1999). NEUROPSI: A brief neuropsychological test battery in Spanish with norms by age and educational level. *Journal of the International Neuropsychological Society, 5,* 413–433. https://doi.org/10.1017/S1355617799555045

Ostrosky-Solís, F., Gómez-Pérez, E., Matute, E., Rosselli, M., Ardila, A., & Pineda, D. A. (2007). NEUROPSI Attention and Memory: A neuropsychological test battery in Spanish with norms by age and educational level. *Applied Neuropsychology, 14,* 156–170. https://doi.org/10.1080/09084280701508655

Paap, K. R. (2016). The neuroanatomy of bilingualism: Will winds of change lift the fog? *Language, Cognition and Neuroscience, 31,* 331–334. https://doi.org/10.1080/23273798.2015.1082607

Peña-Casanova, J., Blesa, R., Aguilard, M., Gramunt-Fombuena, N., Gómez-Ansón, B., Oliva, R., . . . Sol, J. M. (2009). Spanish multicenter normative studies (NEURONORMA Project): Methods and sample characteristics. *Archives of Clinical Neuropsychology, 24,* 307–319. https://doi.org/10.1093/arclin/acp027

Penny, R. (2002). *A history of the Spanish language* (2nd ed.). New York, NY: Cambridge University Press.

Perani, D., Farsad, M., Ballarini, T., Lubian, F., Malpetti, M., Fracchetti, A., Magnani, G., March, A., & Abutalebi, J. (2017). The impact of bilingualism on brain reserve and metabolic connectivity in Alzheimer's dementia. *Proceedings of the National Academy of Sciences, 114,* 1690–1695.

Pew Research Center. (2012, June 19). The rise of Asian Americans (revised July 12, 2012). Retrieved from https://assets.pewresearch.org/wp-content/uploads/sites/3/2013/01/SDT_Rise_of_Asian_Americans.pdf

Pharies, D. A. (2007). *A brief history of the Spanish language.* Chicago, IL: University of Chicago Press.

Pliatsikas, C., Moschopoulou, E., & Saddy, J. D. (2015). The effects of bilingualism on the white matter structure of the brain. *Proceedings of the National Academy of Sciences, 112,* 1334–1337. https://doi.org/10.1073/pnas.1414183112

Pontón, M. O. (2001). Hispanic culture in the United States. In M. O. Pontón, & J. León-Carrión (Eds.), *Neuropsychology and the Hispanic patient: A clinical handbook* (pp. 39–58). Mahwah, NJ: Erlbaum.

Pontón, M. O., Gonzalez, J. J., Hernandez, I., Herrera, L., & Higareda, I. (2000). Factor analysis of the Neuropsychological Screening Battery for Hispanics (NeSBHIS). *Applied Neuropsychology, 7,* 32–39. https://doi.org/10.1207/S15324826AN0701_5

Pontón, M. O., & León-Carrión, J. (Eds.). (2001). *Neuropsychology and the Hispanic patient: A clinical handbook.* Mahwah, NJ: Erlbaum.

Pontón, M. O., Satz, P., Herrera, L., Ortiz, F., Urrutia, C. P., Young, R., . . . Namerow, N. (1996). Normative data stratified by age and education for the Neuropsychological Screening Battery for Hispanics (NeSBHIS): Initial report. *Journal of the International Neuropsychological Society, 2,* 96–104. https://doi.org/10.1017/S1355617700000941

Portes, A., & Rumbaut, R. G. (2006). *Immigrant America: A portrait* (3rd ed.). Berkeley, CA: University of California Press.

Potowski, K., & Carreira, M. (2010). Spanish in the USA. In K. Potowski (Ed.), *Language diversity in the USA* (pp. 66–80). Cambridge, UK: Cambridge University Press.

Pountain, C. J. (1999). Spanish and English in the 21st century. *Donaire, 12,* 33–43.

Puente, A. E., & Ardila, A. (2000). Neuropsychological assessment of Hispanics. In E. Fletcher-Janzen, T. L. Strickland, & C. R. Reynolds (Eds.), *Handbook of cross-cultural neuropsychology* (pp. 87–104). New York, NY: Springer.

Puente, A. E., Zink, D., Hernandez, M., Jackman-Venanzi, T., & Ardila, A. (2013). Bilingualism and its impact on psychological assessment. In L. T. Benuto (Ed.), *Guide to psychological assessment with Hispanics* (pp. 15–31). New York, NY: Springer.

Randolph, C. (1998). *Repeatable battery for the assessment of neuropsychological status (RBANS).* San Antonio, TX: The Psychological Corportation.

Ressel, V., Pallier, C., Ventura-Campos, N., Díaz, B., Roessler, A., Ávila, C., & Sebastián-Gallés, N. (2012). An effect of bilingualism on the auditory cortex. *Journal of Neuroscience, 32,* 16597–16601. https://doi.org/10.1523/JNEUROSCI.1996-12.2012

Roberts, P. M., Garcia, L. J., Desrochers, A., & Hernandez, D. (2002). English performance of proficient bilingual adults on the Boston Naming Test. *Aphasiology, 16,* 635–645. https://doi.org/10.1080/02687030244000220

Romero, H. R., Lageman, S. K., Kamath, V. V., Irani, F., Sim, A., Suarez, P., . . . Attix, D.K. (2009). Challenges in the neuropsychological assessment of ethnic minorities: Summit proceedings. *The Clinical Neuropsychologist, 23,* 761–779. https://doi.org/10.1080/13854040902881958

Rosselli, M., Ardila, A., Jurado, M. B., & Salvatierra, J. L. (2012). Cognate facilitation effect in balanced and non-balanced Spanish-English bilinguals using the Boston Naming Test. *International Journal of Bilingualism, 18,* 649–662. https://doi.org/10.1177/1367006912466313

Ryan, C. (2013, August). Language use in the United States: 2011. *U.S. Census Bureau.* American Community Survey Reports No. 22. Retrieved from https://www2.census.gov/library/publications/2013/acs/acs-22/acs-22.pdf

Saklofske, D. H., Van de Vijver, F. J., Oakland, T., Mpofu, E., & Susuki, L. A. (2015). Intelligence and culture: History and assessment. In S. Goldstein, D. Princiotta, & J. A. Naglieri (Eds.), *Handbook of intelligence: Evolutionary theory, historical perspective, and current concepts* (pp. 341–366). New York, NY: Springer Science+Business Media.

Salazar, G. D., Perez Garcia, M., & Puente, A. E. (2007). Clinical neuropsychology of Spanish speakers: The challenge and pitfalls of a neuropsychology of a heterogeneous population. In B. P. Uzzell, M. O. Pontón, & A. Ardila (Eds.), *International handbood of cross-cultural neuropsychology* (pp. 283–302). Mahwah, NJ: Erlbaum.

Salillas, E., & Wicha, N. Y. (2012). Early learning shapes the memory networks for arithmetic: Evidence from brain potentials in bilinguals. *Psychological Science, 23,* 745–755. https://doi.org/10.1177/0956797612446347

Salinas, C. M., Bordes Edgar, V., & Puente, A. E. (2016). Barriers and practical approaches to neuropsychological assessment of Spanish speakers. In F. R. Ferraro (Ed.), *Minority and cross-cultural aspects of neuropsychological assessment: Enduring and emerging trends* (2nd ed., pp. 229–258). New York, NY: Taylor & Francis.

Sánchez-Benavides, G., Peña-Casanova, J., Casals-Coll, M., Gramunt, N., Molinuevo, J. L., Gómez-Ansón, B., . . . Blesa, R. (2014). Cognitive and neuroimaging profiles in mild cognitive impairment and Alzheimer's disease: Data from the Spanish Multicenter Normative Studies (NEURONORMA) Project. *Journal of Alzheimer's Disease, 41,* 887–901. https://doi.org/10.3233/JAD-132186

Schlueter, J. E., Carlson, J. F., Geisinger, K. F., & Murphy, L. M. (Eds.). (2013). *Pruebas publicadas en Español: An index of Spanish tests in print*. Lincoln, NE: University of Nebraska Press.

Schweizer, T. A., Ware, J., Fischer, C. E., Craik, F. I., & Bialystok, E. (2012). Bilingualism as a contributor to cognitive reserve: Evidence from brain atrophy in Alzheimer's disease. *Cortex, 48,* 991–996. https://doi.org/10.1016/j.cortex.2011.04.009

Searight, H. R., & Searight, B. K. (2009). Working with foreign language interpreters: Recommendations for psychological practice. *Professional Psychology: Research and Practice, 40,* 444–451. https://doi.org/10.1037/a0016788

Sireci, S. G. (2005). Using bilinguals to evaluate the comparability of different language versions of a test. In R. K. Hambleton, P. F. Merenda, & C. D. Spielberger (Eds.), *Adapting educational and psychological tests for cross-cultural assessment* (pp. 117–138). Mahwah, NJ: Erlbaum.

Sternberg, R. (2004). Culture and intelligence. *American Psychologist, 59,* 325–338. https://doi.org/10.1037/0003-066X.59.5.325

Strutt, A. M., Burton, V. J., Resendiz, C. V., & Peery, S. (2016). Neurocognitive assessment of Hispanic individuals residing in the United States: Current issues and potential solutions. In F. R. Ferraro (Ed.), *Minority and cross-cultural aspects of neuropsychological assessment* (2nd ed., pp. 201–228). New York, NY: Taylor & Francis.

Tanzer, N. K. (2005). Developing tests for use in multiple languages and cultures: A plea for simultaneous development. In R. K. Hambleton, P. F. Merenda, & C. D. Spielberger (Eds.), *Adapting educational and psychological tests from cross-cultural assessment* (pp. 235–263). Mahwah, NJ: Erlbaum.

Taussig, I. M., Henderson, V. W., & Mack, W. (1992). Spanish translation and validation of a neuropsychological battery. *Clinical Gerontologist, 11,* 95–108. https://doi.org/10.1300/J018v11n03_07

U.S. Department of Health and Human Services. (2001). *Mental health: Culture, race, and ethnicity—A supplement to* Mental health: A report of the Surgeon General. Rockville, MD: U.S. Department of Health and Human Services, Substance Abuse and Mental Health Services Administration, Center for Mental Health Services.

Uzzell, B. P., Pontón, M. O., & Ardila, A. (Eds.). (2007). *International handbook of cross-cultural neuropsychology*. Mahwah, NJ: Erlbaum.

Van de Vijver, F., & Tanzer, N. K. (2004). Bias and equivalence in cross-cultural assessment: An overview. *Revue européenne de psychologie appliquée, 54,* 119–135. https://doi.org/10.1016/j.erap.2003.12.004

Van de Vijver, F. J. R., & Poortinga, Y. H. (2005). Conceptual and methodological issues in adapting tests. In R. K. Hambleton, P. F. Merenda, & C. D. Spielberger (Eds.), *Adapting educational and psychological tests for cross-cultural assessment* (pp. 39–63). Mahwah, NJ: Lawrence Erlbaum Associates.

Van Hamme, L. (1996). American Indian cultures and the classroom. *Journal of Indian Education, 35,* 21–36.

Veltman, C. (1983). *Language shift in the United States*. Berlin, Germany: Mouton.

Wei, L. (2007). Dimensions of bilingualism. In L. Wei (Ed.), *The bilingualism reader* (2nd ed., pp. 3–22). London, England: Routledge.

Weintraub, S., Salmon, D., Mercaldo, N., Ferris, S., Graff-Radford, N. R., Chui, H., . . . Morris, J. C. (2009). The Alzheimer's Disease Centers' Uniform Data Set (UDS): The neuropsychologic test battery. *Alzheimer Disease & Associated Disorders, 23,* 91–101. https://doi.org/10.1097/WAD.0b013e318191c7dd

9

Neurocognitive Development of Semantics in Chinese- and English-Speaking Children With and Without Autism

TAI-LI CHOU AND JAMES BOOTH

INTRODUCTION

In this chapter, we will review the neurocognitive organization and processing of semantic knowledge. By organization, we are referring to how the actual representations themselves are organized in occipitotemporal cortex. By processing, we are referring to how these representations are accessed and manipulated by frontoparietal cortex. The first section of the paper will review association strength versus categorical relations in semantic knowledge. The second and third sections discuss what is known about the organization and processing of semantic knowledge in typical children. The fourth and fifth sections review the organization and processing of semantic knowledge in autistic children. Finally, we end the chapter with a summary and implications of this work for future research.

ASSOCIATION STRENGTH VERSUS CATEGORICAL RELATIONS
IN SEMANTIC KNOWLEDGE

At least two different kinds of semantic relations have been proposed in the organization of semantic knowledge. One of the semantic relations is association strength for stimulus pairs based on free association norms (e.g., bread–butter; Chou, Booth, Burman et al., 2006; Chou, Chen, Fan, Chen, & Booth, 2009). Another of the semantic relations is categorical relatedness (e.g., bread–cake), which refers to stimulus pairs that have shared features or properties (Nation & Snowling, 1999; Wong, Chen, Chou, & Lee, 2011). However, association strength has not been distinguished from categorical relatedness in neuroimaging work, which causes potential confounds in interpreting developmental differences in semantic knowledge (McRae & Boisvert, 1998). Therefore, it is necessary to systematically manipulate association strength and categorical relatedness to investigate the developmental changes of semantic knowledge in children. For example, umbrella–rain is an associative pair but not a categorical pair. In contrast, bed–desk is a categorical pair but not an associative pair (Nation & Snowling, 1999). A high categorical pair (boat–ship) has more overlapping features than a low categorical pair (bow–arrow; Nation & Snowling, 1999; Plaut, 1995).

Connectionist models use two computational properties to explain the difference between association strength and categorical relatedness (Plaut, 1995). In these models, associatively related words follow each other often during training. Associative priming occurs because the network learns to make a rapid transition from the meaning of the prime to the meaning of its associated word. Thus, it is easier to get out of one attractor basin and into the other. This associative priming behaves differently from categorical priming. Categorically related words are defined to overlap in their semantic features. Categorical priming occurs because the network computes feature similarity between the prime and target. The computation across all features (both overlapping and nonoverlapping) makes categorical priming relatively weaker as compared to associative priming.

For association strength, several previous studies have found higher accuracy on stronger (e.g., sword–knife) than weaker association (cookie–grain) items (Hung, Lee, Chen, & Chou, 2010; Lee, Chen, & Chou, 2009; Wong et al., 2011). However, these studies have not distinguished categorical relatedness from association strength, and, as reviewed previously, this distinction is important on theoretical grounds. To make this distinction clear, three lines of evidence are provided to explain the differences between association strength and categorical relatedness, including modeling studies, stimulus characteristics, and neural substrates. First, modeling studies based on distributed networks have shown that associative relations and categorical relations have different impacts on semantic knowledge (McClelland & Rogers, 2003; Plaut, 1995). Second, word association norms can contain a miscellaneous variety of relations, such as synonyms, antonyms, or same categories (Hue, Gao, & Lo, 2005). Indeed, Hutchison (2003) classified each stimulus and its primary associate from norms into 1 or more of 14 possible relations, showing that stimuli may fall into more than one classification. In fact, previous studies have examined association pairs independent of categorical relatedness and categorical pairs independent of association strength (Nation & Snowling 1999; Wong, Chen, & Chou, 2014). Third, there are different neural substrates corresponding to association strength and categorical relatedness. Detailed findings of brain activation will be discussed later in this chapter. Thus, it is necessary to separate categorical relatedness from association strength to examine the processing and organization of semantic knowledge in children.

For categorical relatedness, computational models suggest that related concepts are represented by similar patterns such that the similarity between two concepts depends on the degree of feature overlap (Plaut, 1995). For example, cows and sheep share many features, such as being warm-blooded, and belong to the category of mammals. Members of the same category are all related to one another by their shared properties (Estes, Golonka, & Jones, 2011). As compared with low categorical pairs, high categorical pairs produce a greater priming effect, suggesting that the high categorical relatedness should produce greater activation of the memory representation of the category (Hines, Czerwinski, Sawyer, & Dwyer, 1986). Similarly, modeling findings demonstrate greater semantic priming for high categorical pairs with more overlapping features as compared to low categorical pairs with fewer overlapping features (Plaut, 1995).

DEVELOPMENTAL CHANGES IN THE ORGANIZATION OF SEMANTIC KNOWLEDGE IN TYPICAL CHILDREN

This section focuses on two different kinds of semantic relations that have been proposed in the organization of semantic knowledge, including association strength and categorical relations. For association strength, a critical region implicated in the organization of semantic knowledge is the left middle temporal gyrus (MTG). Greater activation in this region in adults is thought to be related to the storage of lexical representations (Lau, Phillips, & Poeppel, 2008) and the representation of conceptual contents (Fairhall & Caramazza, 2013). Child studies using association strength have found greater activation for weaker association pairs in the left MTG, indicating more elaborate semantic representations in older children in English (Chou, Booth, Burman et al., 2006; Chou, Booth, Bitan et al., 2006) and in Chinese (Chou et al, 2009; Lee, Booth, Chen, & Chou, 2011; Lee, Chen, & Chou, 2015). Age-related increases in the left MTG have been always found for weak association strength, suggesting that there may be a greater number of potential lexical associates with age in the left MTG. Older children may have more elaborate semantic representations in the temporal cortex.

Using a longitudinal approach, Lee, Booth, and Chou (2016) further explored the organization of semantic knowledge with association strength in the left MTG. They scanned 15 Chinese children (aged 8–11) and 15 adolescents (aged 12–15) to examine whether initial brain measures could predict future behavioral improvement in a semantic judgment task. Their results showed greater developmental changes from Time 1 to Time 2 for weaker association pairs in the left MTG for the children (aged 8–11) as compared to the adolescents (aged 12–15). Moreover, connectivity with the left MTG in weak association pairs was uniquely predictive of behavioral improvement from Time 1 to Time 2 for the children but not the adolescents. Taken together, their results suggest relatively rapid development before adolescence of semantic representations in the left MTG. Moreover, the left MTG is a core node of association strength in the semantic network over development.

For categorical relatedness, a key region implicated in the organization of semantic knowledge is the left occipitotemporal cortex. This region plays a critical role in storing visual-perceptual features contributing to category-specific semantic memory (Binder, Desai, Graves, & Conant, 2009), and greater activation in this region is related to difficulty of access of visual-perceptual features of object knowledge (Grossman et al., 2013). Greater activation in this region is thought to reflect the storage of many features of a general-level category rather than a few features of a specific-level category (Taylor, Devereux, & Tyler, 2011). Thus, high categorical pairs with more overlapping features are hypothesized to be related to the storage of visual features in the occipitotemporal cortex.

With a longitudinal approach, Chou, Wong, Chen, Fan, and Booth (2019) orthogonally manipulated association strength (strong/weak) and categorical relatedness (high/low) to examine the developmental changes in activation of sixteen 10- to 14-year-old Chinese children over a 2-year interval. For Time 2–Time 1, the weak versus strong association strength produced greater activation in the left MTG, suggesting more elaborate semantic representations over development. Moreover, the high versus low categorical relatedness produced greater activation in the left occipitotemporal cortex, suggesting more elaborate features of categorical knowledge over development. In conclusion, developmental changes in the organization of semantic knowledge are fairly cultural-invariant. In other words, age-related increases have been found in the organization of association strength in the left MTG in both English and Chinese. Also, age-related increases have been found in the organization of categorical relatedness in the left occipitotemporal cortex in both English and Chinese.

STRUCTURAL DIFFERENCES IN LINGUISTIC PROPERTIES BETWEEN CHINESE AND ENGLISH

As compared to cultural invariance in developmental changes in the *organization* of semantic knowledge, developmental changes in the *processing* of semantic knowledge seem to be culturally dependent. Processing semantic knowledge may be related to variant brain regions because of prominent linguistic features that differ between Chinese and English. Before discussing developmental changes in the processing of semantic knowledge, we need to mention the linguistic properties of Chinese.

There exist many homophones in Chinese, and thus mapping is particularly unsystematic from spoken to written word forms. Spoken Chinese (Mandarin) uses four tones to differentiate different words with different meanings. Chinese has about 5,000 common words but only about 400 different syllables independent of tone, or around 1,300 if a change in tone is considered to create a different syllable (Reich, Chou, & Patterson, 2003). Therefore, Chinese contains many monosyllabic homophones with different meanings. About 55% of monosyllables correspond to more than five homophones (Lee et al., 2011). When a Chinese word is pronounced without context, it is almost impossible for the listener to know which visual word or meaning is referred to. This is in contrast to English, which does not have a large number of homophones and has a relatively consistent mapping between spoken and written word forms.

Another prominent feature is the mapping from orthography to semantics in Chinese. The relation between form and meaning in English is only partly consistent and admits many exceptions (e.g., –or in *traitor* and *anchor*), reflecting the lack of reliability of semantic information at the sublexical level (Seidenberg &

Gonnerman, 2000). In contrast, Chinese includes greater semantic information at the sublexical level, showing a more direct mapping between orthography and semantics. In terms of the orthographic characteristics of Chinese characters, about 80% of Chinese characters are phonetic compound (phonograms) that consist of a semantic radical and a phonetic radical. These semantic radicals may provide a reliable cue to the semantic category of the character. For instance, all characters that contain the semantic radical 金 (/jin1/, metal), such as 銅(/tong2/, copper), 鐵 (/tie3/, iron), indicate that the characters are related to the category of metal. In addition, many semantic radicals may not stand alone as a character in Chinese. An example is 湖 (/hwu2/, lake), which is composed of a semantic radical of three dots arranged vertically on the left with an associated meaning of *water*. The semantic radical in this example does not correspond to a real character.

DEVELOPMENTAL CHANGES IN THE PROCESSING OF SEMANTIC KNOWLEDGE IN TYPICAL CHILDREN

By processing, we are referring to how lexical representations are accessed and manipulated by frontoparietal cortex. For association strength, the neuroimaging studies show that the left inferior frontal gyrus (IFG; Brodmann area [BA] 45/47) exhibits greater activation for weak association pairs as compared to strong association pairs in English (Booth et al., 2002). Previous work has suggested that greater activation in the left IFG is related to increased demands on retrieval and selection of semantic knowledge (Binder et al., 2009; Badre & Wagner, 2007; Fletcher, Shallice, & Dolan, 2000; Zhu et al., 2009). Therefore, increased left IFG activation due to weaker association strength is likely due to greater demands on these mechanisms (Chou, Booth, Burman et al., 2006; Chou, Booth, Bitan et al., 2006).

In contrast to distantly related pairs, closely related pairs with stronger association produce greater activation in the left inferior parietal lobule (BA 40) in English (Chou, Booth, Burman et al., 2006; Chou, Booth, Bitan et al., 2006) or in the left angular gyrus (BA 39) in Chinese (Chou et al., 2009). The cultural difference in processing strong association strength may be explained by differences in reading for meaning between Chinese and English (Chou et al., 2009). Reading for meaning may show greater engagement of the mapping from orthography to semantics at word level in Chinese. Semantic radicals may allow for greater semantic integration in Chinese reading, which may be related to greater activation in the left angular gyrus. In contrast, the greater involvement of phonological processing may be related to greater activation in the left inferior parietal lobule because the systematic alphabet in English reading may allow for greater involvement of phonology.

Regarding developmental changes in processing association strength in English, the neural correlates of semantic judgments in the auditory modality were explored in a group of 9- to 15-year-old children (Chou, Booth Burman et al., 2006). Children showed greater activations in bilateral IFG (BAs 47, 45) for semantic processing. Moreover, words with weak association strength elicited activation in the left IFG, whereas words with strong association strength elicited significantly greater activation in the left inferior parietal lobule (BA 40). Using the visual modality, another group of 9- to 15-year-old children did the same semantic association task (Chou, Booth, Bitan et al., 2006). Consistent with the findings of the auditory modality (Chou, Booth, Burman et al., 2006), words with weak association strength elicited greater activation in left IFG (BA 45), suggesting more difficult semantic search. In contrast, words with strong association strength elicited greater activation in bilateral inferior parietal lobules (BA 40), which also showed an overall developmental increase, suggesting stronger integration of highly related semantic knowledge. More generally, the findings of these two studies support modality-independent brain regions for semantic processing, including the left IFG with weaker association strength and left inferior parietal lobule with strong association strength in English.

As to weak association strength, Lee et al. (2011) took advantage of mapping from spoken to written word forms in Chinese by using a cross-modal presentation in which the first word was presented visually and the second word was presented auditorily (visual–auditory). Participants were asked to judge if these two sequentially presented words were related in meaning or not in a semantic judgment task. In hearing the sound in the visual–auditory task, the participants may activate all the possible orthographic candidates and meanings (Hung et al., 2010; Booth, Cho, Burman, & Bitan, 2007). Thus, Lee et al. (2011) was able to test whether the multiple homophones may induce greater engagement of selecting an appropriate character and/or meaning among spoken homophones when words are presented auditorily. Indeed, in a group of 10- to15-year-old Chinese children, weaker association pairs produced greater activation in the left IFG (BA 45) for both tasks. However, this effect was stronger for the visual–auditory task than for the visual–visual task and this difference was stronger for older compared to younger children. The findings suggest greater involvement of semantic selection mechanisms in the cross-modal task requiring the access of the appropriate meaning of homophonic spoken words in Chinese, especially for older children (Figure 9.1).

As to strong association strength, a study by Lee et al. (2015) manipulated association strength (i.e., a global reading unit) and semantic radical (i.e., a local reading unit) to explore the interaction of lexical and sublexical semantic information in making semantic judgments in Chinese. In the semantic judgment task, two types of stimuli were used: visually–similar (i.e., shared a semantic radical) versus visually–dissimilar (i.e., did not share a semantic radical) character pairs. Participants were asked to indicate if two Chinese characters, arranged according to association strength, were related in meaning. Lee et al. showed greater developmental increases in activation in left angular gyrus (BA 39) in the visually similar compared to the visually-dissimilar pairs for the strong association, suggesting that shared semantics at the sublexical level facilitates the integration of semantic knowledge at the lexical level in older children (Figure 9.2).

Regarding processing categorical relatedness in English children, Nation and Snowling (1999) orthogonally manipulated categorical relatedness (e.g., cat–dog

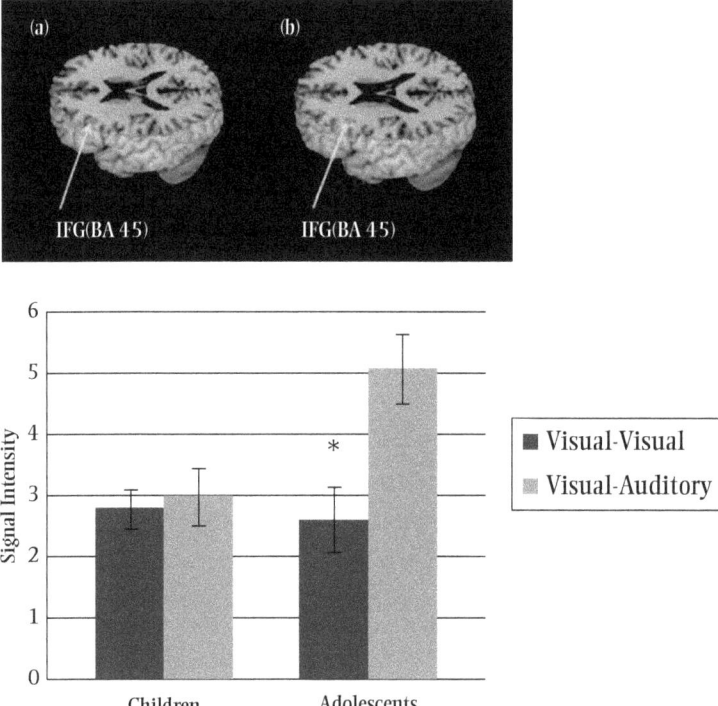

Figure 9.1. (Top panel) Increasing age was correlated with greater activation in the midventral region of left inferior frontal gyrus (IFG, BA 45) with weak association strength in the visual-visual semantic judgment task (in green) and in the visual-auditory semantic judgment task (in red). (Bottom panel) Greater involvement of semantic selection mechanisms in the cross-modal task requiring the access of the appropriate meaning of homophonic spoken words in Chinese for adolescents. (See color plate) From S. H. Lee, J. R. Booth, S. Y. Chen, & T. L. Chou, 2011, Developmental changes in the inferior frontal cortex for selecting semantic representations. *Developmental Cognitive Neuroscience, 1,* 338–350.

for categorical relations; broom–floor for functional relations) and association strength (high/low) to assess children with good and poor reading comprehension, matched for decoding skill. Both groups of children showed priming for function-related pairs. For category-related pairs, poor comprehenders only showed priming if the category pairs also shared high association strength. In contrast, good comprehenders showed priming for category-related pairs, irrespective of the degree of prime-target association. These findings suggests that category knowledge is gradually abstracted and refined from children's event-based knowledge and that poor comprehenders are less sensitive to abstract semantic relations than normal readers (Nation & Snowling, 1999).

Regarding developmental changes in processing categorical relatedness in Chinese, Wong et al. (2011) tested a group of third graders to distinguish functional from categorical relations of words. Their participants performed semantic

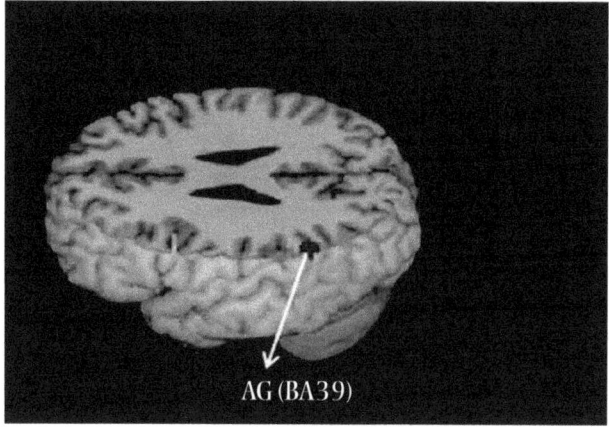

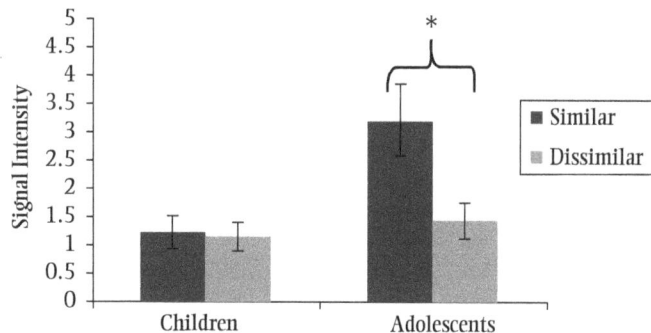

Figure 9.2. (Top panel) Developmental increases in activation in left angular gyrus (AG, BA 39) with strong association strength for both visually-similar and visually-dissimilar pairs in Chinese. (Bottom panel) Greater activation for the similar pairs than the dissimilar pairs in adolescents, but not in children. (See color plate) From S. H. Lee, J. R. Booth, & T. L. Chou, 2015, Developmental changes in the neural influence of sublexical information on semantic processing. *Neuropsychologia, 73,* 25–34.

judgments on word pairs with either the functional or categorical relation. After controlling for word recognition abilities, increased accuracy was found for word pairs with the functional relation as compared to those with the categorical relation. The findings reported by Wong et al. (2011) suggest that the organization of semantic representations may be predominantly based on the functional relation rather than the categorical relation for third graders. A longitudinal finding reported by Wong et al. (2014) further suggests that the organization of semantic representations may be predominantly based on the categorical relation for fifth graders, suggesting a shift from functional to categorical relations along with age (also see Chou, Fan, Lee, & Wong, 2013).

The neural substrate of processing categorical relatedness may be related to the left precuneus in German speakers. Sachs, Weis, Krings, Huber, and Kircher

(2008) examined semantic relations with a category construction task in which participants selected among two options that best went with a target (e.g., car). Their results showed that choosing an option with categorical relatedness (e.g., bus) was associated with increased activation of the left precuneus. When participants processed categorical pairs in a lexical decision task, greater activation was also found in the precuneus (Sachs, Weis, Zellagui et al., 2008). Furthermore, an age-related increase in precuneus activation is related to feature extraction for semantic categorization in a visual working memory task (Ciesielski, Lesnik, Savoy, Grant, & Ahlfors, 2006). Taken together, previous research suggests greater activation in the precuneus to be related to categorical relatedness.

Similar to the finding reported in English, the neural substrate of processing categorical relatedness in Chinese may be related to the left precuneus. With a longitudinal approach, Chou, Wong, Chen, Fan, and Booth (2019) orthogonally manipulated association strength (strong/weak) and categorical relatedness (high/low) to examine the developmental changes in activation of sixteen 10- to 14-year-old children over a 2-year interval. Moreover, they examined whether initial behavioral performance (Time 1) predicted brain activation changes (Time 2–Time 1). For Time 2–Time 1, the high versus low categorical relatedness produced greater activation in the left precuneus and accuracy (Time 1) predicted activation changes in the precuneus, suggesting more complete integration over development.

In summary, the processing of semantic knowledge refers to how representations are accessed and manipulated by frontoparietal cortex. For association strength, age-related increases in IFG activation with weak association strength are related to greater engagement of retrieval/selection of semantic knowledge, and age-related increases in activation in the inferior parietal lobule/angular gyrus with strong association strength are related to more complete semantic integration. For categorical relatedness, age-related increases in precuneus activation with high categorical relations are related to more complete integration of visual-perceptual features.

DEVELOPMENTAL CHANGES IN THE ORGANIZATION OF SEMANTIC KNOWLEDGE IN CHILDREN WITH AUTISM

Children with autism perform more poorly on categorization than typically developing children (Church et al., 2010). Johnson and Rakison (2006) propose categorization and concept formation deficits in children with autism, for example, in the formation of coherent concepts for animates and inanimates. Development of such concepts is thought to be a crucial building block for young children's emerging understanding that different object kinds possess different physical, psychological, biological, and motion-related features. Johnson and Rakison (2006) tested preschoolers with autism for early concept formation in English. Their results suggest that the preschoolers with autism are delayed in the processes by which they form categories but nonetheless possess relevant knowledge about the motion properties of animates and inanimates.

To examine the neural correlates of developmental changes in the organization of semantic knowledge in English, Dunn and Bates (2005) used event-related potential to study children (aged 8–9) and adolescents (aged 11–12) with autism and typically developing controls during semantic categorization. The autistic children showed a greater N4 effect on a wrong categorization as compared to typical children. In contrast, the autistic adolescents showed an attenuated N4 response as compared to typical adolescents. The authors suggest that overall changes in the amplitude of N4 may reflect changes in semantic expectancy and that children with autism may have deficits in semantic classification. The attenuated N4 response in adolescents with autism might be due to the fact that the adolescents have more experience with language use along with increasing age (Dunn and Bates, 2005).

An fMRI study by Wong, Gau, and Chou (in press) orthogonally manipulated association strength and categorical relatedness to test 34 boys with autism (mean age = 12.3 years) and 36 typically developing boys (mean age = 11.9 years) in Chinese during semantic judgments. For the autism group, reduced activation was found in the MTG for weak association strength, and in the left occipitotemporal cortex for high categorical relatedness as compared to the typical group. There may be a different organization of semantic knowledge between the autistic and typical boys. The boys with autism may have less well defined semantic associations and less elaborate features of categorical knowledge.

To further explore developmental changes in the organization of semantic knowledge in Chinese, Chen, Gau, Lee, and Chou (2016) used functional MRI to investigate boys with autism and typically developing boys. The boys were divided into children (aged 8–12) and adolescents (aged 13–17). The behavioral results showed that the autistic boys had lower accuracy in the related condition relative to the typical boys. The neuroimaging results showed greater activation in the cuneus in boys with autism than typical boys. As to developmental changes, autistic children produced greater activation in the cuneus than typical children. The findings of Chen et al.'s (2016) study suggest that boys with autism may rely more on lower-level visual information during semantic judgments (Figure 9.3).

DEVELOPMENTAL CHANGES IN THE PROCESSING OF SEMANTIC KNOWLEDGE IN CHILDREN WITH AUTISM

To explore the neural correlates of processing semantic knowledge, Wong, Gau, and Chou (in press) orthogonally manipulated association strength and categorical relatedness to test 34 Chinese-speaking boys with autism (mean age = 12.3 years) and 36 typically developing boys (mean age = 11.9 years) during semantic judgments. For the autism group, reduced activation was found in the IFG for weak association strength, and in the left precuneus for high categorical relatedness as compared to the typical group. There may be different processes of semantic knowledge between the autistic and typical boys. The autistic boys may have poor ability of retrieving/selecting semantic knowledge and integrating features.

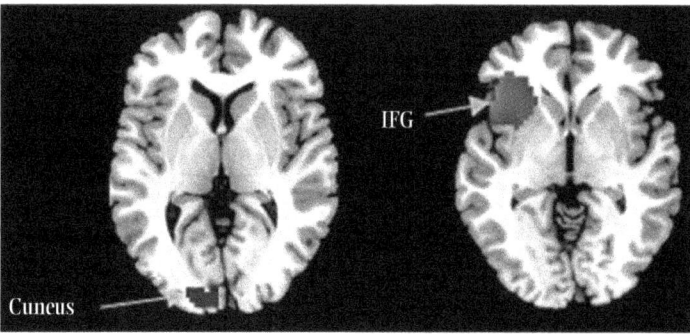

Figure 9.3. Children with autism had greater cuneus activation than typically developing children (in blue) while adolescents with autism showed reduced left inferior frontal gyrus (IFG) activation as compared to typically developing adolescents (in green). (See color plate) From P. J. Chen, S. S. Gau, S. H. Lee, & T. L. Chou, 2016, Differences in age-dependent neural correlates of semantic processing between youths with autism spectrum disorder and typically developing youths. *Autism Research, 2,* 1–11.

Regarding developmental changes of neural activity during semantic processing, Chen et al. (2016) used functional MRI to investigate Chinese boys with autism and typically developing boys. The boys were divided into children (aged 8–12) and adolescents (aged 13–17). The neuroimaging results showed less activation in the left IFG in boys with autism than typical boys. As to developmental changes in the processing of semantic knowledge, adolescents with autism showed reduced left IFG activation as compared to typical adolescents. The findings of Chen et al.'s study suggest that typical boys may engage more in higher-level processing of retrieving or selecting semantic knowledge. The findings imply different semantic processes between these two groups (Figure 9.3).

SUMMARY AND FUTURE DIRECTIONS

We have reviewed the neurocognitive development of semantics at the word level in Chinese and English typical children and those with autism. The organization of semantic knowledge is related to brain activation in occipitotemporal cortex. Developmental increases in activation have been found in the MTG for weak association strength and in the occipitotemporal cortex for high categorical relatedness. As compared to typical children, autistic children produce reduced activation in the MTG for weak association strength and in the occipitotemporal cortex for high categorical relatedness. However, autistic children produce greater activation in the cuneus during semantic judgments. Overall, these results suggest that the *organization* of semantic representations in the occipital and temporal cortex becomes more refined with development and is altered in children with autism.

The processing of semantic knowledge is related to brain activation in frontoparietal cortex. Developmental increases in activation have been found in the IFG for weak association strength, in the inferior parietal lobule/angular gyrus for strong association strength, and in the precuneus for high categorical relatedness. As compared to typical children, autistic children produce reduced activation in the IFG for weak association strength and in the precuneus for high categorical relatedness. Overall, these results suggest that the *processing* of semantic representations becomes more engaged over development, perhaps due the refinement of these representations, and is altered in autistic children.

To further explore semantic knowledge, it is crucial to consider the social-communication aspects of words in children with autism. The organization of semantic knowledge appears to rely on experiences and interaction with the environment during lexical learning (Martin & Chao, 2001). Children with autism lack interactive learning experiences to help them organize the semantic system due to their pragmatic impairments. Previous research has showed that diminished and anomalous child–caregiver interactions are associated with poor language outcomes (Bopp, Mirenda, & Zumbo, 2009; Siller & Sigman, 2008). Also, early word learning is based in part on the apprehension of social cues such as joint attention. Young children with autism are less likely to utilize such social cues to support their word processing (Baron-Cohen, Baldwin, & Crowson, 1997; Parish-Morris et al., 2007). To echo this point, one of the significant changes with autism spectrum disorders (ASD) is that deficits in social interaction and communication, two of the three core features (repetitive/restricted behaviors included) of autism (American Psychiatric Association, 1994), are combined (social-communication deficit) as one of the two core symptoms in the *Diagnostic and Statistical Manual of Mental Disorders*–fifth edition (American Psychiatric Association, 2013) diagnostic criteria for ASD (Greaves-Lord et al., 2013; Huerta, Bishop, Duncan, Hus, & Lord, 2012). However, little is known about the involvement of social deficits on semantic processing in children with ASD. Therefore, in the future it will be important to explore the neural substrates of interaction between social ability and semantic processing in children with ASD as compared to typically developing children.

REFERENCES

American Psychiatric Association. (1994). *Diagnostic and statistical manual of mental disorders* (4th ed.). Washington, DC: American Psychiatric Association.

American Psychiatric Association. (2013). *Diagnostic and statistical manual of mental disorders* (5th ed.). Washington, DC: American Psychiatric Association.

Badre, D., & Wagner, A. D. (2007). Left ventrolateral prefrontal cortex and the cognitive control of memory. *Neuropsychologia, 45,* 2883–2901. https://doi.org/10.1016/j.neuropsychologia.2007.06.015

Baron-Cohen, S., Baldwin, D. A. & Crowson, M. (1997). Do children with autism use the speaker's direction of gaze strategy to crack the code of language? *Child Development, 68,* 48–57. https://doi.org/10.2307/1131924

Binder, J. R., Desai, R. H., Graves, W. W., & Conant, L. L. (2009). Where is the semantic system? A critical review and meta-analysis of 120 functional neuroimaging studies. *Cerebral Cortex, 19,* 2767–2796. https://doi.org/10.1093/cercor/bhp055

Booth, J. R., Burman, D. D., Meyer, J. R., Gitelman, D. R., Parrish, T. B., & Mesulam, M. M. (2002). Modality independence of word comprehension. *Human Brain Mapping, 16,* 251–261. https://doi.org/10.1002/hbm.10054

Booth, J. R., Cho, S., Burman, D. D., & Bitan, T. (2007). Neural correlates of mapping from phonology to orthography in children performing an auditory spelling task. *Developmental Science, 10,* 441–451. https://doi.org/10.1111/j.1467-7687.2007.00598.x

Bopp, K. D., Mirenda, P., & Zumbo, B. D. (2009). Behavior predictors of language development over 2 years in children with autism spectrum disorders. *Journal of Speech, Language, and Hearing Research, 52,* 1106–1120. https://doi.org/10.1044/1092-4388(2009/07-0262)

Chen, P. J. Gau, S. S. Lee, S. H. & Chou, T. L. (2016). Differences in age-dependent neural correlates of semantic processing between youths with autism spectrum disorder and typically developing youths. *Autism Research, 2,* 1–11. https://doi.org/10.1002/aur.1616

Chou, T. L., Booth, J. R., Burman, D. D., Bitan, T., Bigio, J. D., Lu, D., & Cone, N. E. (2006). Developmental changes in the neural correlates of semantic processing. *Neuroimage, 29,* 1141–1149. https://doi.org/10.1002/hbm.20231

Chou, T. L., Booth, J. R., Bitan, T., Burman, D. D., Bigio, J. D., Cone, N. E., . . . Cao, F. (2006). Developmental and skill effects on the neural correlates of semantic processing to visually presented words. *Human Brain Mapping, 27,* 915–924.

Chou, T. L., Chen, C. W., Fan, L. Y., Chen, S. Y., & Booth, J. R. (2009). Testing for a cultural influence on reading for meaning in the developing brain: The neural basis of semantic processing in Chinese children. *Frontiers in Human Neuroscience, 3,* 27. https://doi.org/10.3389/neuro.09.027.2009

Chou, T. L. Fan, L. Y. Lee, S. H. & Wong, C. H. (2013). From semantic processing and representations to semantic development. *Chinese Journal of Psychology, 55,* 277–288. https://doi.org/10.6129/CJP.20130406a

Chou, T. L., Wong, C. H., Chen, S. Y., Fan, L. Y. & Booth, J. R. (2019). Developmental changes of association strength and categorical relatedness on semantic processing in the brain. *Brain and Language, 189,* 10–19. https://doi.org/10.1016/j.bandl.2018.12.006

Church, B. A., Krauss, M. S., Lopata, C., Toomey, J. A., Thomeer, M. L., Coutinho, M. V., . . . Mercado, E. (2010). Atypical categorization in children with high-functioning autism spectrum disorder. *Psychonomic Bulletin & Review, 17,* 862–868. https://doi.org/10.3758/PBR.17.6.862

Ciesielski, K. T., Lesnik, P. G., Savoy, R. L., Grant, E. P., & Ahlfors, S. P. (2006). Developmental neural networks in children performing a Categorical N-Back Task. *Neuroimage, 33,* 980–990. https://doi.org/10.1016/j.neuroimage.2006.07.028

Dunn, M. A., & Bates, J. C. (2005). Developmental change in neutral processing of words by children with autism. *Journal of Autism and Developmental Disorders, 35,* 361–376. https://doi.org/10.1007/s10803-005-3304-3

Estes, Z., Golonka, S., & Jones, L. L. (2011). Thematic thinking: The apprehension and consequences of thematic relations. In B. H. Ross (Ed.), *The psychology of learning and motivation: Advances in research and theory* (Vol. 54, pp. 249–294). San Diego, CA: Academic Press. https://doi.org/10.1016/B978-0-12-385527-5.00008-5

Fairhall, S. L., & Caramazza, A. (2013). Brain regions that represent amodal conceptual knowledge. *Journal of Neuroscience, 33,* 10552–10558. https://doi.org/10.1523/JNEUROSCI.0051-13.2013

Fletcher, P. C., Shallice, T., & Dolan, R. J. (2000). "Sculpting the response space": An account of left prefrontal activation at encoding. *Neuroimage, 12,* 404–417. https://doi.org/10.1006/nimg.2000.0633

Greaves-Lord, K., Eussen, M. L., Verhulst, F. C., Minderaa, R. B., Mandy, W., Hudziak, J. J., ... Hartman, C. A. (2013). Empirically based phenotypic profiles of children with pervasive developmental disorders: Interpretation in the light of the DSM-5. *Journal of Autism and Developmental Disorders, 43,* 1784–1797. https://doi.org/10.1007/s10803-012-1724-4

Grossman, M., Peelle, J. E., Smith, E. E., McMillan, C. T., Cook, P., Powers, J., ... Boller, A. (2013). Category-specific semantic memory: Converging evidence from bold fMRI and Alzheimer's disease. *Neuroimage, 68,* 263–274. https://doi.org/10.1016/j.neuroimage.2012.11.057

Hines, D., Czerwinski, M., Sawyer, P. K., & Dwyer, M. (1986). Automatic semantic priming: Effect of category exemplar level and word association level. *Journal of Experimental Psychology: Human Perception and Performance, 12,* 370-379. https://doi.org/10.1037/0096-1523.12.3.370

Hue, C. W., Gao, C. H., & Lo, M. (2005). *Association norms for 600 Chinese characters.* Taipei, Taiwan: Taiwanese Psychological Association.

Huerta, M., Bishop, S.L., Duncan, A., Hus, V. & Lord, C. (2012). Application of DSM-5 criteria for autism spectrum disorder to three samples of children with DSM-IV diagnoses of pervasive developmental disorders. *American Journal of Psychiatry, 169,* 1056–1064. https://doi.org/10.1176/appi.ajp.2012.12020276

Hung, K. C., Lee, S. H., Chen, S. Y., & Chou, T. L. (2010). Effect of semantic radical and semantic association on semantic processing of Chinese characters for adults and fifth graders. *Chinese Journal of Psychology, 52,* 327–344. https://doi.org/10.6129/CJP.2010.5203.06

Hutchison, K. A. (2003). Is semantic priming due to association strength or feature overlap? A microanalytic review. *Psychonomic Bulletin & Review, 10,* 785–813. https://doi.org/10.3758/BF03196544

Johnson, C. R., & Rakison, D. (2006). Early categorization of animate/inanimate concepts in young children with autism. *Journal of Developmental and Physical Disabilities, 18,* 73–89. https://doi.org/10.1007/s10882-006-9007-7

Lau, E. F., Phillips, C., & Poeppel, D. (2008). A cortical network for semantics: (De)constructing the N400. *Nature Reviews Neuroscience, 9,* 920–933. https://doi.org/10.1038/nrn2532

Lee, S. H., Booth, J. R., Chen, S. Y., & Chou, T. L. (2011). Developmental changes in the inferior frontal cortex for selecting semantic representations. *Developmental Cognitive Neuroscience, 1,* 338–350. https://doi.org/10.1016/j.dcn.2011.01.005

Lee, S. H., Booth, J. R., & Chou, T. L. (2015). Developmental changes in the neural influence of sublexical information on semantic processing. *Neuropsychologia, 73,* 25–34. https://doi.org/10.1016/j.neuropsychologia.2015.05.001

Lee, S. H. Booth, J. R. & Chou, T. L. (2016). Temporo-parietal connectivity uniquely predicts reading change from childhood to adolescence. *NeuroImage, 142,* 126–134. https://doi.org/10.1016/j.neuroimage.2016.06.055

Lee, S. H., Chen, S. Y., & Chou, T. L. (2009). Effect of vocabulary size on semantic processing of Chinese characters for fifth graders and adults. *Formosa Journal of Mental Health, 22,* 354–382. https://doi.org/10.30074/FJMH.200912_22(4).0001

Martin, A., & Chao, L. L. (2001). Semantic memory and the brain: Structure and processes. *Current Opinion in Neurobiology, 11,* 194–201. https://doi.org/10.1016/S0959-4388(00)00196-3

McClelland, J. L., & Rogers, T. T. (2003). The parallel distributed processing approach to semantic cognition. *Nature Reviews Neuroscience, 4,* 310–322. https://doi.org/10.1038/nrn1076

McRae, K., & Boisvert, S. (1998). Automatic semantic similarity priming. *Journal of Experimental Psychology: Learning, Memory, and Cognition, 24,* 558–572. https://doi.org/10.1037/0278-7393.24.3.558

Nation, K., & Snowling, M. J. (1999). Developmental differences in sensitivity to semantic relations among good and poor comprehenders: Evidence from semantic priming. *Cognition, 70,* B1–B13. https://doi.org/10.1016/S0010-0277(99)00004-9

Parish-Morris, J., Hennon, E. A., Hirsh-Pasek, K., Golinkoff, R. M., & Tager-Flusberg, H. (2007). Children with autism illuminate the role of social intention in word learning. *Child development, 78,* 1265–1287. https://doi.org/10.1111/j.1467-8624.2007.01065.x

Plaut, D. C. (1995, July). *Semantic and associative priming in a distributed attractor network.* Paper presented at the 17th Annual Conference of the Cognitive Science Society, Pittsburgh, PA.

Reich, S., Chou, T. L., & Patterson, K. (2003). Acquired dysgraphia in Chinese: Further evidence on the link between phonology and orthography. *Aphasiology, 17,* 585–604. https://doi.org/10.1080/02687030344000049

Sachs, O., Weis, S., Krings, T., Huber, W., & Kircher, T. (2008). Categorical and thematic knowledge representation in the brain: Neural correlates of taxonomic and thematic conceptual relations. *Neuropsychologia, 46,* 409–418. https://doi.org/10.1016/j.brainres.2008.03.045

Sachs, O., Weis, S., Zellagui, N., Huber, W., Zvyagintsev, M., Mathiak, K., & Kircher, T. (2008). Automatic processing of semantic relations in fMRI: Neural activation during semantic priming of taxonomic and thematic categories. *Brain Research, 1218,* 194–205.

Seidenberg, M. S., & Gonnerman, L. M. (2000). Explaining derivational morphology as the convergence of codes. *Trends in Cognitive Neuroscience, 4,* 353–361. https://doi.org/10.1016/S1364-6613(00)01515-1

Siller, M., & Sigman, M. (2008). Modeling longitudinal change in the language abilities of children with autism: Parent behaviors and child characteristics as predictors of change. *Developmental Psychology, 44,* 1691–1704. https://doi.org/10.1037/a0013771

Taylor, K. I., Devereux, B. J., & Tyler, L. K. (2011). Conceptual structure: Towards an integrated neurocognitive account. *Language and Cognitive Processes, 26,* 1368–1401. https://doi.org/10.1080/01690965.2011.568227

Wong, C. H., Chen, S. Y., Chou, T. L., & Lee, S. H. (2011). The impacts of word recognition ability and semantic relation on semantic processing for third graders. *Chinese Journal of Psychology, 53,* 293–307. https://doi.org/10.6129/CJP.2011.5303.03

Wong, C. H., Chen, S. Y., & Chou, T. L. (2014). A longitudinal study of semantic association and categorical relatedness on children's semantic processing of Chinese characters. *Chinese Journal of Psychology, 56,* 65–81. https://doi.org/10.6129/CJP.20131103

Wong, C. H., Gau, S. S., & Chou, T. L. (in press). Neural correlates of processing categorical relatedness in youths with autism spectrum disorder. *Autism Research.*

Zhu, Z., Zhang, J. X., Wang, S., Xiao, Z., Huang, J., & Chen, H. C. (2009). Involvement of left inferior frontal gyrus in sentence-level semantic integration. *Neuroimage, 47,* 756–763. https://doi.org/10.1016/j.neuroimage.2009.04.086

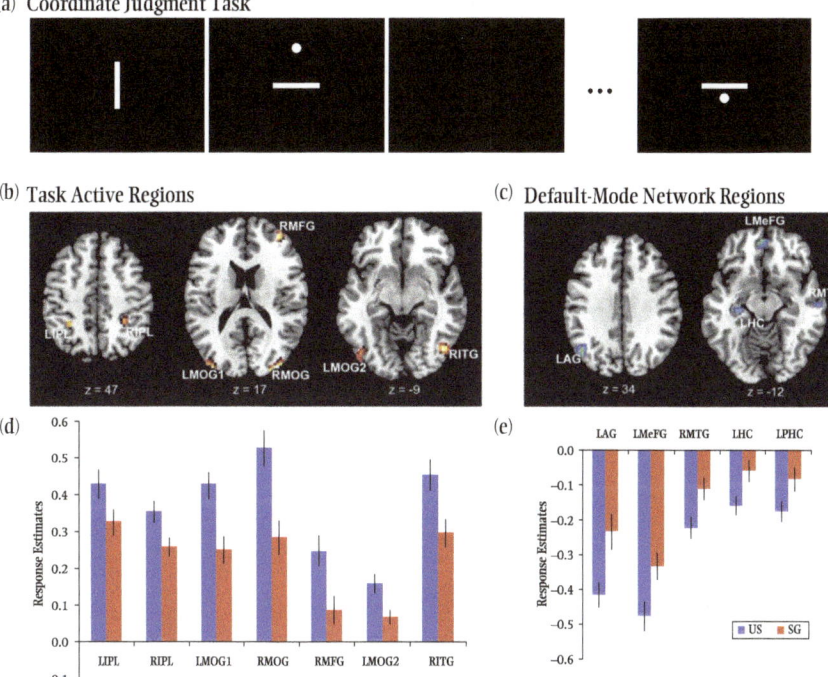

Figure 6.3. Functional neural responses in American (US) Westerners and Singaporean (SG) East Asians when making relative visuospatial judgments. (A) Sample stimuli for the visuospatial Coordinate Judgment Task in which participants were to encode the length of the target vertical bar stimuli (leftmost slide) and decide in subsequent trials whether a dot was farther or nearer from a horizontal bar relative to the length of the vertical bar (right slides; slides were blank during rest trials). (B) Task active and (C) default-mode network regions that significantly modulated their functional responses during the coordinate judgment task across Westerners and East Asians. L = left; R = right; IPL = inferior parietal lobule; MOG = middle occipital gyrus; MFG = Middle Frontal Gyrus; ITG = inferior temporal gyrus, AG = angular gyrus, HC = hippocampus; MeFG = medial frontal gyrus; MTG = middle temporal gyrus. (D) Westerners showed higher functional neural responses than East Asians in task active regions. (E) Westerners showed more suppressed functional neural responses than East Asians in default-mode network regions. Adapted and reproduced with permission from Joshua O. S. Goh et al. Culture differences in neural processing of faces and houses in the ventral visual cortex. J. O.S. Goh, E. D. Leshikar, B. P. Sutton, J. C. Tan, S. K. Y. Sim, A. C. Hebrank, & D. C. Park, 2010, Culture differences in neural processing of faces and houses in the ventral visual cortex. *Social Cognitive and Affective Neuroscience, 5,* 227–235. © The Author (2010). Published by Oxford University Press.

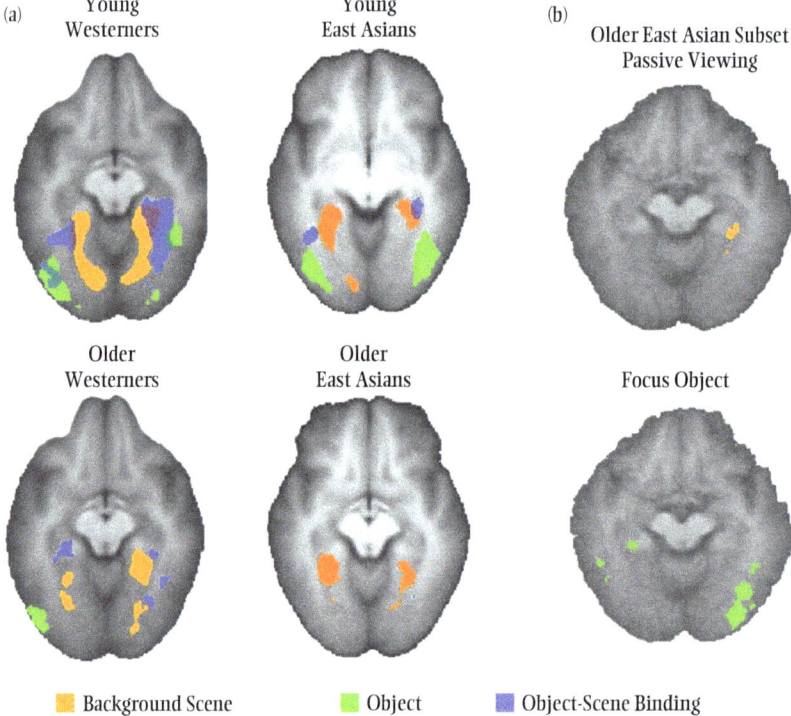

Figure 6.4. (A) Object, background scene, and object-background binding processing brain regions involved during passive viewing of picture quartets (see Figure 6.2). Compared to young Westerners, young East Asians, and older Westerners, older East Asians showed reduced involvement of object processing in the lateral occipital areas and binding processing in the medial temporal areas. (B) Object processing in the lateral occipital areas in a subset of older East Asians whose passive viewing response is shown above is rescued when instructed to focus on the central objects in pictures. Adapted and reproduced with permission from M. W. Chee, J. O. S. Goh, V. Venkatraman, J. C. Tan, A. Gutchess, B. Sutton, ... D. Park, Agerelated changes in object processing and contextual binding revealed using fMR adaptation, *Journal of Cognitive Neuroscience, 18,* 495–507. © 2006 by the Massachusetts Institute of Technology; J. O. S. Goh, S. C. Siong, D. Park, A. Gutchess, A. Hebrank, & M. W. L. Chee, 2004, Cortical areas involved in object, background, and object-background processing revealed with functional magnetic resonance adaptation. *Journal of Neuroscience, 24,* 10223–10228; and J. O.S. Goh, M. W. Chee, J C. Tan, V. Venkatraman, A. Hebrank, E. D. Leshikar, ... D. C. Park, 2007, Age and culture modulate object processing and object–scene binding in the ventral visual area. *Cognitive, Affective, & Behavioral Neuroscience, 7,* 44–52. Copyright © 2007 Springer US. All Rights Reserved.

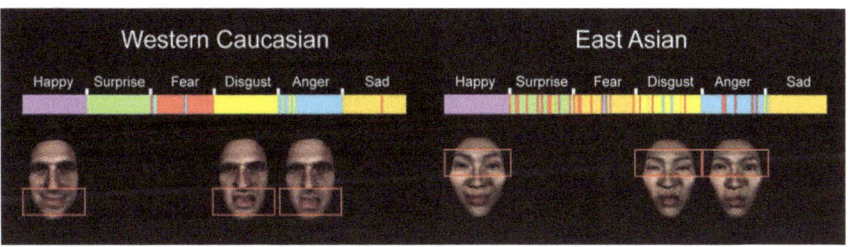

Figure 6.6. Clustering of computer generated emotional faces based on ratings from Western and East Asian participants in relation to the Facial Action Coding System (Ekman & Friesen, 1977; top colored bars). Western ratings yielded homogenous and distinctive clusters for each of the six facial emotion expressions as illustrated by relatively well separated colors for each emotion. By contrast, East Asian ratings yielded more heterogeneous and overlapping categorizations of the faces depicted by the scattered colors across emotions. Also, Westerners weighted the mouth as more important for emotional expression interpretation whereas East Asians weighted more the eyes (bottom faces). Adapted and reproduced with permission from R. E. Jack, O. G. B. Garrod, H. Yu, R. Caldara, & P. G. Schyns, 2012, Facial expressions of emotion are not culturally universal. *Proceedings of the National Academy of Sciences of the United States of America, 109,* 7241–7244.

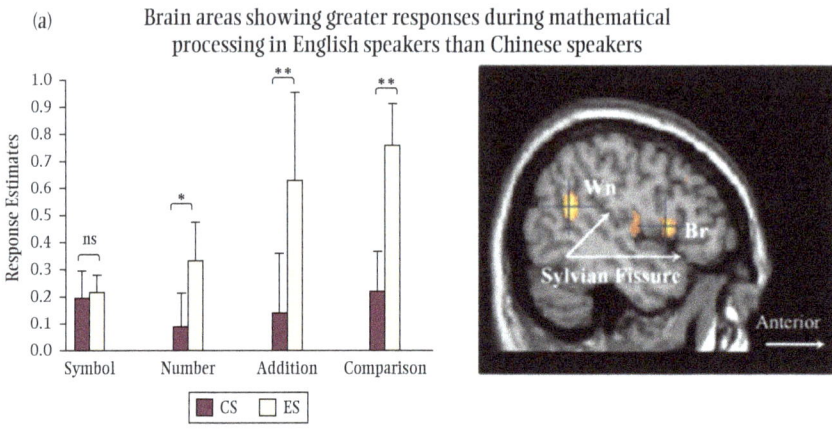

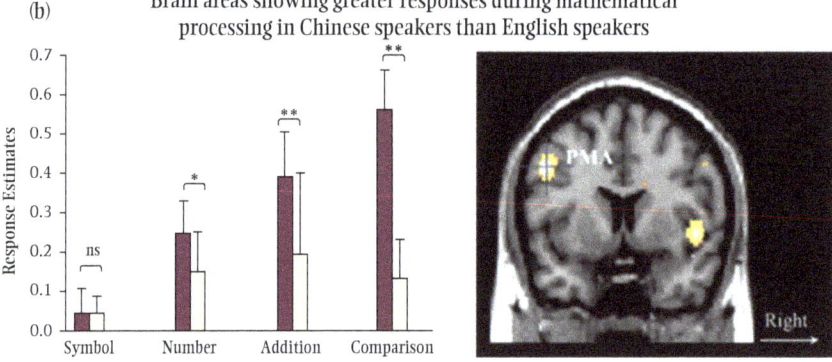

Figure 6.7. (A) The perisylvian area shows higher responses in English (ES) than Chinese (CS) speaking participants across different mathematical tasks. (B) The premotor and visual (not shown) areas show higher neural responses in CS than ES instead during mathematical processing. Adapted and reproduced with permission from Y. Tang, W. Zhang, K. Chen, S. Feng, Y. Ji, J. Shen, J., . . . Y. Liu, 2006, Arithmetic processing in the brain shaped by cultures. *Proceedings of the National Academy of Sciences, 103,* 10775–10780. Copyright (2006) National Academy of Sciences, USA.

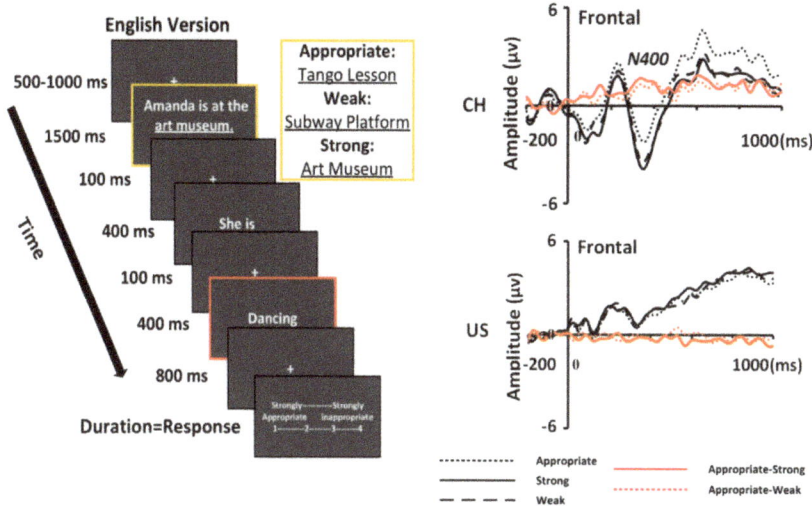

Figure 6.9. Sample stimuli describing behaviors that have different degrees of appropriateness under different contexts. East Asians showed greater N400 responses compared to Westerners to target words (red box) that describe norm violations relative to more appropriate behaviors. Adapted and reproduced with permission from Y. Mu, S. Kitayama, S. Han, & M. J. Gelfand, 2015, How culture gets embrained: Cultural differences in event-related potentials of social norm violations. *Proceedings of the National Academy of Sciences of the United States of America, 112,* 15348–15353.

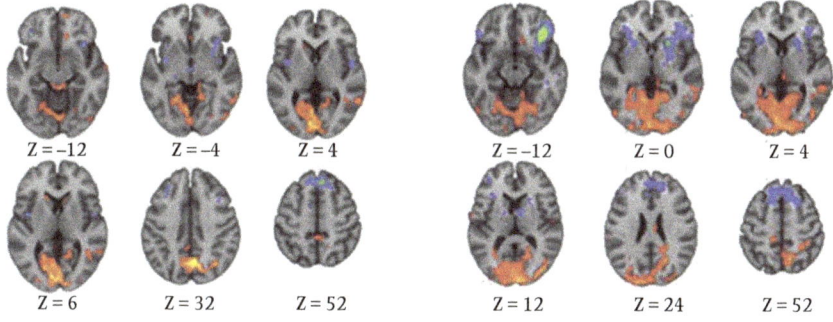

Figure 7.1. Resting-state functional connectivity for monolinguals and bilinguals. Regions showing correlated activity with the inferior frontal seed voxels. Regions of positively correlated activity with the seed were observed in the frontal–posterior network, which was strongly represented in bilinguals (warm-colored voxels). Regions of negatively correlated activity with the seed were observed in the anterior left–right network, which was similar for both monolinguals and bilinguals (cool-colored voxels). Adapted from G. Luk, E. Bialystok, F. I. M. Craik, & C. L. Grady, 2011, Lifelong bilingualism maintains white matter integrity in older adults. *Journal of Neuroscience, 31,* 16808–16813. Used with permission.

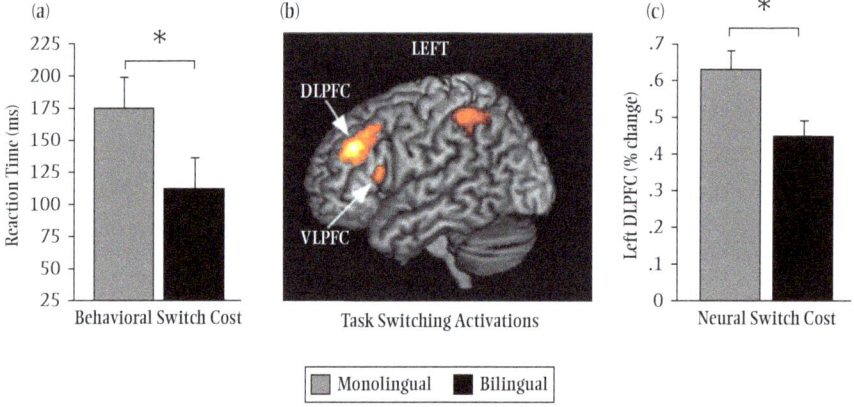

Figure 7.2. (A) Task switching behavioral and brain activation differences for monolinguals and bilinguals. Older bilinguals showed smaller behavioral switch costs than their monolingual peers. (B) Task switching (switch–nonswitch contrast) was associated with the activation of frontostriatal and frontoparietal regions across participants, including left frontal activations seen here in the DLPFC and VLPFC. (C) Older adult bilinguals showed lower task switching fMRI activation (i.e., lower neural switch costs) in several frontal regions, including the left DLPFC. Adapted from B.T. Gold, C. Kim, N. F. Johnson, R. J. Kryscio, & C. D. Smith, 2013, Lifelong bilingualism maintains neural efficiency for cognitive control in aging. *Journal of Neuroscience, 33*, 387–396. Used with permission. *p < 0.05.

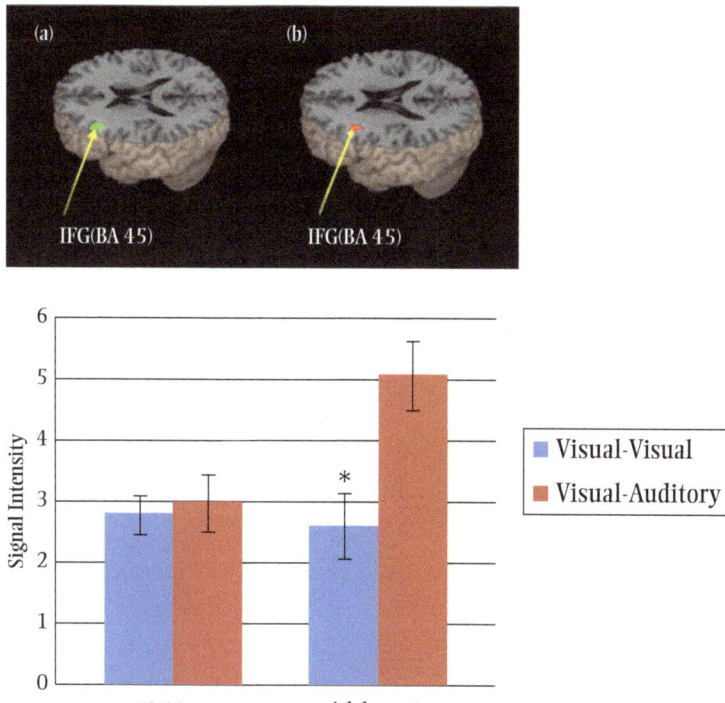

Figure 9.1. (Top panel) Increasing age was correlated with greater activation in the mid-ventral region of left inferior frontal gyrus (IFG, BA 45) with weak association strength in the visual-visual semantic judgment task (in green) and in the visual-auditory semantic judgment task (in red). (Bottom panel) Greater involvement of semantic selection mechanisms in the cross-modal task requiring the access of the appropriate meaning of homophonic spoken words in Chinese for adolescents. From S. H. Lee, J. R. Booth, S. Y. Chen, & T. L. Chou, 2011, Developmental changes in the inferior frontal cortex for selecting semantic representations. *Developmental Cognitive Neuroscience, 1,* 338–350.

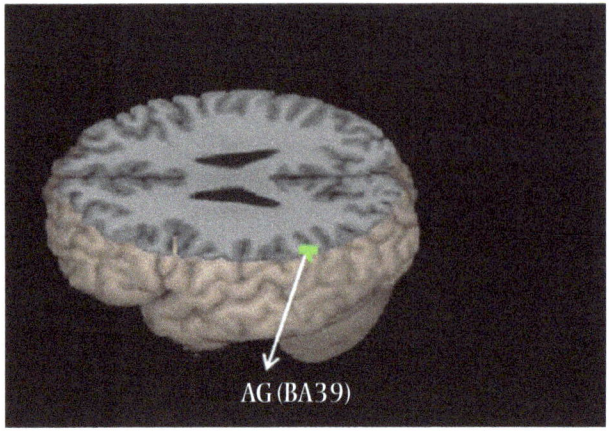
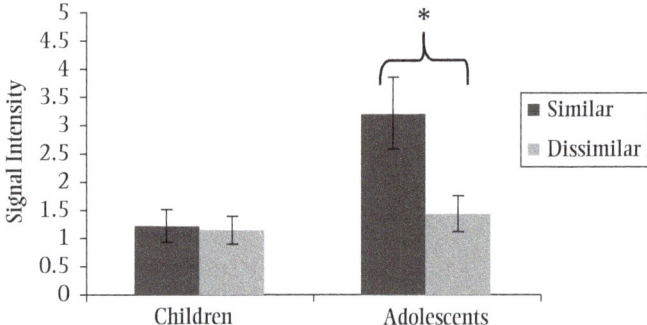

Figure 9.2. (Top panel) Developmental increases in activation in left angular gyrus (AG, BA 39) with strong association strength for both visually-similar and visually-dissimilar pairs in Chinese. (Bottom pane) Greater activation for the similar pairs than the dissimilar pairs in adolescents, but not in children. From S. H. Lee, J. R. Booth, & T. L. Chou, 2015, Developmental changes in the neural influence of sublexical information on semantic processing. *Neuropsychologia, 73,* 25–34.

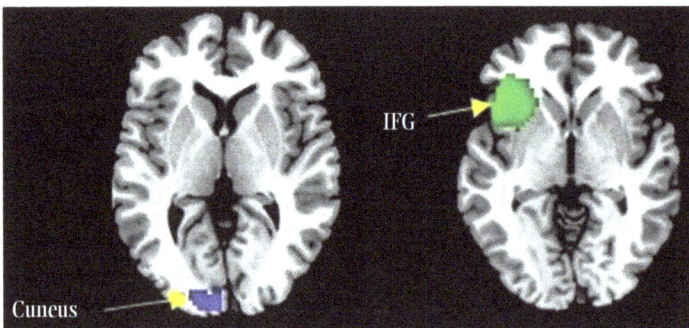

Figure 9.3. Children with autism had greater cuneus activation than typically developing children (in blue) while adolescents with autism showed reduced left inferior frontal gyrus (IFG) activation as compared to typically developing adolescents (in green). From P. J. Chen, S. S. Gau, S. H. Lee, & T. L. Chou, 2016, Differences in age-dependent neural correlates of semantic processing between youths with autism spectrum disorder and typically developing youths. *Autism Research, 2,* 1–11.

10

Culture and Language Diversity in Pediatric Neuropsychology

VERONICA BORDES EDGAR AND REGILDA ANNE ROMERO

INTRODUCTION

There is a growing need for the understanding of cultural and language diversity in the neuropsychological assessment of children. With the rapid demographic shifts in the United States, the need has never been greater. This is evidenced by national movements such as the Relevance 2050 initiative by the American Academy of Clinical Neuropsychology (2016; AACN), which estimates that by 2050, 60% of the American population will be "untestable" with the current toolkit. Although there has been increased diversity in patient populations throughout the United States, no longer limited by geographic area, there are still gaps in the science and practice of clinical pediatric neuropsychology.

This chapter will address cultural and linguistic diversity considerations for the assessment of children with special focus on the impact of bilingualism and education. Potential issues with test bias and norms, test translation, and language proficiency in children will be discussed. Additionally, we will address best practice recommendations in the clinical neuropsychological evaluation of bilingual children.

BACKGROUND

The current U.S. demographics indicate that minorities make up 38% of the population, with 48% of the population younger than 18 years of age. By 2023, minority children will be the majority (U.S. Census 2014). Socioeconomic variables such as poverty and parental education have also been established as having an influence on assessment and intervention outcomes for children. Currently, 21% of children under 18 in the United States live in poverty. This number ranges from 10% of non-Hispanic Whites to 26% of Blacks. When children are living in unrelated families, this figure more than doubles to 27% (U.S. Census, 2014). In terms of parental education, Hispanics/Latinos have a greater likelihood of having a sixth grade education or lower (14%) as compared to the general population (3%) as well as compared with African Americans/Blacks (2%), Asians (4%) and non-Hispanic Whites (3%). They also have a much lower representation of parents with a bachelor's degree (11%) as do African Americans (14%) compared to the general population (21%). Asians have a higher representation in this category (33%; U.S. Census, 2015).

With respect to bilingualism, 20.3% of children ages 5 to 7 enrolled in school speak a language other than English at home. In fact, 10 states had 20% or more of children speaking a language other than English at home and an additional 17 states with 10% to 20% of the population. These figures are similar when looking at young school children enrolled in nursery/kindergarten to fourth grade. These languages include Spanish, Chinese, Arabic, Korean, Vietnamese, and French (Kominski, Shin, & Marotz, 2008). While this continues to establish the growing need for appropriate bilingual assessment, perhaps a more important factor to look at related to assessment is how well these children are speaking English. In this same analysis, researchers found that for children aged 5 to 17 enrolled in school, 5% spoke English at a rating of less than "very well," and five states had a population of 6% or more that spoke English at a level less than "very well." Not surprisingly, these numbers are slightly higher for young school-aged children with a national rate of 6.5% and 11 states having a population of 6% or higher speaking English less than "very well." These figures suggest that the likelihood of practitioners needing to manage a bilingual evaluation is quite high. Of course, it is certainly highest for Spanish with more than 35 million people in the United States speaking Spanish at home (U.S. Census, 2009). However, for Latinos/Hispanics, variability is largely seen dependent on time in the United States. Although one third of foreign-born Hispanics speak English at home or report to speak English "very well," by the second generation only 8% report prefer Spanish, while 50% identify as bilingual (Pew Research Center, 2013). So the question becomes, who is going to test these children?

The underrepresentation of Spanish-speaking neuropsychologists, let alone board-certified neuropsychologists, has been a long-standing issue that continues to be an area of concern (Echemendia, Harris, Congett, Diaz, & Puente, 1997; Romero, et al., 2009; Salinas, Bordes Edgar, & Puente, 2016). Beyond bilingualism, there are not enough Latino psychologists to test all of the Latinos who need testing, and the problem is only continuing to grow. Simply "referring out" is no longer a viable solution. Even then, consideration of within- and between-group differences for each racial and ethnic group is important. Therefore, establishing formal competence in cultural neuropsychology and increased opportunities for training for all practitioners is a priority for the profession as a whole. We can look to organizations such as the Hispanic Neuropsychological Society to lead the way in these endeavors as has been seen in collaboration with the National Academy of Neuropsychology (NAN) in the development of a position paper (Judd et al., 2009) outlining many of the issues and best practice guidelines as well more recent endeavors with AACN and the Relevance 2050 initiative. From the 2009 NAN position paper, qualifications for psychologists engaging in cross-cultural assessment include adequate knowledge of the examinee's culture, understanding of one's own perspective and cultural biases, understanding issues in cross-cultural communication, and awareness of ethical issues and qualifications of interpreters and translators (Judd et al., 2009). Additionally, Puente and McCaffrey (1992) noted that neuropsychological assessments should be based not on only on the tests and brain function, but also on the biopsychosocial aspect of the person, suggesting that it is important to view the person within the context of the person's life, which includes acculturation level and language preference and proficiency. To that end, the next section describes issues impacting cultural neuropsychological assessment including sociocultural factors, values, bilingualism, acculturation/parent education, and neighborhood factors. Consideration of these variables is as starting point for culturally competent practice.

Sociocultural Factors Impacting Effort and Engagement

Cultural diversity and heterogeneity even within cultures can influence rapport, test performance, interpretation, and treatment recommendations. Beginning with rapport, this can often be influenced by the patient's value of *personalismo* or having a personal connection as well as eye contact. There is often a reduction in eye contact as a sign of respect. For many minorities, there are certain beliefs about psychologists or physicians that may begin with a level of mistrust. Often, this is perpetuated by sociohistorical factors as well as fears of what the provider may do. For example, in the current U.S. climate, many Latino families fear deportation if they report their current immigration status. However, this information can impact the types of interventions that they receive, as it limits accessibility of services. Across a number of minority groups, there can also be fear related to provider perspectives on natural or homeopathic therapies. The latter can influence symptom presentation and reporting. Another importance in understanding sociocultural variables is the ability for patients to maneuver within systems, which can subsequently influence diagnostic and treatment choices. Although certain interventions may be recommended, they may not be financially or logistically feasible.

For many minority families, the value of *familismo* or the value and loyalty for family and importance of extended families can also impact assessment and interventions. For example, in some families, this may mean that grandparents are primary caregivers in addition to parents and that behavioral or parenting strategies must also be feasible and agreed to by this extended network. Within the family, gender role expectations may result in greater or fewer adaptive capabilities than might be expected. Conversations with parents regarding these expectations often yield valuable information that can improve interpretation of results. One example of this is in East Indian culture where boys may not be expected to engage in independent self-care (e.g., feeding, dressing) well into the school-age years. Gender differences and the value of *familismo* may also impact the type and quality of information received in a diagnostic interview. For example, a mother may not disclose certain information to a male practitioner nor to a female practitioner through a male interpreter. The practitioner's own values of gender and gender equality may also impact the diagnostic formulation. Considerations of logistics such as financial status, transportation, child care, and participation of extended families in care should be part of the therapeutic framework.

Other behavioral variables such as the use of eye contact and head nodding have been addressed in most multicultural training, but are important to keep in mind for the purposes of assessing effort and understanding. Patients may nod their head to be respectful and appear to agree, but they may not have a full understanding of the instructions or the information being conveyed. Finally, the value of speed and time has significant cultural variability. If not taken into account, this can also impact interpretations of effort as well as ability. Bush et al. (2005) noted that these cultural factors may lead to exaggeration or denial of symptoms without intent or motivation (conscious or unconscious) to deceive.

Familial and Neighborhood Factors

Familial factors such as parent education and acculturation have been well documented to be important areas of consideration within an evaluation (Pontón & Leon-Carrion, 2001). As previously noted, there is significant variability and a trend for undereducation of minorities. When evaluating children, it is important to consider the level of education of their parents as well as their own education. The quality of education they have received as well as trends for persistence in their community should be considered. In some cultures. formal education may end prior to the Western view of high school completion. Such is the case for Old Order Anabaptists who may end formal schooling around age 14 (Kraybill & Gilliam, 2012). Additionally, American Indians, for example, may receive variable quality of education on and off tribal lands, as well as have higher rates of high school completion when living off tribal lands (67% on tribal lands vs. 73% off tribal lands, compared to 80% for the U.S. population; U.S. Census, 2006). There is also a significant amount of history including familial separation and compulsory education by the U.S. government that may contribute to an environment of mistrust (Verney, Bennett, & Hamilton, 2016). Similar to education, acculturation cuts across cultural groups and has been shown to mitigate cultural differences (Ferraro, 2016). Acculturation is most commonly considered with Hispanics, but should also be assessed in other racial minorities. Assessment in first- and second-generation immigrants is also recommended and can be conducted formally through measures as well as informally through interview.

Finally, the concepts of neighborhood poverty and socioeconomic status (SES) need to be addressed. Multiple indicators of SES should be gathered by the provider including knowledge of parental education and family income (Manly, 2008). Nutrition, housing conditions, environmental exposures, and access to healthcare are closely tied with SES and may impact performance on neuropsychological tests (Suzuki, Naqvi, & Hill, 2013). The differential impact of SES on assessment performance and validity has been well studied (Liaw & Brooks-Gunn, 1994; Noble, McCandliss, & Farah, 2007; Noble, Norman, & Farah, 2005) and has been documented for Hispanics (Chin, Negash, & Hamilton, 2011). In the study by Noble et al. (2007), first graders were assessed across a number of neurocognitive domains. Regressions for SES (based on parent education, occupation, and income) revealed significant variance accounted for in the area of language (32%; most notably receptive language), visuospatial skills (17%) and memory (10%). Approximately 6% of the variance in working memory and cognitive control was also accounted for by SES. Further, this relationship was not mediated by an exposure to a second language at home. Although researchers included large numbers of African American (34%) and Latino (23%) children, analyses were not examined across groups. Across the research, little to no information for other ethnic groups that are at high risk, such as American Indians, is available (Verney et al., 2016).

Bilingualism/Multiculturalism

Multilingualism is the norm for many areas of the world. Multilinguals experience cognitive and social advantages such as improved inhibition, working memory, and cognitive flexibility (Barac, Bialystok, Castro & Sanchez, 2014). Research focusing on bilingualism and neurocognitive functioning is increasing in the United States along with the changing demographics. From a developmental perspective, being knowledgeable about theories of language development for bilinguals is critical for pediatric neuropsychologists. It is first important to distinguish three types of bilingual language acquisition described by Mushi (2002)—simultaneous, sequential/successive, and circumstantial. The latter of these includes individuals who learn a second language in passing and are not the focus of this chapter.

The dual language system hypothesis (Genesee, 1989) appears to be well supported by research. This assumes two linguistic systems are established as in the case of simultaneous bilingual acquisition. In simultaneous acquisition, children are exposed to both languages from an early age and acquisition of language, therefore, does not differ from those exposed to a single language (Krashen, 1982). This similarity in language acquisition to monolingual learners has been demonstrated from newborn preference and language discrimination (Byers-Heinlein, Burns, Werker, 2010; Jusczyk, 1997) and the acquisition of early language milestones through the second birthday when children are developing grammar (Genesee, Paradis & Crago, 2004). In fact, research has demonstrated that typically developing bilingual children pass through the stages of language development in the same manner (age) as monolingual peers (Genesee et al., 2004; Paradis & Genesee, 1997; Meisel, 1994) and have similar total (combined-language) vocabularies (Hoff et al., 2011; Pearson, 1998). Studies in infancy have shown that children exposed to two languages from birth can process each language in a native manner phonetically, but this process requires a live person with social interaction. Additionally, babies between 6 and 12 months of age can hear sounds of all languages spoken, but only develop prototypes for the ones they hear frequently (Kuhl, 1993; Werker & Desjardins, 1995). In children with autism spectrum disorder where language delays are often seen, researchers have failed to find differences in vocabulary, receptive and expressive language, and early language milestones between monolingual and bilingual children (Hambly & Fombonne, 2012). The old way of thinking that bilingual children may be delayed is therefore inappropriate and often leads to delayed intervention for children who need it.

In contrast, sequential/successive bilinguals begin learning a single language prior to exposure to a second language. Typically, these are children who acquire their first language in infancy and then begin exposure to a second language after age 3 when they enter preschool, or for some, age 5 when they enter kindergarten. Regardless of the type of bilingual (simultaneous or sequential), individuals have a dominant language. This dominant language is considered their L1 or SL (stronger language) and their nondominant language can also be considered their L2 or WL (weaker language), depending on the terms used by researchers. This is not always stable throughout the lifespan and an individual may shift their L2 to become their L1 as in the case of children raised in Spanish-speaking homes (L1) who are immersed in English-only education (L2) and activities. This can also be seen in immigrant children as their level of acculturation increases. There has also been discussion of possible attrition or reduction in L1 (Artiola i Fortuny, 2009). Overall, it is thought that the age and extent of when L2 exposure begins will impact a child's acquisition.

In the bilingual brain, theories such as the competition model (Bates & MacWhinney; 1989; MacWhinney, 2005) explain how languages can interact, compete and how language is transferred. Translation and code-switching involves inhibition of one language to activate the second (Costa & Santesteban, 2004). There are questions that remain in how environmental influences and exposure to both languages impact lateralization and localization of language (Hull & Vaid, 2007; Vaid, 2008) as well as how language of bilinguals is represented within the brain (Green, 1998; Paradis, 2004) given contradictory findings even in imaging studies. Of special note, Espinosa (2016) found that language development for bilingual and minority individuals is dependent on several factors including country of origin, languages spoken, age of exposure, amount and quality of exposure, socioeconomic variables, and opportunities to use language.

Impact on Education

Bilingualism and the extent of L1 versus L2 development when a child enters the school setting can have lasting impact. Basic interpersonal communication skills (BICS; i.e., peer-appropriate or social language) are typically acquired first (within 2 years) as compared to cognitive and academic language proficiency (CALP), which may take up to 5 to 10 years (Collier, 1995; Cummins, 1984; 1999; French & Llorente, 2008). As such, educators as well as examiners who are unfamiliar with the child's language exposure may falsely conclude that a child is more proficient in the language of instruction or assessment than they may truly be. Dual language versus single language (immersion) education remains a highly contested concern. Research has found that individuals with schooling in their L1 prior to language immersion took less time to gain proficiency than those who had no formal schooling in their L1 (Collier, 1995). This is supported by the developmental notion of scaffolding in that if an individual has the rules for language then he or she is more likely to build on those rules in development of a second language.

Beyond the language in the classroom, there is differential impact on academic achievement based on SES. Bilingual children both living in poverty and with limited access to early education were at increased vulnerability (Castro, Espinosa & Paez, 2011). This trend persists beyond early childhood as bilingual children from low income families are at increased risk for academic underachievement throughout childhood based on state achievement data, high school completion, and college enrollment (Espinosa, 2016; Rumberger, 2004). On the other hand, higher achievement in English was seen for bilingual children who were not low SES (Espinosa, 2016; Espinosa et al., 2017).

Issues with Test Measures

Beyond linguistics, the underlying problem of test bias has been well studied. The appropriateness of a test measure is dependent on different factors including the patient's language preference and proficiency, acculturation level, and *test equivalence* (Groth-Marnat, 2009). Test bias occurs when the measure used is not linguistically (appropriately translated), conceptually (construct validity), or psychometrically equivalent. This can result in either an overestimate or underestimate of what is being measured (Reynolds and Suzuki, 2013). Addressing test bias limits the likelihood of unfair and inaccurate assessment of bilingual individuals, or individuals from diverse background for that matter. Consequently, addressing test bias provides a more accurate interpretation of the results. Some of the test bias that affects bilingual or multicultural assessments are content and construct bias.

In selecting cognitive measures for diverse populations, understanding factors such as *cultural loading* of test and *linguistic demands* are essential, particularly for interpretation of results. Cultural loading refers to tests or items that are culture specific. A test is culturally biased when its items have a high cultural loading and is administered to an individual from a different culture (Reynolds & Suzuki, 2013). Cultural loading factors that may affect testing include a person's level of acculturation and the extent that his or her cultural knowledge (Rhodes, Ochoa, & Ortiz, 2005).

In terms of linguistic demands, it is known that most neuropsychological tests are language-based. Even visual or nonverbal tests tend to be verbally mediated because they are affected, to varying degrees, by language proficiency (Solano-Flores, 2008).The length and wordiness of the instructions and complexity of verbal directions are components of verbal directions that need to be considered or examined (Cormier, McGrew, & Evans, 2011). Hence, individuals with limited English-proficiency or individuals who are bilingual perform significantly lower than native-English speakers on cognitive measures (Nieves-Brull, 2006; Rhodes et al., 2005). Comprehension of instructions and ability to answer items of current measures of intelligence and cognitive abilities are based on age- or grade-level expectations of abilities, which includes language development and proficiency (Rhodes et al., 2005). Further, the quality of exposure to and the age of English learning are thought to be variables that interact with linguistic demands (Ortiz & Dynda, 2005).

Psychological and neuropsychological measures are mainly published in the English language; therefore, performance on these measures depends on the American English language proficiency (Hebben & Milberg, 2002). Similarly, these measures were normed in the United States and based on the cultural aspects of Americans. Consequently, performance on these measures depends on the familiarity of the American cultural context. While results do not necessarily become invalidated, caution in interpreting them is needed when applied outside the norm due to possible underestimation of abilities of bilingual children and possible lack of generalizability of measures.

Of the four pillars of assessment including norm-referenced measures, interviews, behavioral observations, and informal assessment, the ones most viewed as culturally biased are norm-referenced measures (i.e., standardized tests based on a group of individuals that represent demographic characteristics such as gender, age, race/ethnicity, etc.; Sattler, 2008). Most neuropsychological assessments were developed some decades ago, with minimal inclusion of people from the minority groups in design, development, and standardization of these tests (Davies, Strickland, & Cao, 2014). While test developers aim to include minorities in their normative sample, normative data continue to be limited when it comes to within-group sampling. While bilingual and bicultural groups remain to be rarely represented (Hebben & Milberg, 2002), minority representation in the standardization of the norm group is important as it is the basis on which individual test scores are compared to gain meaningful interpretation (Groth-Marnat, 2009). There is a need for modification and accommodation for cultural variability to produce reliable results and accurate diagnosis (Davies et al., 2014).

Test Adaptation/Translations and Validity

In 1992, the International Test Commission, with the help of other agencies, prepared guidelines for test translation and adaptation, which were subsequently field-tested by Hambleton and his colleagues (Hambleton, 2001; 2005; Hambleton, Yu, & Slater, 1999). These guidelines serve as a reference so that the translation and adaptation of assessment tools to a different language are comparable to their original measures. These not only apply to verbal instructions and test items but also to the nonverbal tests and visual/graphic materials. That said, a well-established standard or protocol for cultural consideration in child neuropsychology is still lacking, as compared to the field of adult neuropsychology (Byrd, Arentoft, Scheiner, Westerveld, & Baron, 2008). In recently years, aside from test translations and adaptations, toolkits (i.e., the World Bank toolkit and a toolkit by Gates Foundation) have been developed to provide guidelines for neuropsychological testing in low- and middle income countries (as cited by Semrud-Clikeman et al., 2017). These guidelines may also be useful in assessing children from diverse backgrounds in the United States.

Another important concept in testing is the statistical measurement of performance, namely sensitivity and specificity. *Test sensitivity* measures the proportion of true positives (i.e., correctly identifies as having the condition), while *test specificity* measures the proportion of false negatives (i.e., correctly identifies as not having the condition). In neuropsychological testing, cultural background, particularly one's language, may attribute to false-positive errors (Hebben & Milberg, 2002).

Likewise, English language proficiency may affect symptom validity testing and effort. The lack of command of the English language may affect the child's interpretation of some test items, possibly inaccurately endorsing items. This could result in validity testing results that suggest exaggeration of symptoms, when this is not the case. Language proficiency and other cultural factors such as level of acculturation and collectivistic values may inhibit the child from endorsing symptoms and, therefore, minimalize symptoms. A child struggling to understand English may appear uncooperative or disinterested (Groth-Marnat, 2009), which may seem be mistaken as not putting forth effort.

Construct validity, by definition, is the degree to which a test instrument is measuring what it is intended to measure. Translation of English to another language does not necessarily ensure accurate—or, at least, adequate—transfer of the context to the other language (Davies et al., 2014), which could pose threat to construct validity. Additionally, translation and adaption of neuropsychological tests do not necessarily assume the same sensitivity and specificity levels as its original version. Further, false-positive errors were attributed to reasons why more minorities are placed in special education classes (Hebben & Milberg, 2002).

Now that we have explored background cultural considerations, a more in depth process for bilingual evaluations is reviewed.

NEUROPSYCHOLOGICAL ASSESSMENT OF BILINGUAL CHILDREN

There are many factors or components to the assessment of bilingual children. These include appropriate evaluators, use of interpreters and translators, language of the report, and feedback and recommendations. Bilingual assessment has historically been defined as a bilingual individual being assessed by a bilingual examiner with dual-language methods (Grosjean, 1989), not a monolingual individual assessing a bilingual individual in only one language. More commonly, however, examiners are engaging in cross-cultural assessment where there are significant cultural or language differences between the examiner and examinee, tests, or social context (Judd et al., 2009).

Assessment of language proficiency should be the first step when evaluating children from diverse backgrounds (Groth-Marnat, 2009; Judd et al., 2009; Salinas et al., 2016). This can be done both formally through assessment as well as informally through interviewing the parent and child. It may be necessary to conduct the evaluation in the child's native language if deemed mastery of the English language is found insufficient.

Appropriate Evaluators

An examiner from the same minority group may get the more reliable and valid information from the examinee (Sattler, 2008). If the clinician is sufficiently knowledgeable and is able to conduct the assessment in the child's native language, then the clinician may very well assess the patient (Groth-Marnat, 2009). Linguistic qualifications of professionals participating in English Language Learning (ELL) test development are not usually included in the literature (Solano-Flores, 2008). Merely describing these individuals as bilingual does not usually encompass the level of their linguistic abilities and level of bilingualism. Similarly, research on the characteristics of individuals participating in the process of ELL testing is scare; hence, it is unclear as to their qualifications. However, best practice recommendations are to use a bilingual and bicultural examiner with similar cultural background as the patient who has received training in both bilingual and cultural assessment (Judd et al., 2009; Salinas et al., 2016). A second preference would be for someone with more informal training in these areas with similar linguistic and cultural background, and so forth.

Use of Interpreters and Translators

Despite the preference of an individual with both training and cultural/linguistic similarity to the patient, the use of an interpreter for spoken language and translator for written language may be necessary. Utilizing interpreters poses many difficulties (Sattler, 2008) such as interjecting their personal biases, giving patient context cues, and intentionally omitting or adding information. As such, interpreter selection can greatly impact the assessment and interpretation process (Rhodes et al., 2005).

Things to consider during interpreter/translator selection: (a) The interpreter/translator should have at least commensurate English proficiency and native-language proficiency; (b) family members and friends of the patient should not be chosen to interpret/translate due to ethical reasons; and (c) interpreters/translators who have credentials/certifications or, at the least, have professional training should be sought (Judd et al., 2009; Rhodes et al., 2005; Sattler, 2008). In working with the interpreter, it is essential that they are given instructions prior to the evaluation in terms of the process and expectations (Sattler, 2008). A clinician should ascertain with them the limits of confidentiality and specific issues that may be encountered. For example, it is important to ask interpreters to provide language produced by the child with exact replication and without correcting for grammar or adding interpretation of meaning. Further, technical terms should be discussed prior to the session. Additionally, interpreters should be involved in the assessment process as assistants and not as co-examiners unless they are qualified mental health providers themselves (Sattler, 2008). At the end of the session, the clinician should debrief with the interpreter especially if he or she had to convey difficult information to families. Providing feedback to the interpreter on the process may also be appropriate. Finally, in documentation, clinicians should identify interpreters by name along with their qualifications, as well as other personnel used for interpretive practice (e.g. psychometrists), and any translations used (Judd et al, 2009).

When interpreter/translator services are not provided within a clinician's agency/practice, contracting with a formal agency is best (Rhodes et al., 2005). Additionally, there are other resources that may provide interpreters such as the judicial system, embassy/consulate offices, phone companies, and other professional organizations. When interpretation services are clearly not feasible, modification of testing procedures may be warranted (Rhodes et al., 2005).

Finally, practitioners should be aware of potential problems with interpreters that go beyond the problem of language. Within the authors' own practice, this includes problems with telephonic interpretation (e.g., reduction in personal space) and potential dual relationships when there are small communities. On the other hand, interpreters can also provide a valuable source of cultural information that can assist the examiner in interpretation of performance. For example, the interpreter may be able to notify the clinician that a particular item would be unfamiliar to the child within his or her culture.

Bilingual Assessment

Along with establishing language proficiency and the need for an interpreter, the examiner should assess the child's opportunity to learn English. This will inform the decision toward the child's most proficient language and the one most appropriate for evaluation. As previously discussed, test materials and norms may be limited, but the use of multiple measures to derive at convergent validity may be helpful in this regard. It is also important to know as one assesses language abilities for bilinguals that bilingual children with language impairment are similar in deficit patters and acquisition of language as monolingual children with language impairment (Genesee et al., 2004). Therefore, as previously discussed, merely being bilingual does not automatically mean one is susceptible to language delay; having such delay may be a sign of true language impairment as opposed to a problem with bilingual language acquisition. It is also important for clinicians to identify the proficiency and competency they are trying to evaluate. In some cases, English-only testing is appropriate such as in the case of academic functioning when trying to determine what a child has learned in their English-based classroom. Additionally, single-language testing is most appropriate when conducting a formal assessment of language proficiency. However, in other cases, conceptual scoring may be most appropriate. This includes acceptance of a response in either language as a correct response. In any case, it is important for the clinicians to document the child's choices and supporting reasons. For example, if a child reports preference for English with behavioral dominance although he or she does not have a clearly assessed dominance in that language, primary or initial assessment in English may be appropriate.

Language of Feedback: Verbal and Written

Following the evaluation, examiners should provide information to all involved in their preferred language. In doing so, an interpreter may be necessary, and, as such, there will be the need for discussion of confidentiality and debriefing with the interpreter. During feedback, clinicians are encouraged to use visual explanations such as the bell curve to help convey the concepts of level of performance. Some families may struggle with this concept and therefore age equivalents can often be a helpful added tool. If a clinician is tracking progress as in the case of a reevaluation, a visual explanation of the child's relationship to typically developing peers (such as on a growth curve) has shown to be helpful in discussing changing scores. Finally, clinicians should use caution in their language to avoid idioms and any cultural taboos.

Just like in monolingual assessment, families should also be provided with written feedback in a manner that is accessible to them. There has been debate regarding whether the entire report should be translated or whether there should only be a summary letter. While the authors would advocate that the entire report be translated, this is not always feasible from a financial or time perspective. It has been our experience that this can be not only a costly endeavor depending on the language of translation when using formal translation services, but can also take up to several months in a hospital-based system. The greater the length of time between receiving the English report and the translated report, the less likely the parents understand it is a translated version of the same document and the less likelihood that they can use the document to advocate for interventions for their child as often school placement meetings have passed. An alternative proposed and utilized by the authors is to write a summary or provide a translation of the impressions and recommendations that can be given to parents shortly after the English report is completed with an option for translation of the full report. Ideally, the translated summary and English report can be provided simultaneously if this does not detain the provision of the English report for an extended length of time. This alternative allows for equitable level of care in that the full report can be provided, but it also allows for greater utility of the final product because parents now have a document that they can understand (both in their language and reading level) and have it in a timely manner so that they can use it for child advocacy.

Culturally Appropriate Treatment Recommendations/Interventions

Demographic information and other cultural factors should be considered for understanding the referral question (Davies et al., 2014). These variables should be the basis for assessment method selection and treatment planning and recommendation as these cultural factors likely correlate with test performance. The authors recommend limiting to 8 to 10 recommendations that are organized in order of priority. Language should be simple and concise. Clinicians should provide realistic resources as well as material in a language accessible to the parents. Acculturation and family dynamics (e.g., number of children, involvement of extended family networks, family hierarchy, religion, SES) play a role in responsiveness to interventions and ability to carry out recommendations. For individuals in some cultures (e.g., American Indian, Latino, Old Order Anabaptists, Chinese), acceptance by the clinician of spiritual and alternative healthcare as complementary is important. A challenge for pediatric neuropsychologists especially is accepting that advanced formal education may not be in line with cultural expectations.

With respect to bilingual individuals, learning English is important, but should not come at the expense of development of their home/native language (L1). This is not in the best interest of the child as it may impact communication with family members and increase isolation within their community, and he or she will lose out on the cultural learning and advantages that bilinguals benefit from. Further concern is the impact to the child's ethnic identity development. As discussed by Salinas et al. (2016), improvement in one language can assist in second-language acquisition. There are no studies showing that monolingual intervention is superior to bilingual intervention (French & Llorente, 2008; Thordardottir, 2010). Even for preschoolers, children benefit from speech therapy in English (L2) with support for their L1 at home (Tsybina & Eriks-Brophy; 2010). Further, the recommendation of the major speech-language associations has been to conduct bilingual therapy with a native or near native speaker and to use interpreters if needed (Thordardottir, 2010). Salinas et al. (2016) has proposed that if intervention in only one language is to be carried out, then individual variables such as the age of the child and his or her environment should be carefully evaluated before selecting English as the sole language for intervention.

Strategies for bilingual development include labeling in both languages such as through the use of paired verbal/visual schedules and checklists. Anecdotally, having each parent speak in different languages does not impair a child's ability to develop those language abilities, but rather assists in the development of bilingualism. This is the one parent–one language approach, but can be complicated to carry out on a practical level. Finally, Espinosa (2015) has a list of bilingual academic best practices that include bringing the home language into the classroom and partnering with families; the building of bilingual literacy through books, songs, and visual cues; the improvement of generalizability through practice in a variety of situations; and ongoing assessment of the child's skills.

FUTURE DIRECTIONS

There continues to be a scarcity in assessing bilinguals or individuals from diverse populations in spite of the advances in test development (i.e., methods and procedures, which include participant recruitment, norming, statistical analysis, etc.), test adaptations, and translations (Suzuki & Ponterotto, 2008). Bringing more awareness to bicultural assessment and adding to the existing literature by either conducting new studies or replicating and updating current norms to include more diverse population is warranted. Similarly, further development of alternative methods of assessment may be helpful. Developing standardized and nonstandardized assessment tools are warranted not just for bilingual assessments but also to gather information regarding other diversity issues such effects of SES, gender and transgender identity, etc.

Assessment courses and training programs should incorporate issues in treating and assessing diverse populations. Aside from attending formal courses, consulting with patients and collaborating with minority communities could enrich the assessment process not just on a single case but in terms of resources such as translator/interpreters (Susuki &Ponterotto, 2008). Increased advocacy on the part of clinicians is necessary across systems such as organizational supports in healthcare organizations to ensure culturally and linguistically appropriate services and access to care. This includes appropriate patient paperwork, availability of interpreters, translation support within a timely manner, institutional signage, and community outreach. Advocacy on the state and national levels is also necessary as bilingual assessments can often take longer including building of rapport, assessment of acculturation and language proficiency, assessment in two languages, the length of time that it takes to integrate and translate reports, and the time that it takes to debrief interpreters. This results in evaluations that are more expensive and can impact access to care. As such advocacy with insurance companies is also necessary.

REFERENCES

American Academy of Clinical Neuropsychology. (2016). Relevance 2050 Initiative. https://theaacn.org/relevance-2050-initiative/#gsc.tab=0

Artiola i Fortuny, L. (2009). Research and practice: Ethical issues with immigrant adults and children. In J. E. Morgan & J. H. Ricker (Eds.), *Textbook of clinical neuropsychology* (pp. 960–981). New York, NY: Taylor & Francis.

Bates, E., & MacWhinney, B. (1989). Functionalism and the competition model. In B. MacWhinney & E. Bates (Eds.), *The crosslinguistic study of sentence processing* (pp. 3–73). New York, NY: Cambridge University Press.

Barac, R., Bialystok, E., Castro, D. C., & Sanchez, M. (2014). The cognitive development of young dual language learners: A critical review. *Early Childhood Research Quarterly, 29*, 699–714. https://doi.org/10.1016/j.ecresq.2014.02.003

Bush, S. S., Ruff, R. M., Troster, A. I., Barth, J. T., Koffler, S. P., Pliskin, N. H., . . . Silver, C. H. (2005). Symptom validity assessment: Practice issues and medical necessity. *Archives of Clinical Neuropsychology, 20*, 419–426. https://doi.org/ 10.1016/j.acn.2005.02.002

Byers-Heinlein, K., Burns, T. C., & Werker, J. F. (2010). The roots of bilingualism in newborns. *Psychological Science, 21*, 343–348. https://doi.org/10.1177%2F0956797609360758

Byrd, D., Arentoft, A., Scheiner, D., Westerveld, M., & Baron, I. S. (2008). State of multicultural neuropsychological assessment in children: Current research issues. *Neuropsychology Review, 18,* 214–222. https://doi.org/10.1007/s11065-008-9065-y

Castro, D., Espinosa, L. & Paez, M. (2011). Defining and measuring quality in early childhood practices that promote dual language learners' development and learning. In M. Zaslow, I. Martinez-Beck, K. Tout, & T. Halle (Eds.), *Quality measurement in early childhood settings* (pp. 130–145). Baltimore, MD: Paul H. Brookes.

Chin, A., Negash, S., & Hamilton, R. (2011). Diversity and disparity in dementia: The impact of ethnoracial differences in Alzheimer disease. *Alzheimer Disease and Associated Disorders, 25,* 187–195. https://org/10.1097/WAD.0b013e318211c6c9

Collier, V. P. (1995). Acquiring a second language for school. *Directions in Language and Education, 1,* 1–10.

Cormier, D. C., McGrew, K. S., & Evans, J. J. (2011). Quantifying the "degree of linguistic demand" in spoken intelligence test directions. *Journal of Psychoeducational Assessment, 29,* 515–533. https://doi.org/10.1177/0734282911405962

Costa, A. & Santesteban, M. (2004). Lexical access in bilingual speech production: Evidence from language switching in highly proficient bilinguals and L2 learners. *Journal of Memory and Language, 50,* 491–511. https://doi.org/10.1016/j.jml.2004.02.002

Cummins, J. (1984). *Bilingualism and special education: Issues in assessment and pedagogy.* Clevedon, England: Multilingual Matters.

Cummins, J. (1999). BICS and CALP: Clarifying the distinction (ERIC Document Reproduction Service No. ED438551). Retrieved from https://files.eric.ed.gov/fulltext/ED438551.pdf

Davies, M. S., Strickland, T. L., & Cao, M. (2014). Neuropsychological evaluation of culturally diverse populations. In F. T. L. Leong, L. Comas-Díaz, G. C. Nagayama Hall, V. C. McLoyd, & J. E. Trimble (Eds.), *APA handbook of multicultural psychology: Vol. 2. Applications and training* (pp. 231–251). Washington, DC: American Psychological Association.

Echemendia, R. J., Harris, J. G., Congett, S. M., Diaz, L. M., & Puente, A. (1997). Neuropsychological training and practices with Hispanics: A national survey. *The Clinical Neuropsychologist, 11,* 229–243. https://doi.org/10.1080/13854049708400451

Espinosa, L. (2015). *Getting it right for young children from diverse backgrounds: Applying research to improve practice with a focus on dual language learners* (2nd ed). Upper Saddle River, NJ: Pearson.

Espinosa, L. (2016, May 11). The science of dual language learning for children birth through age five: Effective practices that improve outcome [Webinar]. Early Childhood Investigations. Retrieved from https://www.earlychildhoodwebinars.com/webinars/14008/

Espinosa, L, LaForett, D. R., Burchinal, M., Winsler, A. Tien, H-C., Peisner-Feinberg, E., & Castro, D. (2017). Child care experiences among dual language learners in the United States: Analyses of the early childhood longitudinal survey—Birth cohort. *AERA Open, 3*(2), 1–15. https://doi.org/10.1177/2332858417699380

Ferraro, F. R. (Ed.). (2016). *Minority and cross-cultural aspects of neuropsychological assessment: Enduring and emerging trends* (2nd ed). New York, NY: Taylor and Francis.

French, C., & Llorente, A. M. (2008). Language: Development, bilingualism, and abnormal states (pp. 78–120). In A. M Llorente (Ed.), *Principles of neuropsychological assessment with Hispanics: Theoretical foundations and clinical practice* (pp. 78–120). New York, NY: Springer.

Genesee, F. (1989). Early bilingual development: One language or two? *Journal of Child Language, 16,* 161–180. https://doi.org/10.1017/S0305000900013490

Genesee, F., Paradis, J., & Crago, M. (2004). *Dual language development and disorders: A handbook on bilingualism and second language learning.* Baltimore, MD: Brookes.

Green, D. W. (1998). Mental control of the bilingual lexico-semantic system. *Bilingualism: Language and Cognition, 1,* 67–81. https://doi.org/10.1017/S1366728998000133

Grosjean F. (1989). Neurolinguists, beware! The bilingual is not two monolinguals in one person. *Brain and Language, 36,* 3–15. https://doi.org/10.1016/0093-934X(89)90048-5

Groth-Marnat, G. (2009). *Handbook of Psychological Assessment* (5th ed.). New York, NY: Wiley.

Hambleton, R. K. (2001). The next generation of the ITC Test Translation and Adaptation Guidelines. *European Journal of Psychological Assessment, 17,* 164–172. https://doi.org/10.1027//1015-5759.17.3.164

Hambleton, R. K. (2005). Issues, designs, and technical guidelines for adapting tests into multiple languages and cultures. In R. K. Hambleton, P. F. Merenda, & C. D. Spielberger (Eds.), *Adapting educational and psychological tests for cross-cultural assessment* (pp. 3–38). Mahwah, NJ: Erlbaum.

Hambleton, R. K., Yu, J., & Slater, S. C. (1999). Field test of the ITC guidelines for adapting educational and psychological tests. *European Journal of Psychological Assessment, 15,* 270. https://doi.org/10.1027//1015-5759.15.3.270

Hambly, C., & Fombonne, E. (2012). The impact of bilingual environments on language development in children with autism spectrum disorders. *Journal of Autism and Developmental Disorders, 42,* 1342–1352. https://doi.org/10.1007/s10803-011-1365-z

Hebben, N., & Milberg, W. (2002). *Essentials of neuropsychological assessments.* New York, NY: Wiley.

Hoff, E., Core, C., Place, S., Rumiche, R., Señor, M., & Parra, M. (2011). Dual language exposure and early bilingual development. *Journal of Child Language, 39,* 1–27. https://doi.org/10.1017/S0305000910000759

Hull, R., & Vaid, J. (2007). Bilingual language lateralization: A meta-analytic tale of two hemispheres. *Neuropsychologia, 45,* 1987–2008. https://doi.org/10.1016/j.neuropsychologia.2007.03.002

Judd, T., Capetillo, D., Carrión-Baralt, J., Mármol, L. M., San Miguel-Montes, L., Navarrete, M. G., . . . NAN Policy and Planning Committee. (2009). Professional considerations for improving the neuropsychological evaluation of Hispanics: A National Academy of Neuropsychology Education Paper. *Archives of Clinical Neuropsychology, 24,* 127–135. https://doi.org/10.1093/arclin/acp016

Jusczyk, P. (1997). *The discovery of spoken language.* Cambridge, MA: MIT Press.

Kominski, R. A., Shin, H. B., & Marotz, K. (2008). Language needs of school-age children [Poster]. *U.S. Census Bureau.* Retrieved from http://www2.census.gov/library/working-papers/2008/demo/2008_Kominski-Shin-Marotz-poster.pdf

Krashen, S. D. (Ed.). (1982). *Child–adult differences in second language acquisition.* Rowley, MA: Newbury House.

Kraybill, D. B., & Gilliam, J. M. (2012). Culturally competent safety interventions for children in Old Order Anabaptist communities. *Journal of Agromedicine, 17,* 247–250. https://doi.org/10.1080/1059924X.2012.658303

Kuhl, P. K. (1993). Developmental speech perception: Implications for models of language impairment. In P. Tallal, A. M. Galaburda, R. R. Llinás, & C. von Euler (Eds.), *Temporal information processing in the nervous system.* (pp. 248–263). New York: The New York Academy of Sciences. https://doi.org/10.1111/j.1749-6632.1993.tb22973.x

Liaw, F. R., & Brooks-Gunn, J. (1994). Cumulative familial risk and low-birthweight children's cognitive and behavioral development. *Journal of Clinical Child Psychology, 23,* 360–372. https://doi.org/10.1207/s15374424jccp2304_2

MacWhinney, B. (2005). A unified model of language acquisition. In J. F Kroll & A. M. B. de Groot (Eds.), *Handbook of bilingualism: Psycholinguistic approaches* (pp. 49–67). New York, NY: Oxford University Press.

Manly, J. J. (2008). Critical issues in cultural neuropsychology: Profit from diversity. *Neuropsychological Review, 18,* 179–183. https://doi.org/10.1007/s11065-008-9068-8

Meisel, J. M. (1994). Code-switching in young bilingual children: The acquisition of grammatical constraints. *Studies in Second Language Acquisition, 16,* 413–441. https://doi.org/10.1017/S0272263100013449

Mushi, S. L. (2002). Acquisition of multiple languages among children of immigrant families: Parents' role in the home-school language pendulum. *Early Child Development and Care, 172,* 517–530. https://doi.org/10.1080/03004430214546

Nieves-Brull, A. I. (2006). *Evaluation of the culture-language matrix: A validation study of test performance in monolingual English speaking and bilingual English/Spanish speaking populations* (Unpublished doctoral dissertation). St. John's University, New York, NY.

Noble, K. G, McCandliss, B. D., & Farah, M. J. (2007). Socioeconomic gradients predict individual differences in neurocognitive abilities. *Developmental Science, 10,* 464–480. https://doi.org/10.1111/j.1467-7687.2007.00600.x

Noble, K. G., Norman, M. F., & Farah, M. J. (2005). Neurocognitive correlates of socioeconomic status in kindergarten children. *Developmental Science, 8,* 74–87. https://doi.org/10.1111/j.1467-7687.2005.00394.x

Ortiz, S. O., & Dynda, A. M. (2005). Use of intelligence tests with culturally and linguistically diverse populations. In D. P. Flanagan & P. H. Harrison (Eds.), *Contemporary intellectual assessment: Theories, tests, and issues* (2nd ed., pp. 545–556). New York, NY: Guilford.

Paradis, M. (2004). *An integrated neurolinguistic theory of bilingualism*. Philadelphia, PA: John Benjamins.

Paradis, J., & Genesee, F. (1997). On continuity and the emergence of functional categories in bilingual first-language acquisition. *Language Acquisition, 6,* 91–124. https://doi.org/10.1207/s15327817la0602_1

Pearson, B. Z. (1998). Assessing lexical development in bilingual babies and toddlers. *International Journal of Bilingualism, 2,* 347–372. https://doi.org/10.1177/136700699800200305

Pew Research Center. (2013, September 15). 2013 survey of Hispanics [Data file]. Retrieved from https://www.pewhispanic.org/2015/09/15/2013-survey-of-hispanics/

Pontón, M. O., & Leon-Carrion, J. (Eds.). (2001). *Neuropsychology and the Hispanic patient: A clinical handbook*. Mahwah, NJ: Erlbaum.

Puente, A. E., & McCaffrey, R. J. (1992). *Psychobiological variables in neuropsychological assessment*. New York, NY: Plenum.

Reynolds, C. R., & Suzuki, L. (2013). Bias in psychological assessment: An empirical review and recommendations. In J. R. Graham, J. A. Naglieri, & I. B. Weiner (Eds.), *Handbook of psychology: Vol. 10. Assessment psychology.* (2nd ed., pp. 82–113). Hoboken, NJ: Wiley.

Rhodes, R. L, Ochoa, S. H., & Ortiz, S. O. (2005). *Assessing culturally and linguistically diverse students: A practical guide*. New York, NY: Guilford.

Romero, H., Lageman, S., K., Kamath, W., Irani, F., Sim, A., Suarez, P., . . . Summit Participants. (2009). Challenges in the neuropsychological assessment of ethnic minorities: Summit proceedings. *The Clinical Neuropsychologist, 23,* 761–769. https://doi.org/10.1080/13854040902881958

Rumberger, R. W. (2004). What can be done to reduce school dropouts? In G. Orfied (Ed.), *Dropouts in America: Confronting the graduation rate crisis* (pp.243–254). Cambridge, MA: Harvard Education Press.

Salinas, C, M., Bordes Edgar, V., & Puente, A. (2016). Barriers and practical approaches to neuropsychological assessment of Spanish speakers. In F. R. Ferraro (Ed.), *Minority and cross-cultural aspects of neuropsychological assessment: Enduring and emerging trends* (2nd ed., pp. 229–258). New York, NY: Taylor and Francis.

Sattler, J. M. (2008). *Assessment of children: Cognitive foundations* (5th ed.). San Diego, CA: Author.

Semrud-Clikeman, M., Romero, R. A. A., Prado, E. L., Shapiro, E. G., Bangirana, P., & John, C. C. (2017). Selecting measures for the neurodevelopmental assessment of children in low-and middle-income countries. *Child Neuropsychology, 23,* 761–802. https://doi.org/10.1080/09297049.2016.1216536

Solano-Flores, G. (2008). Who is given tests in what language by whom, when, and where? The need for probabilistic views of language in the testing of English language learners. *Educational Researcher, 37,* 189–199. https://doi.org/10.3102/0013189X08319569

Suzuki, L. A., Naqvi, S., & Hill, J. S. (2013). Assessing intelligence in a cultural context. In F. T. L. Leong (Ed.), *APA handbook of multicultural psychology: Theory and research* (Vol. 1, pp. 247–266). Washington, DC: American Psychological Association.

Suzuki, L. A., & Ponterotto, J. G. (Eds.). (2008). *Handbook of multicultural assessment: Clinical, psychological, and educational applications.* New York, NY: Wiley.

Thordardottir, E. (2010). Towards evidence-based practice in language intervention for bilingual children. *Journal of Communication Disorders, 43,* 523–537. https://doi.org/10.1016/j.jcomdis.2010.06.001

Tsybina, I., & Eriks-Brophy, A. (2010). Bilingual dialogic book-reading intervention for preschoolers with slow expressive vocabulary development. *Journal of Communication Disorders, 43,* 538–556. https://doi.org/10.1016/j.jcomdis.2010.05.006

U.S. Census Bureau. (2006). *We the people: American Indians and Alaska Natives in the United States.* Washington, DC: Author.

U.S. Census Bureau. (2009). [Data file]. Retrieved from http://www.census.gov/data/tables/2013/demo/2009-2013-lang-tables.html

U.S. Census Bureau. (2014). Table B16001: Language spoken at home by ability to speak English for the population 5 years and over, 2010–2014. *American Community Survey 5 Year Estimates.* Retrieved from https://factfinder.census.gov/faces/tableservices/jsf/pages/productview.xhtml?src=bkmk/

U.S. Census Bureau. (2015). Current population survey: Annual social and economic (ASEC) supplement. Retrieved from https://www2.census.gov/programs-surveys/cps/techdocs/cpsmar15.pdf

Vaid, J. (2008). The bilingual brain: What is right and what is left? In J. Altarriba & R. R. Heredia (Eds.), *An introduction to bilingualism: Principles and processes* (pp. 129–141). New York, NY: Psychology Press.

Verney, S. P., Bennett, J., & Hamilton, J. M. (2016). Cultural considerations in the neuropsychological assessment of American Indians/Alaska natives. In F. R. Ferraro (Ed.), *Minority and cross-cultural aspects of neuropsychological assessment: Enduring and emerging trends* (2nd ed., pp. 115–158). New York, NY: Taylor and Francis.

Werker, J. F., & Desjardins, R. N. (1995). Listening to speech in the first year of life: Experiential influences on phoneme perception. *Current Directions in Psychological Science, 4,* 76–81. https://doi.org/10.1111/1467-8721.ep10772323

11

Racial Disparities in Alzheimer's Disease

Biological, Social, and Methodological Considerations

MEGAN ZUELSDORFF, LISA L. BARNES, AND
OZIOMA C. OKONKWO

INTRODUCTION

Alzheimer's disease (AD) represents a pressing public health problem not only for the older adults most likely to experience it, but also for caregivers and families, healthcare providers, and policymakers. The population of the United States is aging as the baby boom generation matures. Currently there are approximately 46 million Americans over the age of 65; by 2050, that number will nearly double, to an estimated 83.7 million people. The proportion of those over the age of 85 will also swell, from just 2% of the population as a whole to approximately 4.5%. An aging population brings with it complex medical, economic, and social issues. Increasingly prevalent neurodegenerative diseases like AD and chronic multimorbidities will be associated with rising healthcare utilization and costs as well as new formal and informal caregiving needs. These challenges will be acutely felt, and likely exacerbated, among minority groups within the United States, many of whom are arguably underserved by the healthcare systems currently in place despite representing an increasing segment of the population as the United States grows more ethnically diverse. While 80% of those currently over the age of 65 are non-Hispanic White, by 2050 that proportion will shrink to 60%. Among those aged 85 and older, a third will be a racial or ethnic minority.

The demographic shifts in both age and racial composition bring critical challenges for minority populations who are at increased risk for many of the common chronic conditions affecting older adults, including AD. Such shifts also represent a crucial moment for researchers seeking to understand and maximize healthy aging trajectories across not only racial and ethnic groups but socioeconomic strata as well; in the United States particularly, race and socioeconomic status remain deeply and nearly inextricably intertwined. Because historically disadvantaged populations are at increased risk for poor health, they are more likely to bear a substantially greater portion of the associated economic and social burden. For example, several studies demonstrate that African Americans show increased vulnerability to cognitive impairment, but access treatment and formal caregiving less frequently than do other populations.

Although evidence for racial disparities in AD is accumulating, most of the research that informs current knowledge of diagnosis, risk and protective mechanisms, and disease management has been done in overwhelmingly non-Hispanic White samples. In the current chapter, we present a summary of the growing body of evidence for racial disparities in AD and the biological and social conditions that drive them and provide a discussion on the key methodological challenges facing the field in this area.

DESCRIPTIVE EPIDEMIOLOGY OF ALZHEIMER'S DISEASE

Definitions and Diagnosis

AD accounts for 60% to 80% of diagnosed dementia cases. While the most current diagnostic criteria incorporate AD biomarkers (Jack et al., 2011), it is important to recognize that such additions are intended primarily for research purposes. The public health problem of AD remains centered around the clinical syndrome of impairment that ultimately presents itself and is often, but not always, correlated with underlying pathology. Dementia diagnosis requires cognitive impairment in at least two cognitive domains, and cognition may be measured by either sensitive measures such as a neuropsychological test battery or by briefer screening tools such as the Mini-Mental Status Exam (MMSE). For a clinical diagnosis of AD, impairment must be severe enough to cause functional limitations in daily activities. Diagnosis and subsequent management of AD can occur in one of several settings: by a multidisciplinary team in a comprehensive AD center, by a specialist such as a neurologist or geriatrician, or, increasingly, by a primary care physician (Cho et al., 2014).

Prevalence and Incidence Rates by Ethnicity

An estimated 5.4 million Americans, or one of every nine adults over the age of 65, have a diagnosis of AD; approximately 475,000 new cases are identified each year. While studies have varied in rigor and estimates have varied in magnitude, it is widely accepted that prevalence of AD is approximately two times higher in African Americans than in non-Hispanic Whites (Demirovic et al., 2003). The evidence on incidence rates is also mixed. A seminal study of AD incidence and ethnicity found that in a large sample of Medicare recipients in New York City, seven-year risk of developing AD was approximately twice as high in non-Hispanic African American elders as in non-Hispanic Whites (Tang et al., 2001), and this estimate remains the most widely cited. More recently, some research groups have reported risk differences of similar magnitude (Potter et al., 2009) while others have reported fewer, although still notable, disparities (Mayeda, Glymour, Quesenberry, & Whitmer, 2016; Shadlen et al., 2006). African Americans are also at increased risk for mild cognitive impairment (Katz et al., 2012; Manly et al., 2008), and conversion from mild cognitive impairment to AD, as compared to non-Hispanic Whites (Manly et al., 2008). Notably, a few studies have found that such differences in dementia risk are completely explained by population-level ethnic disparities in socioeconomic factors such as educational attainment (Fillenbaum et al., 1998; Katz et al., 2012; Yaffe et al., 2013). While the varying sampling, measurement, and analytic strategies employed may contribute to mixed findings on the magnitude of between-group differences and the individual contribution of race as a construct, there is a general consensus that African Americans bear a greater burden of AD and related dementias than any other racial group.

DISPARITIES IN COGNITIVE FUNCTION AND DECLINE

Because AD is generally diagnosed by cognitive test performance below a diagnostic threshold, either in an extensive neuropsychological battery or on a brief screener, the epidemiology of cognitive function and age-related change that does not yet meet diagnostic criteria is relevant in understanding risk and disparities in AD and other dementias. Accordingly, many observational studies have examined cognitive test performance and decline over time, rather than presence or absence of disease, as an outcome of interest.

Fluid cognitive abilities, such as those in the domains of episodic memory and executive function, are particularly vulnerable to change during the latter half of the life course, while crystallized abilities remain stable until late in life or in the more advanced stages of dementia (Harada, Natelson Love, & Triebel, 2013). Unlike crystallized abilities, fluid cognitive abilities are believed to be more innate and less influenced by cultural factors and education (Horn & Cattell, 1966). Despite this assumption, there are clear disparities in cognitive testing of these abilities between persons in different socioeconomic and racial strata (Karlamangla et al., 2009; Yaffe et al., 2013). In studies of between-group differences in cognition, African Americans perform more poorly than do non-Hispanic Whites on various cognitive measures and across the majority of established cognitive domains (Gross et al., 2015). This disparity can be seen at all ages, although some studies have found that cross-sectional differences do become attenuated in late life (Karlamangla et al., 2009; Zahodne, Manly, Azar, Brickman, & Glymour, 2016).

Clinical definitions of AD emphasize gradual within-individual cognitive declines over time. Thus, longitudinal research, where change can be directly observed and where participants serve as their own controls, is key in characterizing cognitive health disparities. While there are consistent cross-sectional race-related differences in cognitive test performance and risk for AD, longitudinal studies provide mixed evidence regarding racial differences in *rates* of cognitive decline (Gross et al., 2015). A number of studies have found there to be no difference at all (Castora-Binkley, Peronto, Edwards, & Small, 2015; Early et al., 2013; Masel & Peek, 2009), potentially highlighting the importance of peak midlife cognitive function in determining the age at which a diagnostic threshold is met (Karlamangla et al., 2009). Some studies have found that rate of decline is actually slower among African Americans than among non-Hispanic White participants (Early et al., 2013; Karlamangla et al., 2009; Wilson, Capuano, Sytsma, Bennett, & Barnes, 2015). Despite expectations, given the increased risk for incident disease, only a limited number of studies have found faster rates of decline among racial minorities (Lee et al., 2012; Sawyer, Sachs-Ericsson, Preacher, & Blazer, 2009; Wolinsky et al., 2011). Effect sizes for race, where significant differences have been found, have been quite small as compared to the large effects found for cross-sectional function. As a whole, available evidence suggests that there are probably not strong differences by race in rate of cognitive decline (Barnes & Bennett, 2014).

Efforts to explain differentiating cognitive trajectories, and the path to dementia diagnosis, continue. Some of the discrepant findings on rates of decline may be explained by significant methodological challenges in disparities research, discussed later in this chapter. Further, even subtle differences in neuropathological profiles may have implications for dementia presentation and prognosis.

RACIAL DIFFERENCES IN AD PATHOLOGY AND CLINICOPATHOLOGICAL ASSOCIATIONS

Within the extensive body of research on the neuropathological underpinnings of AD and other dementias, racial minorities remain underrepresented in studies utilizing neuroimaging, biomarker, and postmortem data. As such, exploration of a role for race in clinico-pathological relationships is still relatively nascent. Very early autopsy studies of neuropathological differences by race were severely limited by their small sample sizes, and varying sample characteristics made comparability of findings difficult. One of the first studies, in a convenience sample of brains taken from older patients who died at a large hospital with unknown dementia status, found no significant differences in either type of AD lesions, although they did find that plaques and tangles were slightly more likely to co-occur in the brains of African American patients (Miller, Hicks, D'Amato, & Landis, 1984). A study of 100 elderly deceased persons, limited to those without dementia at the time of death, found that AD lesions were more prevalent in non-Hispanic White patients than in African American patients (de la Monte, Hutchins, & Moore, 1989). Slightly later studies, moving beyond hospital-based samples to explore presence of Alzheimer's lesions among forensic autopsy brains, also found no evidence for racial difference (Riudavets et al., 2006; Sandberg, Stewart, Smialek, & Troncoso, 2001). Notably, these null findings for community-based populations were echoed in one of the first Alzheimer's Disease Research Center-based autopsy studies exploring racial discrepancies in rigorously diagnosed and profiled AD patients (Wilkins, Grant, Schmitt, McKeel, & Morris, 2006).

The early autopsy study evidence demonstrating that few if any ethnoracial differences in neuropathology existed was corroborated by early neuroimaging data. In a study of African Americans with diagnosed dementia, magnetic resonance imaging (MRI) and computerized tomography revealed cerebral atrophy of similar severity to that previously seen in predominantly White samples (Charletta, Gorelick, Dollear, Freels, & Harris, 1995); investigators on an MRI study comparing hippocampal volumes in healthy versus demented African American elders came to similar conclusions (Sencakova et al., 2001). And while these studies were somewhat limited by the same small, nongeneralizable samples as concurrent autopsy studies, their findings are bolstered by a much larger imaging study of a community-based population that revealed comparable ventricle-to-brain ratios in African American and non-Hispanic White participants (Longstreth et al., 2000).

The general consensus that neuropathological differences were not behind racial disparities in cognitive impairment and AD-related dementia has slowly evolved in the wake of increasing recognition of the role for mixed dementia pathologies, particularly infarcts, in many clinically diagnosed cases of AD (Pytel et al., 2006; Schneider, Arvanitakis, Bang, & Bennett, 2007; White et al., 2005). There is evidence that mixed pathology is more strongly associated with clinical manifestation of cognitive impairment and dementia than are singular pathologies (Lo, Jagust, & Alzheimer's Disease Neuroimaging, 2012) and that mixed pathology is more common in African Americans (Barnes et al., 2015). Over time, autopsy and neuroimaging studies have demonstrated racial differences in cardiovascular pathologies, with African American race most consistently associated with the presence of infarcts and white matter hyperintensities (Brickman et al., 2008; Gottesman, Fornage, Knopman, & Mosley, 2015) in both demented and healthy brains. In one recent autopsy study that thoroughly characterized pathology profiles of demographically matched African American and non-Hispanic White AD patients, African American decedents were far more likely to show mixed pathology, as expected; interestingly, the most common mixed pathology was not AD with infarcts, but AD with Lewy bodies, followed by AD with Lewy bodies and infarcts (Barnes et al., 2015).

Racial disparities in diagnosed AD may also be in part due to between-group differences in clinicopathological associations, with pathological brain changes more strongly related to symptom manifestation among African Americans. Although within-group examination of brain–cognition relationships have been difficult due to power limitations stemming from the lack of ethnic diversity in neuropathology study samples, there is preliminary direct evidence that race modifies the association between AD-related brain changes and cognitive decline. One recent analysis within the Washington Heights–Inwood Columbia Aging Project (WHICAP) sample, for instance, found that larger hippocampal volumes were associated with better memory in non-Hispanic White but not African American participants and that white matter hyperintensities were not only more common but also more strongly associated with executive function and language impairment in African Americans (Zahodne et al., 2015). Similarly, in another large, nondemented northern Manhattan sample, amyloid-β (Aβ) burden was associated with accelerated executive function declines over the previous decade among African American but not non-Hispanic White participants (Gu et al., 2015). Notably, these newer findings contradict one earlier study, which failed to find any race–neuropathology interactions (DeCarli et al., 2008). The findings on clinicopathological associations by racial group highlight the need to not only replicate within-group analyses in additional diverse, adequately powered samples but also to more rigorously explore the differential risk profiles that underpin racial effects.

RACE AND RISK FACTORS IN AD AND BRAIN AGING

Race is a commonly assessed variable in medical research, and its effects on health outcomes often examined with an implication of direct biological significance. In reality, however, race is a largely social construct; in African Americans particularly, the influence of ancestry is arguably eclipsed by the influence of lived experience (Williams, 1998). Race, when used to predict complex, later-life health outcomes like AD, is better characterized as a risk marker than a risk factor (Kaplan & Bennett, 2003); that is, it is *associated* at a population level with a host of risk factors that over the life course will impact incidence and progression of impairment but has little independent direct effect on brain and cognitive changes. In this section, we describe a number of those potentially race-associated risk factors. When considering the implications of race as a risk marker, it is important to recognize that even seemingly unrelated factors are rarely operating independently. Instead, they cluster in the most vulnerable communities, accumulating and interacting to potentiate damage and risk. Ultimately, disease delay and prevention is likely to be the most critical component in reducing ethnoracial disparities in AD and related dementias; understanding the impact and interplay of established risk factors within African American populations, then, is key for developing strategies to modify and mitigate that risk burden.

Genetics

Other than age, the most well-established and nonmodifiable risk factor for AD is a genetic polymorphism involving the apolipoprotein E (*APOE*) gene, located on chromosome 19. There are three alleles associated with *APOE*: ε2, ε3, and ε4. Numerous studies across many samples have demonstrated that the presence of one or more *APOE* ε4 alleles is associated with AD and earlier cognitive declines (Evans et al., 1997; Scarabino, Gambina, Broggio, Pelliccia, & Corbo, 2016); there is good evidence that an ε4-related acceleration of amyloid plaque and neurofibrillary tangle accumulation may be behind this association (Yu, Tan, & Hardy, 2014). However, genetic risk for AD has been established using predominantly or exclusively White samples, and the relatively scarce data available for non-White samples have been much more mixed. The *APOE* ε4 allele appears to be more prevalent in African Americans than in non-Hispanic Whites. However, despite evidence from non-Hispanic White cohorts demonstrating that the ε4 allele potentiates neurodegenerative pathologies (Tsuang et al., 2005; Wang et al., 2015), it has been only inconsistently linked with cognitive outcomes in African American populations (Evans et al., 2003; Murrell et al., 2006). The most recent evidence, from a genome-wide association study in a large African American sample, suggests that a second gene, *ABCA7*, may be more strongly associated with AD risk in African Americans (Reitz et al., 2013). While more research on genetic determinants of cognition and clinicopathologic relationships in non-White populations is certainly needed, there is general consensus that differential genetic risk patterns do not fully account for the increased dementia burden borne by African Americans.

Vascular Conditions

As noted, many studies have found greater prevalence of mixed AD and vascular pathologies among older African Americans than their White counterparts. This is not surprising given evidence that African Americans are more than twice as likely as non-Hispanic Whites to experience stroke (Trimble & Morgenstern, 2008). Stroke risk is important in the exploration of AD disparities both because resultant vascular dementia may be misdiagnosed as AD and because damage sustained from a stroke or from additional silent infarcts may diminish the brain's ability to compensate for existing or future AD pathologies (Moroney et al., 1997). Hypertension has been associated with vascular pathologies, AD pathologies, and cognitive dysfunction including dementia (Perrotta, Lembo, & Carnevale, 2016) and is also generally believed to be more common among African Americans than Whites (Redmond, Baer, & Hicks, 2011). However, blood pressure trajectories change over the course of aging, and in older samples, differences in hypertensive status by race have not always been seen (Tang et al., 2001). One recent, large-scale observational study demonstrated that midlife hypertension was associated with faster rates of cognitive decline over the ensuing 20-year period and, further, that the association was weaker among African Americans (Gottesman et al., 2014). Other recent studies in multiethnic clinical samples have reported that hypercholesterolemia, but not hypertension, is associated with poorer cognitive function (Carvalho et al., 2015) and rate of AD progression (Helzner et al., 2009).

Metabolic dysfunction is receiving increased attention as a potential factor in cognitive health disparities. Type 2 diabetes mellitus (DM) is an established risk factor for cognitive decline (Rajan et al., 2016) and AD (Ott et al., 1999). Recent studies have additionally demonstrated that insulin resistance, a physiological precursor to DM that often goes undetected, is associated with poorer cognitive function and cerebrospinal fluid markers of preclinical AD in healthy middle-aged adults (Hoscheidt et al., 2016) as well as risk for AD in older adults (Hildreth, Van Pelt, & Schwartz, 2012). African Americans are at notably increased risk for both DM (Menke, Casagrande, Geiss, & Cowie, 2015; Romero, Romero, Shlay, Ogden, & Dabelea, 2012) and insulin resistance (Hasson, Apovian, & Istfan,

2015) as compared to their non-Hispanic White counterparts. Additionally, at a population level, they develop the conditions earlier and are less likely to achieve optimal management (Menke et al., 2015), and this may be particularly relevant for cognitive aging in African American populations. Studies have shown that while there are no race-based differences in rate of cognitive change following incident DM, it is midlife rather than late-life metabolic disorder that most strongly predicts later cognitive dysfunction (Rajan et al., 2016), and poorer glucose control is independently associated with accelerated decline (Yaffe et al., 2012). Interestingly, a recent systematic review of the literature revealed that in study samples of patients with diagnosed AD, comorbid DM is often associated with a slower rate of cognitive decline (Li, Cesari, Liu, Dong, & Vellas, 2017). Unexpected findings such as this may reflect heterogeneous pathological profiles cloaked by a homogenizing clinical AD diagnosis. Indeed, available evidence from autopsy studies suggests that DM is associated with cerebral infarcts but not typical AD pathology (Arvanitakis et al., 2006; Beeri et al., 2005), and some researchers have theorized that underlying mixed pathology explains decreased rates of decline seen in research participants with AD and DM (Sanz et al., 2009). If, as available evidence suggests, metabolic disorders play a specific role in clinical presentation of mixed pathology in AD, future research on such pathways in larger African American samples may ultimately be crucial in efforts to understand cognitive health disparities.

Educational Attainment

While progressive neuropathological burden is broadly associated with worsening cognitive impairment, there is much heterogeneity within this relationship at an individual level; in fact, the presence of neuropathology is not invariably associated with a decline in cognitive function. This functional resilience is known as *cognitive reserve* and may be an important determinant of ethnoracial differences in later-life cognitive trajectories. The cognitive reserve hypothesis posits that sets of learned skills and behavioral patterns developed over the lifespan may allow some individuals to better cope with age-related damage in the brain (Stern, 2009). Cognitive reserve is conceptual in nature and cannot be measured directly, but a substantial body of literature has established several strong, commonly assessed reserve markers including high educational attainment (Almeida et al., 2015; Bennett et al., 2003), increased occupational complexity (Boots et al., 2015), and engagement in mentally stimulating leisure activities (Schultz et al., 2014). Notably, many traditional markers of cognitive reserve are also markers of socioeconomic advantage. Furthermore, in the United States, markers of socioeconomic status, including educational and occupational attainment, are very strongly associated with race and ethnicity (Glymour & Manly, 2008). African Americans are more likely than non-Hispanic Whites to leave the educational system without a high school diploma and are less likely to attend a four-year college (McDaniel, DiPrete, Buchmann, & Shwed, 2011); accordingly, they are also less likely to enter into high-status occupations (Pais, 2014).

If cognitive reserve is an important determinant of cognitive function in late life, racial differences in reserve-building opportunities, particularly in higher education access, might very well underlie the observed racial disparities in cognitive outcomes even where neuropathological profiles are similar. Interestingly, however, the relationship between race and cognitive dysfunction often withstands control for not only the cardiovascular and other health factors described earlier in this chapter, but also for years of education (Mayeda et al., 2014; Schwartz et al., 2004). This finding may reflect the inadequacy of using markers validated in predominantly White samples to understand health outcomes in racial minorities, or it may highlight differences in educational quality between African Americans and non-Hispanic Whites. It has been well-documented that receipt of low-quality education with few resources limits future access to mainstream resources needed for success, like higher paying occupations and admission to good colleges. In fact, careful analysis of the education–dementia relationship in large, truly biracial samples has shown that educational *quality*, as measured by literacy level or geographic characteristics of the schooling environment, is a much better predictor of cognitive trajectory than is educational *quantity* (Sisco et al., 2015). While it is still unclear whether the role for educational quality supersedes that of cerebrovascular risk in fomenting cognitive health disparities, as some researchers have claimed (Carvalho et al., 2015), the findings on quality versus quantity serve as an important reminder that social structures that differentially impact particular subpopulations—not only the Jim Crow South, but more current inequities in educational resources as well—can greatly shift or nullify the meaning of even the strongest traditional risk factor.

Psychosocial Factors

There is a substantial body of research on psychosocial risk and protective factors for cognitive aging. Most of these studies have been done in predominantly White samples, and few have focused on within-group analyses designed to acknowledge between-group heterogeneity in relationships. Among those studies that have explored race-specific relationships, findings have varied by the psychosocial factor in question. Depressive symptomatology (Wilson, Mendes De Leon, Bennett, Bienias, & Evans, 2004) and perceived stress (Aggarwal et al., 2014) are associated with faster cognitive decline to a similar degree in both African Americans and non-Hispanic Whites. One study utilizing a large biracial sample showed that larger social networks are protective against cognitive decline regardless of race, but in the same sample, researchers demonstrated that self-reported social engagement was only protective in non-Hispanic Whites (Barnes, Mendes de Leon, Wilson, Bienias, & Evans, 2004). And while it is generally accepted that early life adversity increases risk for cognitive dysfunction and accelerated brain aging, two studies that rigorously examined race-specific effects found unexpected relationships: while in both studies African Americans reported greater number of early adverse experiences, one group found that adversity did, in fact, partially mediate the race-cognitive impairment relationship (Zhang, Hayward, &

Yu, 2016), while the other found that in African Americans such adversity was actually associated with slowed decline (Barnes et al., 2012). The available evidence suggests that many more within-group analyses, ideally including histological and biomarker data, will be necessary before we fully understand the role for psychosocial factors in ethnicity-based cognitive health disparities.

METHODOLOGICAL CHALLENGES

In an exploration of racial differences in cognitive aging and dementia, the importance of persisting methodological challenges cannot be overstated. Descriptive and analytical epidemiology of a given health outcome relies upon accurate and equitable measurement of both that outcome and its relevant risk factors. Disparities in cognitive health are particularly challenging in this crucial regard, with obstacles to a more complete understanding arising in both the clinical and research settings.

Sociocultural factors are crucial determinants of AD diagnosis and management, and the disease process itself has been accurately described as a complex interaction between biology and culture that differs immensely across ethnic and racial groups (Chin, Negash, & Hamilton, 2011). Overall, African Americans are less likely to access specialized healthcare than are non-Hispanic Whites (Bellinger et al., 2010); they may also be less likely to regard memory impairment as necessitating clinical diagnosis and oversight until dysfunction is quite severe or is attended by prominent behavioral symptoms (Barnes et al., 2015). Taken together, such facts might suggest that typical or pure AD should be underdiagnosed in African American populations and that disparities are being underestimated.

However, that assumption is greatly complicated by the potential for misdiagnosis of cognitive impairment among racial minorities who do enter the healthcare system. Diagnosis of AD and related dementias is most reliable when assessment involves a neurologist or geriatrician (Cho et al., 2014) subsequent to comprehensive and longitudinal neuropsychological assessment. However, in the absence of specialized care, African American older adults in the primary care setting are likely to be diagnosed using screening instruments such as the MMSE, despite research demonstrating that standardized MMSE cut-off scores are associated with overdiagnosis of AD in African American populations (Bohnstedt, Fox, & Kohatsu, 1994). Importantly, even more rigorous assessment tools may lack validity in non-White populations. The cross-sectional differences in cognitive test performance seen in African Americans across the latter half of the life course may be partially explained by differential item functioning (DIF), wherein test wiseness, which varies by group based upon within-group educational and cultural experiences. influences response to test items and results in test *score* disparity despite similar underlying *ability*. DIF in later-life cognitive testing can create cultural bias very similar to that which has been established for IQ and academic readiness assessment earlier in life (Pedraza et al., 2009). Taking into account the probability of both missed cases in the community and misclassified cases in the clinic, what is most certain is that non-Hispanic White and African American AD patient populations differ from one another in demographic, symptom, and pathology profiles, and this represents a major challenge for clinical researchers seeking to develop the most broadly effective intervention strategies.

Difficulties with clinical assessment of cognitive outcomes among African Americans persist in clinical and epidemiological research on cognitive aging. Large studies often use medical records to determine outcomes; others use cognitive tests prone to the same DIF and cultural bias discussed previously. Further, additional problems arise in the research setting due to challenges in measuring predictive risk and protective factors of interest. Research on risk factors for cognitive aging, particularly modifiable risk factors that require self-reported subjective data from participants, is prone to measurement error due to the potential for cultural differences in response patterns (Woods-Giscombe & Lobel, 2008). Error in measured variables is compounded by the issue of residual confounding, which arises most notably with unmeasured or inadequately measured race-associated socioeconomic or other environmental variables. The problem of residual confounding is well exemplified by the previously discussed discovery that educational quality can account for far more racial disparity than the traditionally measured years of education (Sisco et al., 2015). However, that discovery, while key to subsequent progress in disparities research, is unlikely to be the last example of important residual confounders acting within models of demographics and cognitive health. As rigorous research continues, analogous inadequacies in other variables will likely emerge, and measurements should ideally improve.

Improving measures and fine-tuning modeling strategies, however, will not solve the methodological challenge at the heart of cognitive health disparities research. To move forward, researchers must understand, and commit to addressing, fundamental sampling challenges that inevitably arise from historically entrenched obstacles to research participation among African Americans. In literature addressing racial differences in cognitive risk, the issue of insufficient power frequently comes up. To detect meaningful relationships between given risk factors and cognitive or brain outcomes within specific ethnoracial groups, recruitment and retention of larger samples of traditionally underrepresented groups will be crucial. However, an increase in raw numbers does not represent a complete solution. Selection bias arises from the very different sociodemographic and health profiles of African Americans and non-Hispanic Whites who participate in cognitive aging research as either healthy controls or as patients, and the problem escalates with selective attrition as those with the fewest resources, or the poorest health—often the same populations at highest risk for adverse brain outcomes—become unable or unwilling to continue their participation (Gottesman et al., 2015). Subsamples of underrepresented groups who continue participating in research studies over time, either because they are fundamentally healthier or more resilient than similarly aged peers or because they possess internal or external resources that allow them to compensate for challenges that arise, become progressively less representative of the larger population they are ostensibly drawn from. Inadequately powered studies and the presence of selection bias represent serious threats to internal and external validity of research findings, and significantly hamper progress toward addressing cognitive impairment and AD in the broader African American population.

CONCLUSIONS AND FUTURE DIRECTIONS

There is overwhelming evidence for racial disparities in cognitive health during the latter half of the life course, with African American communities bearing a substantially greater burden of disease and dysfunction than their non-Hispanic White counterparts. Race-associated risk for AD and cognitive impairment is not explained by genetic factors; rather, race is most likely a marker for a population-level clustering of complex health factors and differential exposures to educational, occupational, psychosocial, and environmental conditions that increase vulnerability to adverse brain and cognitive outcomes. Risk pathways in African Americans, including neuropathological points along those pathways, may differ from those established in predominantly non-Hispanic White cohorts, with distinct implications for preventive and pharmacological intervention. However, understanding these pathways and tailoring interventions accordingly will require studies that engage larger and more representative samples of African Americans.

To this end, researchers must work actively within African American communities to communicate the purpose, intent, and value of cognitive aging research for individuals and for the community and establish long-term relationships that fulfill the expectations created and accordingly dispel mistrust of institutional research. Priority must be placed on acknowledging structural barriers to participation: transportation difficulties, unstable housing arrangements, comorbid health problems, or competing time commitments for work or caregiving that can lead to attrition of even very dedicated participants. Clear alternative accommodations, such as off-hours or in-home visits, must be in place to minimize such barriers. In the laboratory, measurement error should be acknowledged and instruments or items with demonstrable cultural bias should be avoided if possible; selective attrition should be carefully assessed, acknowledged, and ideally accounted for at the analytic stage. And finally, resultant findings must be disseminated immediately and accessibly within those communities that are participating. The challenges of an inclusive, engaged research model for cognitive health disparities are significant, but meeting them will lead to substantial progress in the field and to provision of resources and cognitive risk management across ethnoracial and socioeconomic strata.

REFERENCES

Aggarwal, N. T., Wilson, R. S., Beck, T. L., Rajan, K. B., Mendes de Leon, C. F., Evans, D. A., & Everson-Rose, S. A. (2014). Perceived stress and change in cognitive function among adults 65 years and older. *Psychosomatic Medicine, 76,* 80–85. https://doi.org/10.1097/PSY.0000000000000016

Almeida, R. P., Schultz, S. A., Austin, B. P., Boots, E. A., Dowling, N. M., Gleason, C. E., ... Okonkwo, O. C. (2015). Effect of cognitive reserve on age-related changes in cerebrospinal fluid biomarkers of Alzheimer disease. *JAMA Neurology, 72*, 699–706. https://doi.org/10.1001/jamaneurol.2015.0098

Arvanitakis, Z., Schneider, J. A., Wilson, R. S., Li, Y., Arnold, S. E., Wang, Z., & Bennett, D. A. (2006). Diabetes is related to cerebral infarction but not to AD pathology in older persons. *Neurology, 67*, 1960–1965. https://doi.org/10.1212/01.wnl.0000247053.45483.4e

Barnes, L. L., & Bennett, D. A. (2014). Alzheimer's disease in African Americans: Risk factors and challenges for the future. *Health Affairs, 33*, 580–586. https://doi.org/10.1377/hlthaff.2013.1353

Barnes, L. L., Leurgans, S., Aggarwal, N. T., Shah, R. C., Arvanitakis, Z., James, B. D. ... Schneider, J. A. (2015). Mixed pathology is more likely in Black than White decedents with Alzheimer dementia. *Neurology, 85*, 528–534. https://doi.org/10.1212/WNL.0000000000001834

Barnes, L. L., Mendes de Leon, C. F., Wilson, R. S., Bienias, J. L., & Evans, D. A. (2004). Social resources and cognitive decline in a population of older African Americans and Whites. *Neurology, 63*, 2322–2326. https://doi.org/10.1212/01.wnl.0000147473.04043.b3

Barnes, L. L., Wilson, R. S., Everson-Rose, S. A., Hayward, M. D., Evans, D. A., & Mendes de Leon, C. F. (2012). Effects of early-life adversity on cognitive decline in older African Americans and Whites. *Neurology, 79*, 2321–2327. https://doi.org/10.1212/WNL.0b013e318278b607

Beeri, M. S., Silverman, J. M., Davis, K. L., Marin, D., Grossman, H. Z., Schmeidler, J., ... Haroutunian, V. (2005). Type 2 diabetes is negatively associated with Alzheimer's disease neuropathology. *Journals of Gerontology Series A: Biological Sciences and Medical Sciences, 60*, 471–475. https://doi.org/10.1093/gerona/60.4.471

Bellinger, J. D., Hassan, R. M., Rivers, P. A., Cheng, Q., Williams, E., & Glover, S. H. (2010). Specialty care use in US patients with chronic diseases. *International Journal of Environmental Research and Public Health, 7*, 975–990. https://doi.org/10.3390/ijerph7030975

Bennett, D. A., Wilson, R. S., Schneider, J. A., Evans, D. A., Mendes de Leon, C. F., Arnold, S. E., ... Bienias, J. L. (2003). Education modifies the relation of AD pathology to level of cognitive function in older persons. *Neurology, 60*, 1909–1915. https://doi.org/10.1212/01.wnl.0000069923.64550.9f

Bohnstedt, M., Fox, P. J., & Kohatsu, N. D. (1994). Correlates of Mini-Mental Status Examination scores among elderly demented patients: The influence of race-ethnicity. *Journal of Clinical Epidemiology, 47*, 1381–1387. https://doi.org/10.1016/0895-4356(94)90082-5

Boots, E. A., Schultz, S. A., Almeida, R. P., Oh, J. M., Koscik, R. L., Dowling, M. N. ... Okonkwo, O. C. (2015). Occupational complexity and cognitive reserve in a middle-aged cohort at risk for Alzheimer's disease. *Archives of Clinical Neuropsychology, 30*, 634–642. https://doi.org/10.1093/arclin/acv041

Brickman, A. M., Schupf, N., Manly, J. J., Luchsinger, J. A., Andrews, H., Tang, M. X., ... Brown, T. R. (2008). Brain morphology in older African Americans, Caribbean Hispanics, and Whites from northern Manhattan. *Archives of Neurology, 65*, 1053–1061. https://doi.org/10.1001/archneur.65.8.1053

Carvalho, J. O., Tommet, D., Crane, P. K., Thomas, M. L., Claxton, A., Habeck, C., ... Romero, H. R. (2015). Deconstructing racial differences: The effects of quality of education and cerebrovascular risk factors. *Journals of Gerontology Series B: Psychological Sciences and Social Sciences, 70,* 545–556. https://doi.org/10.1093/geronb/gbu086

Castora-Binkley, M., Peronto, C. L., Edwards, J. D., & Small, B. J. (2015). A longitudinal analysis of the influence of race on cognitive performance. *Journals of Gerontology Series B: Psychological Sciences and Soc Sciences, 70,* 512–518. https://doi.org/10.1093/geronb/gbt112

Charletta, D., Gorelick, P. B., Dollear, T. J., Freels, S., & Harris, Y. (1995). CT and MRI findings among African-Americans with Alzheimer's disease, vascular dementia, and stroke without dementia. *Neurology, 45,* 1456–1461. https://doi.org/10.1212/wnl.45.8.1456

Chin, A. L., Negash, S., & Hamilton, R. (2011). Diversity and disparity in dementia: The impact of ethnoracial differences in Alzheimer disease. *Alzheimer Disease and Associated Disorders, 25,* 187–195. https://doi.org/10.1097/WAD.0b013e318211c6c9

Cho, K., Gagnon, D. R., Driver, J. A., Altincatal, A., Kosik, N., Lanes, S., & Lawler, E. V. (2014). Dementia coding, workup, and treatment in the VA New England healthcare system. *International Journal of Alzheimer's Disease, 2014,* 821894. https://doi.org/10.1155/2014/821894

de la Monte, S. M., Hutchins, G. M., & Moore, G. W. (1989). Racial differences in the etiology of dementia and frequency of Alzheimer lesions in the brain. *Journal of the Nationall Medical Association, 81,* 644–652.

DeCarli, C., Reed, B. R., Jagust, W., Martinez, O., Ortega, M., & Mungas, D. (2008). Brain behavior relationships among African Americans, Whites, and Hispanics. *Alzheimer Disease and Associated Disorders, 22,* 382–391. https://doi.org/10.1097/WAD.0b013e318185e7fe

Demirovic, J., Prineas, R., Loewenstein, D., Bean, J., Duara, R., Sevush, S., & Szapocznik, J. (2003). Prevalence of dementia in three ethnic groups: The South Florida program on aging and health. *Annals of Epidemiology, 13,* 472–478. https://doi.org/10.1016/S1047-2797(02)00437-4

Early, D. R., Widaman, K. F., Harvey, D., Beckett, L., Park, L. Q., Farias, S. T., . . . Mungas, D. (2013). Demographic predictors of cognitive change in ethnically diverse older persons. *Psychology of Aging, 28,* 633–645. https://doi.org/10.1037/a0031645

Evans, D. A., Beckett, L. A., Field, T. S., Feng, L., Albert, M. S., Bennett, D. A., . . . Mayeux, R. (1997). Apolipoprotein E epsilon4 and incidence of Alzheimer disease in a community population of older persons. *Journal of the American Medical Association, 277,* 822–824. https://doi.org/10.1001/jama.1997.03540340056033

Evans, D. A., Bennett, D. A., Wilson, R. S., Bienias, J. L., Morris, M. C., Scherr, P. A., . . . Schneider, J. (2003). Incidence of Alzheimer disease in a biracial urban community: Relation to apolipoprotein E allele status. *Archives of Neurology, 60,* 185–189. https://doi.org/10.1001/archneur.60.2.185

Fillenbaum, G. G., Heyman, A., Huber, M. S., Woodbury, M. A., Leiss, J., Schmader, K. E., . . . Trapp-Moen, B. (1998). The prevalence and 3-year incidence of dementia in older Black and White community residents. *Journal of Clinical Epidemiology, 51,* 587–595. https://doi.org/10.1016/S0895-4356(98)00024-9

Glymour, M. M., & Manly, J. J. (2008). Lifecourse social conditions and racial and ethnic patterns of cognitive aging. *Neuropsychology Review, 18,* 223–254. https://doi.org/10.1007/s11065-008-9064-z

Gottesman, R. F., Fornage, M., Knopman, D. S., & Mosley, T. H. (2015). Brain aging in African-Americans: The Atherosclerosis Risk in Communities (ARIC)

experience. *Current Alzheimer Research, 12,* 607–613. https://doi.org/10.2174/1567205012666150701102445

Gottesman, R. F., Schneider, A. L., Albert, M., Alonso, A., Bandeen-Roche, K., Coker, L., ... Mosley, T. H. (2014). Midlife hypertension and 20-year cognitive change: The atherosclerosis risk in communities neurocognitive study. *JAMA Neurology, 71,* 1218–1227. https://doi.org/10.1001/jamaneurol.2014.1646

Gross, A. L., Mungas, D. M., Crane, P. K., Gibbons, L. E., MacKay-Brandt, A., Manly, J. J., ... Jones, R. N. (2015). Effects of education and race on cognitive decline: An integrative study of generalizability versus study-specific results. *Psychology of Aging, 30,* 863–880. https://doi.org/10.1037/pag0000032

Gu, Y., Razlighi, Q. R., Zahodne, L. B., Janicki, S. C., Ichise, M., Manly, J. J., ... Stern, Y. (2015). Brain amyloid deposition and longitudinal cognitive decline in nondemented older subjects: Results from a multi-ethnic population. *PLoS One, 10*(7), e0123743. https://doi.org/10.1371/journal.pone.0123743

Harada, C. N., Natelson Love, M. C., & Triebel, K. L. (2013). Normal cognitive aging. *Clinical Geriatric Medicine, 29,* 737–752. https://doi.org/10.1016/j.cger.2013.07.002

Hasson, B. R., Apovian, C., & Istfan, N. (2015). Racial/ethnic differences in insulin resistance and beta cell function: Relationship to racial disparities in Type 2 diabetes among African Americans versus Caucasians. *Current Obesity Reports, 4,* 241–249. https://doi.org/10.1007/s13679-015-0150-2

Helzner, E. P., Luchsinger, J. A., Scarmeas, N., Cosentino, S., Brickman, A. M., Glymour, M. M., & Stern, Y. (2009). Contribution of vascular risk factors to the progression in Alzheimer disease. *Archives of Neurology, 66,* 343–348. https://doi.org/10.1001/archneur.66.3.343

Hildreth, K. L., Van Pelt, R. E., & Schwartz, R. S. (2012). Obesity, insulin resistance, and Alzheimer's disease. *Obesity (Silver Spring), 20,* 1549–1557. https://doi.org/10.1038/oby.2012.19

Horn, J. L., & Cattell, R. B. (1966). Refinement and test of the theory of fluid and crystallized general intelligences. *Journal of Educational Psychology, 57,* 253–270. https://doi.org/10.1037/h0023816

Hoscheidt, S. M., Starks, E. J., Oh, J. M., Zetterberg, H., Blennow, K., Krause, R. A., ... Bendlin, B. B. (2016). Insulin resistance is associated with increased levels of cerebrospinal fluid biomarkers of Alzheimer's disease and reduced memory function in at-risk healthy middle-aged adults. *Journal of Alzheimer's Disease, 52,* 1373–1383. https://doi.org/10.3233/JAD-160110

Jack, C. R., Jr., Albert, M. S., Knopman, D. S., McKhann, G. M., Sperling, R. A., Carrillo, M. C., ... Phelps, C. H. (2011). Introduction to the recommendations from the National Institute on Aging-Alzheimer's Association workgroups on diagnostic guidelines for Alzheimer's disease. *Alzheimer's & Dementia, 7,* 257–262. https://doi.org/10.1016/j.jalz.2011.03.004

Kaplan, J. B., & Bennett, T. (2003). Use of race and ethnicity in biomedical publication. *Journal of the American Medical Association, 289,* 2709–2716. https://doi.org/10.1001/jama.289.20.2709

Karlamangla, A. S., Miller-Martinez, D., Aneshensel, C. S., Seeman, T. E., Wight, R. G., & Chodosh, J. (2009). Trajectories of cognitive function in late life in the United States: Demographic and socioeconomic predictors. *American Journal of Epidemiology, 170,* 331–342. https://doi.org/10.1093/aje/kwp154

Katz, M. J., Lipton, R. B., Hall, C. B., Zimmerman, M. E., Sanders, A. E., Verghese, J., ... Derby, C. A. (2012). Age-specific and sex-specific prevalence and incidence of mild cognitive impairment, dementia, and Alzheimer dementia in Blacks and Whites: A report from the Einstein Aging Study. *Alzheimer Disease and Associated Disorders, 26,* 335–343. https://doi.org/10.1097/WAD.0b013e31823dbcfc

Lee, H. B., Richardson, A. K., Black, B. S., Shore, A. D., Kasper, J. D., & Rabins, P. V. (2012). Race and cognitive decline among community-dwelling elders with mild cognitive impairment: Findings from the Memory and Medical Care Study. *Aging and Mental Health, 16,* 372–377. https://doi.org/10.1080/13607863.2011.609533

Li, J., Cesari, M., Liu, F., Dong, B., & Vellas, B. (2017). Effects of diabetes mellitus on cognitive decline in patients with Alzheimer disease: A systematic review. *Canadian Journal of Diabetes, 41,* 114–119. https://doi.org/10.1016/j.jcjd.2016.07.003

Lo, R. Y., Jagust, W. J., & Alzheimer's Disease Neuroimaging Initiative. (2012). Vascular burden and Alzheimer disease pathologic progression. *Neurology, 79,* 1349–1355. https://doi.org/10.1212/WNL.0b013e31826c1b9d

Longstreth, W. T., Jr., Arnold, A. M., Manolio, T. A., Burke, G. L., Bryan, N., Jungreis, C. A., . . . Fried, L. (2000). Clinical correlates of ventricular and sulcal size on cranial magnetic resonance imaging of 3,301 elderly people. The Cardiovascular Health Study. *Neuroepidemiology, 19,* 30–42. https://doi.org/10.1159/000026235

Manly, J. J., Tang, M. X., Schupf, N., Stern, Y., Vonsattel, J. P., & Mayeux, R. (2008). Frequency and course of mild cognitive impairment in a multiethnic community. *Annals of Neurology, 63,* 494–506. https://doi.org/10.1002/ana.21326

Masel, M. C., & Peek, M. K. (2009). Ethnic differences in cognitive function over time. *Annals of Epidemiology, 19,* 778–783. https://doi.org/10.1016/j.annepidem.2009.06.008

Mayeda, E. R., Glymour, M. M., Quesenberry, C. P., & Whitmer, R. A. (2016). Inequalities in dementia incidence between six racial and ethnic groups over 14 years. *Alzheimers & Dementia, 12,* 216–224. https://doi.org/10.1016/j.jalz.2015.12.007

Mayeda, E. R., Karter, A. J., Huang, E. S., Moffet, H. H., Haan, M. N., & Whitmer, R. A. (2014). Racial/ethnic differences in dementia risk among older type 2 diabetic patients: The diabetes and aging study. *Diabetes Care, 37,* 1009–1015. https://doi.org/10.2337/dc13-0215

McDaniel, A., DiPrete, T. A., Buchmann, C., & Shwed, U. (2011). The Black gender gap in educational attainment: Historical trends and racial comparisons. *Demography, 48,* 889–914. https://doi.org/10.1007/s13524-011-0037-0

Menke, A., Casagrande, S., Geiss, L., & Cowie, C. C. (2015). Prevalence of and trends in diabetes among adults in the United States, 1988–2012. *Journal of the American Medical Association, 314,* 1021–1029. https://doi.org/10.1001/jama.2015.10029

Miller, F. D., Hicks, S. P., D'Amato, C. J., & Landis, J. R. (1984). A descriptive study of neuritic plaques and neurofibrillary tangles in an autopsy population. *American Journal of Epidemiology, 120,* 331–341. https://doi.org/10.1093/oxfordjournals.aje.a113897

Moroney, J. T., Bagiella, E., Hachinski, V. C., Molsa, P. K., Gustafson, L., Brun, A., . . . Desmond, D. W. (1997). Misclassification of dementia subtype using the Hachinski Ischemic Score: Results of a meta-analysis of patients with pathologically verified dementias. *Annals of the New York Academy of Sciences, 826,* 490–492. https://doi.org/10.1111/j.1749-6632.1997.tb48510.x

Murrell, J. R., Price, B., Lane, K. A., Baiyewu, O., Gureje, O., Ogunniyi, A., . . . Hall, K. S. (2006). Association of apolipoprotein E genotype and Alzheimer disease in African Americans. *Archives of Neurology, 63,* 431–434. https://doi.org/10.1001/archneur.63.3.431

Ott, A., Stolk, R. P., van Harskamp, F., Pols, H. A., Hofman, A., & Breteler, M. M. (1999). Diabetes mellitus and the risk of dementia: The Rotterdam Study. *Neurology, 53,* 1937-1942. https://doi.org/10.1212/wnl.53.9.1937

Pais, J. (2014). Cumulative structural disadvantage and racial health disparities: The pathways of childhood socioeconomic influence. *Demography, 51,* 1729-1753. https://doi.org/10.1007/s13524-014-0330-9

Pedraza, O., Graff-Radford, N. R., Smith, G. E., Ivnik, R. J., Willis, F. B., Petersen, R. C., & Lucas, J. A. (2009). Differential item functioning of the Boston Naming Test in cognitively normal African American and Caucasian older adults. *Journal of the International Neuropsychological Society, 15,* 758-768. https://doi.org/10.1017/S1355617709990361

Perrotta, M., Lembo, G., & Carnevale, D. (2016). Hypertension and dementia: Epidemiological and experimental evidence revealing a detrimental relationship. *International Journal of Molecular Science, 17*(3), 347. https://doi.org/10.3390/ijms17030347

Potter, G. G., Plassman, B. L., Burke, J. R., Kabeto, M. U., Langa, K. M., Llewellyn, D. J., ... Steffens, D. C. (2009). Cognitive performance and informant reports in the diagnosis of cognitive impairment and dementia in African Americans and Whites. *Alzheimer's & Dementia, 5,* 445-453. https://doi.org/10.1016/j.jalz.2009.04.1234

Pytel, P., Cochran, E. J., Bonner, G., Nyenhuis, D. L., Thomas, C., & Gorelick, P. B. (2006). Vascular and Alzheimer-type pathology in an autopsy study of African-Americans. *Neurology, 66,* 433-435. https://doi.org/10.1212/01.wnl.0000196472.93744.57

Rajan, K. B., Arvanitakis, Z., Lynch, E. B., McAninch, E. A., Wilson, R. S., Weuve, J., ... Evans, D. A. (2016). Cognitive decline following incident and preexisting diabetes mellitus in a population sample. *Neurology, 87,* 1681-1687. https://doi.org/10.1212/WNL.0000000000003226

Redmond, N., Baer, H. J., & Hicks, L. S. (2011). Health behaviors and racial disparity in blood pressure control in the national health and nutrition examination survey. *Hypertension, 57,* 383-389. https://doi.org/10.1161/HYPERTENSIONAHA.110.161950

Reitz, C., Jun, G., Naj, A., Rajbhandary, R., Vardarajan, B. N., Wang, L. S., ... Alzheimer Disease Genetics Consortium. (2013). Variants in the ATP-binding cassette transporter (ABCA7), apolipoprotein E 4,and the risk of late-onset Alzheimer disease in African Americans. *Journal of the American Medical Association, 309,* 1483-1492. https://doi.org/10.1001/jama.2013.2973

Riudavets, M. A., Rubio, A., Cox, C., Rudow, G., Fowler, D., & Troncoso, J. C. (2006). The prevalence of Alzheimer neuropathologic lesions is similar in Blacks and Whites. *Journal of Neuropathology and Experimental Neurology, 65,* 1143-1148. https://doi.org/10.1097/01.jnen.0000248548.20799.a3

Romero, C. X., Romero, T. E., Shlay, J. C., Ogden, L. G., & Dabelea, D. (2012). Changing trends in the prevalence and disparities of obesity and other cardiovascular disease risk factors in three racial/ethnic groups of USA adults. *Advances in Preventive Medicine, vol. 2012,* Article ID 172423, 8 pages, 2012. https://doi.org/10.1155/2012/172423

Sandberg, G., Stewart, W., Smialek, J., & Troncoso, J. C. (2001). The prevalence of the neuropathological lesions of Alzheimer's disease is independent of race and gender. *Neurobiology of Aging, 22,* 169-175. https://doi.org/10.1155/2012/172423

Sanz, C., Andrieu, S., Sinclair, A., Hanaire, H., Vellas, B., & Group, R. F. S. (2009). Diabetes is associated with a slower rate of cognitive decline in Alzheimer disease. *Neurology, 73,* 1359-1366. https://doi.org/10.1212/WNL.0b013e3181bd80e9

Sawyer, K., Sachs-Ericsson, N., Preacher, K. J., & Blazer, D. G. (2009). Racial differences in the influence of the APOE epsilon 4 allele on cognitive decline in a sample of community-dwelling older adults. *Gerontology, 55,* 32–40. https://doi.org/10.1159/000137666

Scarabino, D., Gambina, G., Broggio, E., Pelliccia, F., & Corbo, R. M. (2016). Influence of family history of dementia in the development and progression of late-onset Alzheimer's disease. *American Journal of Medical Genetics Part B: Neuropsychiatric Genetics, 171B,* 250–256. https://doi.org/10.1002/ajmg.b.32399

Schneider, J. A., Arvanitakis, Z., Bang, W., & Bennett, D. A. (2007). Mixed brain pathologies account for most dementia cases in community-dwelling older persons. *Neurology, 69,* 2197–2204. https://doi.org/10.1212/01.wnl.0000271090.28148.24

Schultz, S. A., Larson, J., Oh, J., Koscik, R., Dowling, M. N., Gallagher, C. L., . . . Okonkwo, O. C. (2014). Participation in cognitively-stimulating activities is associated with brain structure and cognitive function in preclinical Alzheimer's disease. *Brain Imaging and Behavior, 9,* 729–736. https://doi.org/10.1007/s11682-014-9329-5

Schwartz, B. S., Glass, T. A., Bolla, K. I., Stewart, W. F., Glass, G., Rasmussen, M., . . . Bandeen-Roche, K. (2004). Disparities in cognitive functioning by race/ethnicity in the Baltimore Memory Study. *Environmental Health Perspectives, 112,* 314–320. https://doi.org/10.1289/ehp.6727

Sencakova, D., Graff-Radford, N. R., Willis, F. B., Lucas, J. A., Parfitt, F., Cha, R. H., . . . Jack, C. R., Jr. (2001). Hippocampal atrophy correlates with clinical features of Alzheimer disease in African Americans. *Archives of Neurology, 58,* 1593–1597. https://doi.org/10.1001/archneur.58.10.1593

Shadlen, M. F., Siscovick, D., Fitzpatrick, A. L., Dulberg, C., Kuller, L. H., & Jackson, S. (2006). Education, cognitive test scores, and Black–White differences in dementia risk. *Journal of the American Geriatric Society, 54,* 898–905. https://doi.org/10.1111/j.1532-5415.2006.00747.x

Sisco, S., Gross, A. L., Shih, R. A., Sachs, B. C., Glymour, M. M., Bangen, K. J., . . . Manly, J. J. (2015). The role of early-life educational quality and literacy in explaining racial disparities in cognition in late life. *Journals of Gerontology Series B: Psychological Sciences and Social Sciences, 70,* 557–567. https://doi.org/10.1093/geronb/gbt133

Stern, Y. (2009). Cognitive reserve. *Neuropsychologia, 47,* 2015–2028. https://doi.org/10.1016/j.neuropsychologia.2009.03.004

Tang, M. X., Cross, P., Andrews, H., Jacobs, D. M., Small, S., Bell, K., . . . Mayeux, R. (2001). Incidence of AD in African-Americans, Caribbean Hispanics, and Caucasians in northern Manhattan. *Neurology, 56,* 49–56. https://doi.org/10.1212/wnl.56.1.49

Trimble, B., & Morgenstern, L. B. (2008). Stroke in minorities. *Neurologic Clinics, 26,* 1177–1190. https://doi.org/10.1016/j.ncl.2008.05.010

Tsuang, D. W., Wilson, R. K., Lopez, O. L., Luedecking-Zimmer, E. K., Leverenz, J. B., DeKosky, S. T., . . . Hamilton, R. L. (2005). Genetic association between the APOE*4 allele and Lewy bodies in Alzheimer disease. *Neurology, 64,* 509–513. https://doi.org/10.1212/01.WNL.0000150892.81839.D1

Wang, R., Fratiglioni, L., Laukka, E. J., Lovden, M., Kalpouzos, G., Keller, L., . . . Qiu, C. (2015). Effects of vascular risk factors and APOE epsilon4 on white matter integrity and cognitive decline. *Neurology, 84,* 1128–1135. https://doi.org/10.1212/WNL.0000000000001379

White, L., Small, B. J., Petrovitch, H., Ross, G. W., Masaki, K., Abbott, R. D., . . . Markesbery, W. (2005). Recent clinical-pathologic research on the causes of dementia in late life: Update from the Honolulu–Asia Aging Study. *Journal of Geriatric Psychiatry and Neurology, 18,* 224–227. https://doi.org/10.1177/0891988705281872

Wilkins, C. H., Grant, E. A., Schmitt, S. E., McKeel, D. W., & Morris, J. C. (2006). The neuropathology of Alzheimer disease in African American and White individuals. *Archives of Neurology, 63,* 87–90. https://doi.org/10.1001/archneur.63.1.87

Williams, D. R. (1998). African-American health: The role of the social environment. *Journal of Urban Health, 75,* 300–321. https://doi.org/10.1007/BF02345099

Wilson, R. S., Capuano, A. W., Sytsma, J., Bennett, D. A., & Barnes, L. L. (2015). Cognitive aging in older Black and White persons. *Psychology of Aging, 30,* 279–285. https://doi.org/10.1037/pag0000024

Wilson, R. S., Mendes De Leon, C. F., Bennett, D. A., Bienias, J. L., & Evans, D. A. (2004). Depressive symptoms and cognitive decline in a community population of older persons. *Journal of Neurology, Neurosurgery, & Psychiatry, 75,* 126–129.

Wolinsky, F. D., Bentler, S. E., Hockenberry, J., Jones, M. P., Weigel, P. A., Kaskie, B., & Wallace, R. B. (2011). A prospective cohort study of long-term cognitive changes in older Medicare beneficiaries. *BMC Public Health, 11,* 710. https://doi.org/10.1186/1471-2458-11-710

Woods-Giscombe, C. L., & Lobel, M. (2008). Race and gender matter: A multidimensional approach to conceptualizing and measuring stress in African American women. *Cultural Diversity & Ethnic Minority Psychology, 14,* 173–182. https://doi.org/10.1037/1099-9809.14.3.173

Yaffe, K., Falvey, C., Hamilton, N., Schwartz, A. V., Simonsick, E. M., Satterfield, S., . . . Harris, T. B. (2012). Diabetes, glucose control, and 9-year cognitive decline among older adults without dementia. *Archives of Neurology, 69,* 1170–1175. https://doi.org/10.1001/archneurol.2012.1117

Yaffe, K., Falvey, C., Harris, T. B., Newman, A., Satterfield, S., Koster, A., . . . Simonsick, E. (2013). Effect of socioeconomic disparities on incidence of dementia among biracial older adults: Prospective study. *British Medical Journal, 347,* f7051. https://doi.org/10.1136/bmj.f7051

Yu, J. T., Tan, L., & Hardy, J. (2014). Apolipoprotein E in Alzheimer's disease: An update. *Annual Review of Neuroscience, 37,* 79–100. https://doi.org/10.1146/annurev-neuro-071013-014300

Zahodne, L. B., Manly, J. J., Azar, M., Brickman, A. M., & Glymour, M. M. (2016). Racial disparities in cognitive performance in mid- and late adulthood: Analyses of two cohort studies. *Journal of the American Geriatrics Society, 64,* 959–964. https://doi.org/10.1111/jgs.14113

Zahodne, L. B., Manly, J. J., Narkhede, A., Griffith, E. Y., DeCarli, C., Schupf, N. S., . . . Brickman, A. M. (2015). Structural MRI predictors of late-life cognition differ across African Americans, Hispanics, and Whites. *Current Alzheimer Research, 12,* 632–639. https://doi.org/10.2174/1567205012666150530203214

Zhang, Z., Hayward, M. D., & Yu, Y. L. (2016). Life course pathways to racial disparities in cognitive impairment among older Americans. *Journal of Health and Social Behavior, 57,* 184–199. https://doi.org/10.1177/0022146516645925

12

Bias, Equivalence, and Fairness

OTTO PEDRAZA AND FONS J. R. VAN DE VIJVER

INTRODUCTION

In an increasingly connected and multicultural world, clinicians and investigators are becoming more cognizant of the hazards involved with the indiscreet use of assessment instruments, test stimuli, and experimental paradigms developed in one language or culture to infer cognitive or behavioral states among individuals from another language or culture. At its core, cross-cultural research is nonexperimental because individuals cannot be assigned randomly to a particular cultural group. Consequently, confounding or nuisance variables are not equally distributed among the groups, which can lead to invalid inferences about the construct being studied (van de Vijver & Leung, 2011). Clinicians and investigators therefore must tread carefully when designing, applying, or interpreting assessment instruments across cultures, with the goals of minimizing the impact of bias and enhancing measurement equivalence.

To that end, the present chapter reviews the taxonomy of bias and equivalence, focusing on the methodological and measurement aspects that can impact the validity of cross-cultural assessment. We consider briefly the concept of test fairness and why it differs from bias and nonequivalence. Lastly, we discuss ways to minimize the risk of bias and nonequivalence when developing new assessment instruments, as well as ways to mitigate the cost of bias and nonequivalence in existing clinical and experimental instruments.

BIAS

Imagine the following scenario, modified slightly from van de Vijver and Tanzer (2004): You take a geography test that includes the question "What is the capital of Argentina?" The test is administered to a large international cohort of adults throughout North, Central, and South America. If the intended purpose is to measure knowledge of this specific topic, the test item will provide a valid comparison across adults from the sampled countries. However, if the item is intended to assess the broader construct of knowledge of geography, an examinee's distance to the country of Argentina will become a nuisance variable. Respondents from neighboring countries such as Chile and Uruguay, and those from South America generally, will be at a relative advantage compared with respondents from North America. In this context, the item is considered biased because the probability of a correct response is not determined solely by the examinee's knowledge of geography, but also by his or her relative proximity to the target country.

Now let us consider another example, this one more germane to cross-cultural neuropsychology and neuroscience. Suppose that the Wechsler Digit Span test (a short-term memory task in which a sequence of digits of increasing length must be repeated to the examiner) is administered in English to a group of monolingual adults and a group of English–Spanish bilingual adults. If the test score is interpreted to reflect the number of digits that an individual can repeat instantly, the test may be unbiased. However, if the score is meant to reflect the construct of short-term memory capacity, the person's primary language is likely to constitute a nuisance variable. In the English language, the numbers 1 to 9 are monosyllabic except for 7, whereas in Spanish only three of those digits are monosyllabic (2, 3, 6). Bilingual adults often rely on their primary language for mathematical operations (Salillas & Wicha, 2012). Therefore, some individuals in the bilingual group will process the digit sequence in Spanish and the additional number of syllables presumably will load their short-term memory capacity earlier. Indeed, numerous studies have shown that Spanish speakers tend to perform worse on digit span tasks compared with English speakers (Funes, Rodriguez, & Lopez, 2016; Gasquoine, Croyle, Cavazos-Gonzalez, & Sandoval, 2007). To the extent that the test is considered a measure of short-term memory, the language in which the responses are provided becomes a nuisance factor that biases our interpretation of between-group differences in the underlying ability.

DEFINITIONS OF BIAS

The definition of bias has evolved from a description of the relationship between an item or test score and a predicted criterion to a description of the relationship between an item or test score and a latent attribute, trait, or ability (Borsboom, Romeijn, & Wicherts, 2008). In the initial formulation, most notably found in Cleary (1968), test bias is present whenever the regression of a criterion on a test score is unequal across groups. For instance, an IQ score would be biased if the regression function linking IQ with a criterion (e.g., job success) is unequal across groups (e.g., men and women). The *Standards for Educational and Psychological Testing*, published jointly by the American Education Research Association, the American Psychological Association, and the National Council on Measurement in Education, designates this form of bias as *predictive bias* and notes that it is generally considered in the context of employment or academic success (American Education Research Association, 1999). The Cleary model of predictive bias is equivalent to testing the null hypothesis that there is no significant between-group difference in the slopes and intercepts of the linear regressions. Jensen

(1980) modified this model to indicate that a test should be considered a biased predictor if there is a significant between-group difference in slope, intercepts, or standard error of the estimates.

Advances in psychometric theory and methods have led to another formulation of bias—the unequal relationship between a score and the latent construct due to the presence of nuisance variables (Poortinga, 1989; van de Vijver & Leung, 2011). In a biased instrument, between-group differences are engendered not only by the construct it purports to measure, but also (partly or entirely) by unwanted factors. In one of the examples presented earlier, bias exists because differences in item score (i.e., the percentage of individuals who correctly identify Buenos Aires as Argentina's capital) do not correspond to differences in the underlying trait or ability (i.e., knowledge of geography) and instead reflect a nuisance variable (i.e., respondents' distance to Argentina). The example also illustrates another important quality of bias. It is not intrinsic to the item or test instrument, but arises from the cross-cultural application of the instrument (van de Vijver & Tanzer, 2004). The item from this example may demonstrate substantial bias in a comparison of adults from Argentina and Canada, but minimal to zero bias when adults from Argentina and neighboring Uruguay, or between residents from cities within Argentina, are compared. In the latter comparison, the nuisance factor of proximity has largely become irrelevant. Stated differently, scores from a biased instrument may yield a valid assessment of the construct *within* a cultural group, but an invalid assessment when used for a comparison *between* cultures (van de Vijver & Leung, 2011).

The definition of bias that considers the presence of nuisance factors, and that also incorporates the related concept of measurement equivalence or invariance, has become the predominant formulation (Borsboom, Romeijn, & Wicherts, 2008) and will be adopted throughout the remainder of this chapter. Next, we will examine three categories of bias that are crucial in cross-cultural neuropsychology and neuroscience: construct bias, method bias, and item bias.

CONSTRUCT BIAS

Construct bias is present whenever the measured construct is not identical across cultural groups (van de Vijver & Poortinga, 2005). It constitutes incomplete construct overlap due to (a) dissimilarity in the definition of the construct across cultures, (b) poor sampling of all the relevant behaviors associated with the construct, or (c) differential appropriateness of the behaviors associated with the construct (van de Vijver & Leung, 2011). Construct underrepresentation (poor sampling) often results from a narrow selection of test item content and is particularly problematic when brief instruments are used (Messick, 1995). In contrast, construct irrelevance occurs when there are factors extraneous to the construct under study, particularly if the item content of the assessment tool is too broad.

An oft-cited example of construct bias is found in cross-cultural studies of intelligence (Saklofske, van de Vijver, Oakland, Mpofu, & Susuki, 2015). Intelligence, as defined in Western cultures and measured through a battery of cognitive tasks, typically emphasizes scholastic domains such as reasoning, verbal skills, and processing speed. This notion of intelligence is often labeled "analytical" or "academic" intelligence (Sternberg et al., 2001). However, non-Western cultures additionally value social relations and self-effacement in their conceptualization of intelligence, sometimes at the expense of scholastic abilities (Yang & Sternberg, 1997). As an illustration, the Luo people of rural Kenya have four broad concepts of intelligence: *rieko*, which corresponds to the Western notion of analytical intelligence and includes knowledge, skills, and competence; *luoro*, which always has a positive connotation and corresponds to social qualities such as respect for others and obedience; *paro*, which consists of practical thinking, particularly when solving social problems, and thus occupies an intermediate position between *rieko* and *luoro*; and *winjo*, which refers to comprehension (Grigorenko et al., 2001). These notions of intelligence stem from two latent factors: social-emotional competence and cognitive competence. However, only the notion of *rieko* and the latent factor of cognitive competence correspond with analytical intelligence and correlate with scores on Western intelligence tests.

As this example illustrates, what is measured and described as intelligence in one culture may represent a dissimilar construct in another culture. For psychologists, anthropologists, and neuroscientists who hold a relativistic view of culture, the presence of construct bias is therefore the rule and not the exception (van de Vijver & Poortinga, 2005). The approach proposed in this chapter is an attempt to remove the ideological undertones of this discussion about relativism by emphasizing empirical tests of bias and equivalence.

METHOD BIAS

Method bias is present when nuisance variance is due to methodological issues (van de Vijver & Leung, 2011). Method bias can lead to devastating consequences for the validity of cross-cultural assessment, as it often leads to a shift in mean test scores that obscures potentially valid differences between the groups (van de Vijver & Tanzer, 2004). As a result, the clinician or researcher who compares scores across cultures will face the difficult task of choosing among rival explanations: a valid cross-cultural difference or an invalid contrast due to method bias (van de Vijver & Poortinga, 2005). The most common sources of method bias consist of sample bias, instrument bias, and administration bias.

Sample Bias

Sample bias arises whenever the groups are not comparable in characteristics other than the target variable (e.g., between-group differences in quantity or quality of education, political orientation, or religious affiliation). For instance, Fernandez and Marcopulos (2008) compared normative samples from 10 different countries for the Trail Making Test (a task of divided attention in which the examinee connects a series of alternating numbers and letters). They found large score differences across the samples that reflected incomparability with regard to age, gender, education, occupation, and estimated intelligence. Van de Vijver and Leung (2011) observe that as the degree of cultural distance increases between groups, the greater the number of sample-dependent characteristics that must be considered.

Instrument Bias

This source of bias refers to the properties of an instrument that are unrelated to the target of study, yet contribute to group differences in item or test scores (van de Vijver & Poortinga, 2005). For example, if a computer is used to measure reaction times in children from families with low versus high socioeconomic status, the dissimilar familiarity with computers by virtue of socioeconomic status is expected to influence the obtained results, regardless of the construct being investigated.

In questionnaires and inventories that assess personality, mood, or other traits, instrument bias can be manifested as differential response styles between groups. Response style refers to a systematic tendency to endorse test or scale items on a basis other than the target construct. It includes *acquiescent response style*, in which respondents tend to agree with most of the propositions; *socially desirable response style*, in which respondents tend to select answers that yield a favorable image of themselves; and *extreme response style*, in which respondents tend to select the extreme endpoints for items on a scale (Johnson, Shavitt, & Holbrook, 2011). Socially desirable responding comprises the tendency to engage in impression management (intentional misrepresentation of oneself to appear more favorable) and self-deceptive enhancement (genuinely held inflated beliefs of oneself). Discordant response styles across cultural groups may lead to the misinterpretation of test results as a reflection of group differences instead of variation due to systematic error. For instance, numerous studies suggest that Hispanic adults demonstrate acquiescent and extreme response styles to a greater extent than non-Hispanic whites, a finding that appears to be moderated by years of education and the language-related aspects of acculturation (Davis, Resnicow, & Couper, 2011; Marin, Gamba, & Marin, 1992). If response style is not taken into consideration, a clinician or researcher may inadvertently attribute the between-group discrepancy in scores to differences in the construct measured by the questionnaire.

Administration Bias

The third source of method bias arises from differences in the procedures used to administer a test, research protocol, or other assessment instrument. Sources of administration bias include the physical environment (e.g., test administered at a person's house vs. a memory disorders clinic, differences in school class size); communication barriers (e.g., ineffective communication between examiner and examinee due to language differences, visual or hearing impairment, poor translation or interpretation, or cultural norm violations); unequal opportunity to familiarize oneself with the test format, technical instruments, or recording devices (e.g., experienced versus naïve subjects in a functional magnetic resonance imaging [fMRI] scanner); mode of data collection (e.g., online survey, paper-and-pencil test); and ambiguity in the test or protocol instructions (van de Vijver & Leung, 2011; van de Vijver & Tanzer, 2004).

ITEM BIAS

Whereas construct and method bias jeopardize validity at the test or scale level, item bias is due to anomalies at the item level. Specifically, an item is biased when persons with the same standing on a trait or ability and who belong to different groups do not have the same probability to provide a correct response (van de Vijver & Leung, 2011). The most common sources of bias for an item include poor translation, ambiguity in meaning, diverse word connotations, and low familiarity with the item in a particular culture.

Let us return to the example presented earlier that asks, "What is the capital of Argentina?" Suppose that this question is part of a comprehensive geography test administered to students in Argentina and Canada under equivalent testing conditions. Students from Argentina presumably will obtain a higher score on this item compared with their Canadian counterparts, even if both groups have a comparable level of geography knowledge and obtain similar means in the total test score. As noted by van de Vijver and Leung (2011), such an item would be biased because it favors one cultural group over the other *across all test score levels*. The probability of a correct response is higher for Argentinian students despite their standing on ability; that is, they have a better chance of knowing the name of their capital city regardless of excellent or poor general knowledge of world geography. Unbiased items should be of equal difficulty for all persons with the same underlying ability. In recent decades, the term *item bias* largely has been replaced with *differential item functioning* (DIF), although DIF strictly refers to the statistical procedures used to detect potential item bias. DIF is a necessary but insufficient condition for item bias. Bias is present only when DIF is due to attributes irrelevant to the target construct (Clauser & Mazor, 1998).

EQUIVALENCE

Bias and equivalence are related concepts. Whereas bias refers to the differences in meaning between cultural groups due to nuisance factors, equivalence refers to the measurement level at which scores from different groups can be compared (He & van de Vijver, 2012). A biased instrument yields inequivalent scores, and cross-cultural equivalence requires the absence of bias (van de Vijver & Leung, 2011; van de Vijver & Poortinga, 2005).

Definitions of Equivalence

In a review of the literature, Johnson (2006) lists 62 available definitions of the concept of equivalence, which he distills into two broad categories: interpretive and procedural. Interpretive forms of equivalence focus on similarities in the meaning of concepts and constructs, with an emphasis on how measures are interpreted across cultural groups. Interpretive equivalence reflects *shared meaning*. On the other hand, procedural forms of equivalence focus on the technical problems of cross-cultural measurement, with an emphasis on the factorial, structural, metric, or scalar aspects of the instruments. Procedural equivalence reflects *shared method*. This diversity in definitions and associated tests is rooted in the parallel historical development of cross-cultural studies in fields such as personality psychology, survey research, education, and psychometrics (van de Vijver & Poortinga, 2016).

These taxonomies of equivalence can be delineated further by considering four hierarchically nested levels: construct, structural or functional, metric or measurement unit, and scalar or full score equivalence (van de Vijver & Leung, 2011).

Construct Equivalence

At the basic level of analysis, constructs must be equivalent across cultures if there is any possibility of a valid comparison. Without construct equivalence, there is no foundation for cross-cultural comparisons as it "amounts to comparing apples and oranges" (van de Vijver & Poortinga, 2005, p. 47). Nonequivalent constructs lack shared meaning, and this incomplete overlap in concepts suggests that the clinician or researcher must constrain the breadth of measurement to those subfacets of the construct that are shared (He & van de Vijver, 2012). A quintessential example was introduced earlier, albeit from the perspective of bias: Western notions of intelligence accentuate scholastic and analytical skills, whereas non-Western cultures may further emphasize societal concerns and responsibilities. The Luo people of Kenya regard intelligence to encompass analytical skills (*rieko*) plus notions of social respect (*luoro*), social problem solving (*paro*), and comprehension (*winjo*). If the construct of intelligence is considered without regard to these cultural differences, the comparison of test scores across groups will be biased due to the dissimilarity in the definition of the construct. At the measurement level, construct inequivalence is present because the instruments developed in Western cultures to assess intelligence will fail to capture these additional culturally bound components. To mitigate this inequivalence, the clinician or researcher may constrain the assessment to the subfacet of intelligence that is shared (i.e., *rieko*) and acknowledge the incomplete coverage of the construct in one of the cultures (He & van de Vijver, 2012).

Structural Equivalence

Once construct equivalence is established, it is necessary to determine if the constructs share the same underlying dimensions across groups (van de Vijver & Leung, 2011). Structural or functional equivalence thus extends the level of analysis by identifying the number of latent factors extracted from the measurement instrument. Structural equivalence is usually established through exploratory or confirmatory factor analyses and hence is often labeled *factorial invariance*. Using the framework of confirmatory factor analysis (CFA), Meredith (1993) outlined three forms of factorial invariance (i.e., equivalence) that serve as a useful hierarchy, such that each sequential step places additional constraints on the initial factor model (see also Meredith & Teresi, 2006). *Weak factorial invariance* requires only that the factor loadings be equivalent across groups. If the factor loadings are not equivalent across groups, the indicators (i.e., test items) are not measuring the factors similarly across groups. *Strong (scalar) factorial invariance* places an additional constraint, namely that the factor loadings and indicator intercepts be equal across groups. This implies that individuals with the same factor scores have the same expected scores on the indicators. Finally, *strict factorial invariance* requires that factor loadings, intercepts, and residual variances be equal across groups. This degree of factorial invariance is seldom achieved and presently there is no clear consensus on the necessity of strict invariance versus the sufficiency of strong invariance (Deshon, 2004; Vandenberg & Lance, 2000).

Metric Equivalence

In metric or measurement unit equivalence, assessment instruments have the same interval or ratio unit of measurement despite a different origin. The absence of equal origins means that the scale of one instrument is transposed by a constant value when compared with the other instrument (van de Vijver & Tanzer, 2004). The typical example consists of the Celsius and Kelvin temperature scales, which share the same one degree/kelvin unit of measurement, but their origins are offset by 273 units (e.g., 0 kelvins is equivalent to −273 °C). Scores obtained with a Kelvin electric device and a Celsius thermometer cannot be compared directly, but knowing the value of the origin allows for a straightforward mathematical conversion. In cross-cultural studies, whenever the origin of the assessment scale or instrument is unknown, the scores between groups cannot be compared directly. Nonetheless, within-group differences can be compared between groups because the magnitude of those differences will be expressed using the same unit of measurement.

Scalar Equivalence

Scalar or full score is the highest level of equivalence and is evident when instruments with the same measurement unit (metric equivalence) also have the same origin across groups. With scalar equivalence, scores obtained from persons who belong to different cultural groups can be compared in a valid manner—it represents measurement free of bias (van de Vijver & Tanzer, 2004). However, the presence of scalar equivalence may be tenuous and difficult to achieve due to the numerous sources of nuisance variables. It may be noted that any comparison of ethnic groups on some test score, such as a digit span score, using a *t*-test or analysis of variance assumes that the scores have scalar invariance.

FAIRNESS

Fairness is intrinsically linked to the concepts of equivalence and (absence of) bias just discussed, but it transcends those attributes to involve social, historical, and legal ramifications. It is a notion that has evolved with changes in the sociocultural landscape and, perhaps for that reason, lacks any universal definition (Dorans & Cook, 2016). Fairness refers to how psychologists use tests to ensure that they are administered in ways that are appropriate for all examinees and have adequate item content. Psychologists should not (dis)favor any ethnic group in their practices. In addition, fairness refers to how test scores are used to predict future performance, such as success in school, occupation, or therapy. As explained in the following discussion, this meaning of fairness that is related to equity is more contentious.

The *Standards for Educational and Psychological Testing* (American Education Research Association, 1999) defines fairness in four principal ways. The first two definitions are broadly supported whereas the third and fourth definitions are problematic and will be presented only for the sake of completeness. First, fairness represents the *absence of bias*. An unfair test yields different meanings for scores obtained by members of different groups. Within the taxonomy of bias presented earlier, this definition of fairness overlaps with and is operationally comparable to the absence of construct and item bias. There is a tacit assumption that comparable rates of passing scores or endorsement responses across groups are necessary for an instrument to be considered fair. Although this notion has been rejected by professionals in education and psychology, Zieky (2002) notes that the belief that different mean scores between groups reflects unfairness is so widespread that the existence of those differences is likely to trigger charges of bias. In statistical analyses of bias, there is no assumption that the removal of bias should make means equal across groups. There is indeed empirical evidence that the removal of biased items often does not reduce the size of ethnic differences. For example, Breslau, Javaras, Blacker, Murphy, and Normand (2008) compared European Americans, African Americans, and Hispanics on the prevalence of depression using the Composite International Diagnostic Interview. Bias was found in various items, such as items dealing with self-reproach and suicidality, which European Americans endorsed more frequently. However, conclusions about the prevalence of depression were not affected by the removal of biased items.

Second, fairness refers to *equitable treatment* throughout the testing process. Fair treatment of all examinees requires consideration of the context, purpose, and manner in which a test is administered, scored, interpreted, and reported. Factors such as lack of standardized test administration, unequal opportunity to familiarize oneself with the test format, and unequal exposure to practice materials may contribute to unfair measurement. This definition is operationally comparable to the absence of method bias. The *Standards* additionally clarifies that in circumstances in which special accommodations may be necessary (e.g., due to disability), those accommodations are expected to result in more comparable test results than if standardized procedures were left unmodified. In those circumstances, accommodations would not be considered unfair treatment and, in fact, could be required by law.

Third, fairness may refer to *comparable opportunity to learn*. Under this definition, an examinee's low achievement score in a particular educational domain may be viewed as unfair if the opportunities to learn the subject matter were insufficient or inadequate. This conceptualization of fairness lacks universal agreement, partly due to the significant obstacles in the operationalization and estimation of such opportunities for learning.

Finally, fairness may refer to *testing outcomes*, more specifically to the use of test scores to make statements about examinees from different ethnic groups. This type of fairness implies that no groups are (dis)favored by the use of test scores. A good example can be found in the public debate about the use of psychological tests in South Africa in the mid-1990s. When apartheid was abolished and the first free elections were held in 1994, there was a strong feeling in the country among the victims of apartheid that psychological assessment should be prohibited because assessment was used in the apartheid days to legitimize the favorable treatment of Whites and discrimination of other ethnic groups in South Africa (Laher & Cockroft, 2014). Whites had access to better education and scored better on cognitive tests. Psychological test scores reflected this dissimilarity in educational access. The higher scores on tests were used to exclude non-Whites from jobs and education. The debate was not so much about the psychometric properties of any specific tests, but more about the historically dubious role of psychological assessment, including its quasi-objectivity and legitimization of discrimination; in short, the debate was about perceived unfairness (Foxcroft, 1997). In the United States, fairness in testing outcomes has been mainly studied in the domain of schooling and job selection. The question then addressed is whether the link between test scores and outcome, such as school or job success, is the same in all ethnic groups. Linear regression has often been used to examine this link. There is differential prediction (pointing to the absence of fairness) if the regression coefficient of the test to predict job or school success and/or its intercept differ across ethnic groups. The received wisdom in the United States has been for decades that comparisons of regression lines of European Americans, African Americans, and Hispanics do not show sizeable differences and that there is no differential validity if psychometrically adequate tests are used (e.g., Schmidt, Pearlman, & Hunter, 1980). In recent years, however, there is mounting evidence that the identity of regression lines may have been too easily taken for granted and that the early studies that supported the validity of instruments across ethnic groups suffered from weaknesses, notably small sample size (Aguinis, Culpepper, & Pierce, 2010). Studies with small sample sizes have low power, which means that differences in regression coefficients or intercepts have to become large before these are significant. A recent meta-analysis of the correlation between cognitive ability test scores and performance in a school or job context was highest for European and Asian Americans and lowest for Hispanics and African Americans (Berry, Clark, & McClure, 2011). These data suggest that tests are less valid for the latter two groups than for the former two groups.

An integrative perspective on fairness is provided by the Educational Testing Service (ETS), a private organization that develops, administers, and scores tests in over 180 countries. ETS (2014) defines fairness as "the extent to which the inferences made on the basis of test scores are valid for different groups of test takers" and not influenced by construct-irrelevant score variance (p. 19). To maximize fairness, ETS recommends a review of the appropriateness of test materials, evaluation of linguistic demands, and consideration of DIF or other empirical procedures (pp. 20–21). This definition of fairness overlaps with the concepts of construct bias, method bias, and item bias and also considers differential prediction; therefore, it reflects a psychometric and operational construal that is comparable to the first two definitions provided by the *Standards*.

In general, then, clinicians and investigators involved in cross-cultural research will be justified to encapsulate concerns about fairness within the framework of construct, method, and item bias.

STRATEGIES TO IDENTIFY AND REDUCE BIAS AND NONEQUIVALENCE

Numerous strategies exist to minimize the risk of bias at the test development stage and to mitigate the cost of bias for tests already in existence (van de Vijver & Leung, 2011; van de Vijver & Tanzer, 2004).

Minimizing the Risk of Bias

Strategies to minimize the risk of bias are deployed before data collection and target one or more types of bias (Figure 12.1). Two classical procedures to minimize construct bias consist of cultural decentering and convergence. Decentering refers to the simultaneous development of an instrument across several cultures, with the subsequent removal of culture-specific items and retention of the items and concepts that are shared across the sampled cultures. In this manner, the remaining item pool has few cultural specifics (e.g., names of local places, historical references, national heroes) and hence less likelihood of construct bias. This approach is commonly applied in large-scale assessment such as the Programme for International Student Assessment (PISA) where experts from multiple groups are involved in item development (OECD, 2014). These experts evaluate whether items can be translated in their language and whether the item covers the same psychological meaning as in the original English version. Alternatively, convergence entails the independent development of an instrument within each culture and the subsequent administration of all versions of the instrument to individuals in all cultures. As an illustration, a French and a Japanese neuroscience research team wish to conduct a study in which subjects read a series of emotional vignettes and then complete a fMRI task. Each team translates the vignettes into the other language, and both versions of the instrument are then administered to all subjects in each country. By evaluating the different versions it may be possible

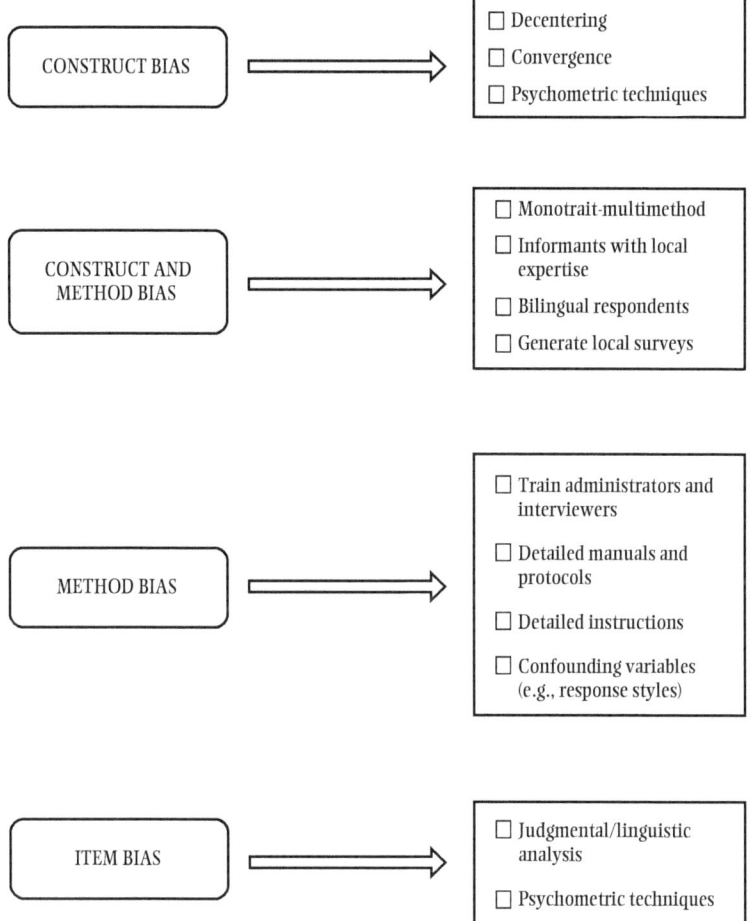

Figure 12.1. Strategies to identify and decrease bias in cross-cultural assessments.

to discern words, phrases, and ideas specific to one culture or shared across the cultures.

Additional strategies have been advanced to minimize the joint risk of construct and method bias. One of these strategies permits the simultaneous evaluation of convergent and discriminant validity by exploring the matrix of correlations across various instruments. In this monotrait-multimethod approach, the correlations among different instruments purported to measure the same trait should be different from zero and sufficiently large to suggest convergent validation. A frequent criticism of this approach, however, is the ambiguity inherent in determining what constitutes a satisfactory result. Further strategies employed to minimize the risk of construct and method bias include the use of informants who are knowledgeable about the local culture and the use of bilingual respondents who can aid in the development and revision of relevant instruments. A related approach is to conduct local surveys that can generate culturally relevant content. Strategies that emphasize procedural standardization across cultural groups further minimize the risk of method bias. These strategies include the extensive training of test administrators and interviewers, the use of detailed manuals and scoring criteria, and the availability of detailed instructions. Other strategies to minimize method bias employ statistical techniques to account for confounding variables (e.g., quantity or quality of education, response styles) that can impact the valid interpretation of results across groups. If a confounding variable has been measured, the opportunity exists to evaluate its influence through covariance or regression analyses.

Strategies to minimize the risk of item bias fall under two kinds of procedures: judgmental and psychometric. Judgmental procedures generally involve linguistic analyses to identify language-related sources of bias, including the examination of grammatical form and structure, meaning and connotation, item format, and words and structures that are prone to be misread (Arffman, 2012). Judgmental reviews additionally consider the impact of inaccurate translations due to omissions or additions as well as semantic, syntactic, or orthographic errors. Arffman (2012) notes that surprisingly few judgmental reviews are conducted to identify language-related sources of bias. Psychometric strategies such as DIF can be deployed at the test development stage, when pilot or normative data are collected or a posteriori for tests already in existence. Thus, psychometric strategies will be presented in the next section.

Mitigating the Cost of Bias

Strategies to mitigate the cost of bias rely on statistical approaches applied once the pertinent data have been collected and are available for analysis. These strategies typically consist of observed or latent structural models to evaluate the presence of construct bias and DIF to evaluate item bias. Structural models test whether the internal structure of the measurement instrument is equivalent between cultural groups and include multidimensional scaling (MDS), exploratory factor analysis (EFA), and CFA (Fischer & Fontaine, 2011).

MULTIDIMENSIONAL SCALING

In MDS, variables are represented as points in geometric space, with the distance between any two items denoting their relatedness (Borg & Groenen, 2005). Items that are more positively correlated tend to share a smaller distance than items that are more negatively correlated. In this manner, the full configuration of all pairwise comparisons of (dis)similarity can be displayed graphically and analyzed visually in multidimensional space or analyzed with mathematical models to evaluate the dimensionality of the data. Dimensionality in MDS refers to the number of coordinate axes that are necessary for the best fit of the data points. A unidimensional solution is thus a single line, a two-dimensional solution is a plane with an x-axis and y-axis and a three-dimensional solution is a sphere with an x-axis, y-axis, and z-axis.

For cross-cultural comparisons on a given instrument, a distance matrix of (dis)similarities is prepared for the reference group. Items that are more closely related are placed near each other and those that are very different are placed far away from each other. The investigator then applies either metric (interval) or nonmetric (ordinal) scaling based on the level of measurement. The number of dimensions or factors extracted from the reference cultural group is determined using the stress goodness-of-fit measure, the scree test, or interpretability. Stress values range from zero to 1, with smaller values reflecting a good configural fit to the data and larger values indicating worse fit. Stress, however, is a poor indicator for solutions in which the number of items approaches the number of dimensions. In a scree test, stress values are plotted on the vertical axis against an ascending number of dimensional solutions on the horizontal axis. Typically, the magnitude of stress values declines precipitously after a few dimensions and then levels off. To select the appropriate number of dimensions, the investigator identifies the "elbow" or point at which the descending point line begins to inflect, suggesting a discontinuity in stress values. In practice, however, clear elbows are seldom evident and the investigator may need to rely on heuristic rules to identify the optimal number of dimensions. Interpretability refers to a subjective judgment about which choice of dimensional configurations is preferable. In certain circumstances, two separate configurations may have comparable stress values, but if one configuration has three dimensions that are easily interpreted and another configuration has two dimensions yet one of them cannot be interpreted, the solution with three dimensions would be preferred. The last step when using MDS for cross-cultural comparisons consists of eliciting a configuration for the nonreference group based on the dimensionality of the reference group. Several procedures, such as generalized Procrustes analysis, are used to align the coordinate systems between the groups, correlate their fit, and determine their congruence (Fischer & Fontaine, 2011). Although multigroup MDS offers flexibility in its handling of a broad range of data, the weakness at its core lies in its descriptive approach, lack of sophisticated statistical tests to evaluate the structural differences between groups, and limited scope for testing specific models.

EXPLORATORY FACTOR ANALYSIS

In multigroup EFA, variables that are highly correlated with each other are combined into linear vectors called factors. The first factor accounts for most of the shared or common variance in item responses, with each subsequent uncorrelated factor accounting for less and less of the remaining variance. The main goal of EFA is to obtain the fewest number of factors that are interpretable and account for most of the common variance. (A related procedure, principal component analysis, seeks to maximize the amount of total variance with the fewest number of linear vectors called components. The two procedures are mathematically comparable in producing linear combinations that summarize the observed correlation matrix, are implemented in an analogous manner in statistical software packages, and frequently yield equivalent results.) To decide how many factors are optimal in the reference group, the investigator can choose among several strategies. One approach is to consider the eigenvalues or the amount of variance across all observed variables accounted for by the factor. The Kaiser criterion suggests that factors with eigenvalues greater than 1 should be retained (Kaiser, 1960). However, this heuristic often yields too many factors (Costello & Osborne, 2005). Alternatively, a scree test can be used to plot the eigenvalues on the vertical axis against the ascending number of factors on the horizontal axis. As in MDS, the eigenvalues tend to drop off sharply and the identification of the inflection point where the descending line changes slope hints at the number of factors to be retained. If this criterion does not produce a clear solution, a parallel analysis can be conducted (Horn, 1965). Random data using the sample size and number of variables in the EFA are generated. A scree test of these random data is compared with the scree test of the real data. The number of factors in the real data is then taken to be the number where the eigenvalues of the random data become smaller than the eigenvalues of the real data. If the investigator has an a priori theoretical expectation for the number of factors, another approach is to select the number of factors that matches this expectation. Lastly, the investigator may choose to select enough factors to reach a predetermined absolute percentage of variance (e.g., select as many factors as possible until 75% of the variance is explained).

To achieve simple structure, in which the indicators have large loadings on one factor but small loadings on other factors, the initial solution may need to be rotated. Two types of rotation are possible: orthogonal or oblique. In an orthogonal rotation, the factors are assumed to be uncorrelated with each other whereas in oblique rotation the factors can correlate. In many situations in the social sciences, oblique rotation is more reasonable due to the greater likelihood that factors will be associated with each other. Once the factors have been rotated to achieve simple structure, the factor loading matrix is examined to interpret the solution. As a convention, factor loading coefficients greater than 0.30 suggest that the indicator loads onto that factor. Variables with cross-factor loadings suggest that the item contributes information to more than one factor. If factor loadings appear to be uninterpretable, it may be necessary to eliminate those items with poor fit and rerun the analysis or consider whether more critical problems exist with item construction or scale design.

For the nonreference cultural group, EFA seeks to extract the same number of factors obtained in the reference group. An orthogonal Procrustes rotation may be necessary to achieve simple structure in the nonreference group (Fischer & Fontaine, 2011). Next, the congruence between the two factorial structures is evaluated through one or more coefficients of fit. The most commonly used of these coefficients of congruence is Tucker's phi (Tucker, 1951), with values in the range of 0.85 to 0.94 corresponding to a fair degree of similarity and values higher than 0.95 implying that the multigroup factorial structures are nearly equal (Lorenzo-Seva & ten Berge, 2006). In all, the strength of the EFA approach is that it can be implemented with accessible software packages (Note: syntax for the Procrustean rotation is provided by Fischer and Fontaine, 2011.) However, the evaluation of multigroup congruence is aimed at the factorial level and does not allow for the examination of item-level incongruence.

CONFIRMATORY FACTOR ANALYSIS

CFA is more flexible than EFA insofar as it permits the investigator to specify one or more factorial models a priori and test their fit to the data (i.e., an estimated covariance matrix is compared with the observed covariance matrix). The first step in multigroup CFA consists of specifying the measurement model in the cultural reference group. Each indicator or item score is considered to be caused by the combination of a latent factor and measurement error, with a minimum of two indicators necessary for each factor. If more than one factor is hypothesized, it is possible to specify in the model if those factors are independent or correlated with one another. Next, one of several estimation algorithms is used to fit the predicted covariance matrix to the sample covariance matrix. Among these, maximum likelihood estimation is the most commonly used method due to its robustness against violations of normality (Chou & Bentler, 1995). Numerous goodness-of-fit indices are available and can be leveraged to evaluate model fit. Among these, the classic chi-square statistic is usually reported yet it is sensitive to sample size and departures from normality (Hu & Bentler, 1995). Consequently, model fit tends to be evaluated by supplementing chi-square with one or more of the following indices: relative chi-square (χ^2/df; Wheaton, Muthen, Alwin, & Summers, 1977), goodness-of-fit index (GFI; Bentler, 1983), adjusted GFI (AGFI; Bentler, 1983), standardized root mean squared residual (SRMR; Bentler, 1995), comparative fit index (CFI; Bentler, 1990), Tucker–Lewis index (TLI; Tucker & Lewis, 1973), and expected cross-validation index (ECVI; Browne & Cudeck, 1993). No established guidelines exist for the relative chi-square index, but usually a value less than 2 or 3 is considered acceptable. In both the GFI and AGFI, values equal to or greater than 0.90 suggest adequate model fit. Hu and Bentler (1998, 1999) recommend a two-index strategy consisting of the SRMR with one of several incremental fit indices, such as the CFI or TLI. A good fit would be indicated by an SRMR less than or equal to 0.08 and a CFI or TLI value close to or above 0.95. Finally, ECVI can be used to determine if the model cross-validates in an independent sample of comparable size, with smaller ECVI values suggesting better model fit (MacCallum & Austin, 2000). If adequate fit is not established, it is possible to compare the original model against more restrictive (nested) models to determine if one is significantly better than the others. Additionally, post hoc modifications may improve model fit, but these modifications often capitalize on chance and should not be pursued if they are not theoretically substantiated (MacCallum, Roznowski, & Necowitz, 1992).

If model fit cannot be established in the reference cultural group despite consideration of nested models and modifications, multigroup CFA should be discontinued (Fischer & Fontaine, 2011). However, if model fit is established, then CFA allows for the evaluation of factorial invariance between the cultural reference and nonreference groups. As discussed earlier, the first form of invariance requires that the factor loadings be similar between the groups (weak invariance). If weak invariance is not established, it may be possible to consider partial invariance, in which invariance is tested for a subset of items or subtests (Byrne, Shavelson, & Muthen, 1989). The next level of constraints is strong or scalar invariance, which requires similar factor loadings and intercepts. If these constraints are met, it may be possible to test for invariant factor loadings, intercepts, and residuals between the groups (strict invariance).

For instance, Bowden, Lange, Weiss, and Saklofske (2008) investigated the invariance of the Wechsler Adult Intelligence Scale-III (WAIS-III) between the U.S. and Canadian standardization samples. The authors first examined the invariant properties of the WAIS-III in a subsample from both countries who completed all 13 subtests, including letter-number sequencing, followed by an analysis of the full standardization samples. Using maximum likelihood CFA, the WAIS-III showed strict factorial invariance between the U.S. and Canadian subsamples and full standardization samples. Another example is provided by Blankson and McArdle (2013), who investigated the invariance of episodic memory and mental status indicators across ethnicity and gender in over 15,000 adults from the Asset and Health Dynamics Among the Oldest Old (AHEAD) study and Health and Retirement Study (HRS). The authors evaluated the fit of a two-factor model (memory and mental status) versus a one-factor model (mental status) on each ethnic and gender group and tested the invariance of the two-factor model (memory and mental status) versus the one-factor model (mental status) across each ethnic and gender group. The results showed weak invariance across White, Black, and Hispanic ethnicity for the two-factor model and the mental status factor, although a reasonable case for strict invariance could be advanced based on practical fit indices. Regarding gender, results differed for the two-factor model versus the mental status factor. Strong invariance held for the two-factor model, although, again, a reasonable case could be made for strict invariance. In contrast, only weak invariance across gender was evident for the mental status factor.

DIFFERENTIAL ITEM FUNCTIONING

There are two types of DIF: uniform and nonuniform (Hambleton, Swaminathan, & Rogers, 1991). Uniform DIF occurs when the probability of a correct response is greater for one group than another across all levels of the ability spectrum. In contrast, nonuniform DIF occurs when the probability of a correct response varies across the ability spectrum (i.e., an interaction exists between group membership and ability). A visual depiction of lack of bias (no DIF), uniform DIF, and nonuniform DIF in a hypothetical item is shown in Figure 12.2. Pedraza et al. (2009) evaluated the presence of uniform and nonuniform DIF between Caucasian and African American older adults on items from the Boston Naming Test (an instrument in which the examinee must name a series of object drawings). Under comparable testing conditions, a randomly selected Caucasian adult and a randomly selected African American adult with equal naming ability should have a similar probability to provide correct item responses. If the conditional probability of a correct response differs between the two ability-matched groups, then the item demonstrates DIF. Nonuniform DIF would be present, for instance, if the probability of a correct item response is higher for Caucasians than for African Americans at the lower levels of naming ability, but higher for African Americans than for Caucasians at the higher naming ability levels. Using methods from item response theory, Pedraza et al. found that 12 items demonstrated DIF, of which 9 were due to uniform DIF and 3 were due to nonuniform DIF.

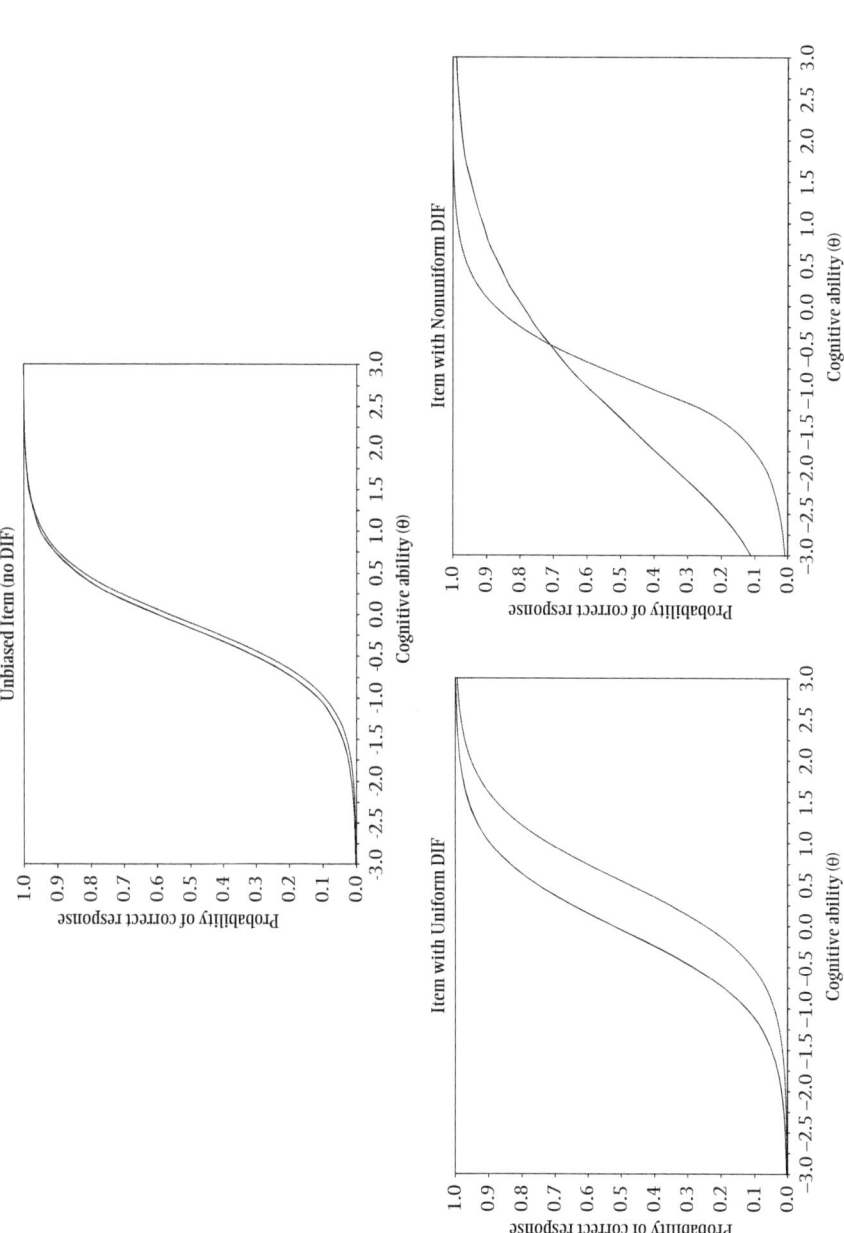

Figure 12.2. Examples of items without bias and with uniform and nonuniform bias (DIF).

Numerous statistical procedures have been developed to detect DIF in dichotomous or polytomous models, including the Mantel–Haenszel method, logistic regression, and item response theory (IRT; Sireci, Patsula, & Hambleton, 2005; Teresi, 2002; van de Vijver & Leung, 2011). The Mantel–Haenszel method consists of stratified 2×2 contingency tables used to calculate a weighted odds ratio of the correct item response for one group versus another (Holland & Thayer, 1988). A key advantage of the Mantel–Haenszel method is that it can be used with small samples. The logistic regression approach predicts performance on a specific item based on score and group membership (Rogers & Swaminathan, 1993; Swaminathan & Rogers, 1990). It is more powerful to detect nonuniform DIF and equally powerful to detect uniform DIF but yields a higher rate of false positives. In contrast to these methods, an IRT model expresses the probabilistic association between a person's observable item responses and their unobservable but estimated ability level (Hambleton & Swaminathan, 1985; van der Linden & Hambleton, 1997). Two fundamental parameters in IRT models are item discrimination and item difficulty. Item discrimination (α) represents the degree to which the item can distinguish individuals with higher ability from those with lower ability. Item difficulty (β) represents the ability level at which a person has a 50% chance of responding correctly to an item. Under a two-parameter logistic IRT model, latent ability level (θ) is thus estimated based upon the person's pattern of passed versus failed items when considering each item's discrimination and difficulty parameters. For this reason, IRT provides an attractive framework for the investigation of uniform and nonuniform DIF (Teresi, Kleinman, & Ocepek-Welikson, 2000). DIF in linear data is often studied with CFA, in which a difference in loadings is associated with non-uniform DIF and a difference in intercept with uniform DIF.

SUMMARY AND CONCLUSIONS

Clinical and experimental instruments used across cultures are vulnerable to threats from nuisance variables at the construct (construct bias), administration (method bias), and item (item bias) level of analysis. These sources of bias can have a devastating effect on the equivalence of test scores and their meaningful interpretation across cultural groups. Consequently, the evaluation of bias and measurement invariance may be as crucial in cross-cultural studies as the demonstration of validity and reliability is for all research involving psychometric tests. Invariance is an implicit assumption in the application of assessment instruments across groups, yet it is seldom demonstrated empirically. At a minimum, strong (scalar) invariance allows for the valid comparison of the observed test means and implies that meaningful differences in latent means exist between the groups.

The demonstration of cross-cultural invariance for established tests is important, but it could be argued that a more critical step is the maximization of validity at the initial test design and development phase. Decentering and convergence allow the investigator to pare down the presence of cultural particulars and lessen the risk of construct bias. This process can be enhanced further by engaging local experts and bilingual respondents and conducting pilot studies in each cultural group. The risk of method bias can be minimized through extensive training of test administrators, interviewers, and other key personnel, as well as the provision of detailed manuals, scoring criteria, and standardized administration instructions. Linguistic-judgmental strategies can be leveraged to explore item bias, with DIF applied to the pilot version of a test to identify additional problems.

Because bias is not an intrinsic property of items or tests, but an emergent characteristic of the cross-cultural application of those items or tests, it may be difficult to design an instrument that is fully immune from bias. Nonetheless, by considering these cultural, linguistic, and psychometric considerations, it is possible to design assessment instruments with reduced likelihood of bias. Modern psychometric procedures then can be deployed upon existing instruments to establish their equivalence and facilitate the quality of cross-cultural measurement, thereby enhancing the validity of our inferences and the usefulness of our work .

REFERENCES

Aguinis, H., Culpepper, S. A., & Pierce, C. A. (2010). Revival of test bias research in preemployment testing. *Journal of Applied Psychology, 95,* 648–680. doi:10.1037/a0018714

American Education Research Association, American Psychological Association, and the National Council on Measurement in Education. (1999). *Standards for educational and psychological testing.* Washington, DC: American Education Research Association.

Arffman, I. (2012). International education studies: Increasing their linguistic comparability by developing judgmental reviews. *International Scholarly Research Network Education, 12,* 1–11. doi:10.5402/2012/179824

Bentler, P. M. (1983). Some contributions to efficient statistics for structural models: Specification and estimation of moment structures. *Psychometrika, 48,* 493–571.

Bentler, P. M. (1990). Comparative fit indexes in structural models. *Psychological Bulletin, 107,* 238–246. https://doi.org/10.1037/0033-2909.107.2.238

Bentler, P. M. (1995). *EQS structural programming manual.* Encino, CA: Multivariate Software.

Berry, C. M., Clark, M. A., & McClure, T. K. (2011). Racial/ethnic differences in the criterion-related validity of cognitive ability tests: A qualitative and quantitative review. *Journal of Applied Psychology, 96,* 881–906. doi:10.1037/a0023222

Blankson, A. N., & McArdle, J. J. (2013). Measurement invariance of cognitive abilities across ethnicity, gender, and time among older Americans. *Journals of Gerontology, Series B: Psychological Sciences and Social Sciences, 70,* 386–397. doi:10.1093/geronb/gbt106

Borg, I., & Groenen, P. J. F. (2005). *Modern multidimensional scaling: Theory and applications* (2nd ed.). New York, NY: Springer.

Borsboom, D., Romeijn, J. W., & Wicherts, J. M. (2008). Measurement invariance versus selection invariance: Is fair selection possible? *Psychological Methods, 13,* 75–98. doi:10.1037/1082-989X.13.2.75

Bowden, S. C., Lange, R. T., Weiss, L. G., & Saklofske, D. (2008). Invariance of the measurement model underlying the Wechsler Adult Intelligence Scale-III in the United States and Canada. *Educational and Psychological Measurement, 68,* 1024–1040. doi:10.1177/0013164408318769

Breslau, J., Javaras, K. N., Blacker, D., Murphy, J. M., & Normand, S. L. T. (2008). Differential item functioning between ethnic groups in the epidemiological assessment of depression. *Journal of Nervous and Mental Disease, 196,* 297–306. doi:NMD.0b013e31816a490e

Browne, M. W., & Cudeck, R. (1993). Alternate ways of assessing model fit. In K. A. Bollen & J. S. Long (Eds.), *Testing structural equation models* (pp. 136–162). Newbury Park, CA: SAGE.

Byrne, B. M., Shavelson, R. J., & Muthén, B. (1989). Testing for the equivalence of factor covariance and mean structures: The issue of partial measurement invariance. *Psychological Bulletin, 105,* 456–466.

Chou, C., & Bentler, P. M. (1995). Estimates and tests in structural equation modeling. In R. H. Hoyle (Ed.), *Structural equation modeling: Concepts, issues, and application* (pp. 37–55). Thousand Oaks, CA: SAGE.

Clauser, B. E., & Mazor, K. M. (1998). Using statistical procedures to identify differentially functioning test items. *Educational Measurement: Issues and Practice, 17,* 31–44.

Cleary, T. A. (1968). Test bias: Prediction of grades of Negro and White students in integrated colleges. *Journal of Educational Measurement, 5,* 115–124.

Costello, A. B., & Osborne, J. W. (2005). Best practices in exploratory factor analysis: Four recommendations for getting the most from your analysis. *Practical Assessment, Research & Evaluation, 10*(7), 1–9. Retrieved from hhttp://pareonline.net/pdf/v10n7.pdf

Davis, R. E., Resnicow, K., & Couper, M. P. (2011). Survey response styles, acculturation, and culture among a sample of Mexican American adults. *Journal of Cross-Cultural Psychology, 42,* 1219–1236. doi:10.1177/0022022110383317

Deshon, R. P. (2004). Measures are not invariant across groups without error variance homogeneity. *Psychology Science, 46,* 137–149.

Dorans, N. J., & Cook, L., L. (2016). *Fairness in educational assessment and methods.* New York, NY: Taylor & Francis.

Educational Testing Service. (2014). *ETS standards for quality and fairness.* Princeton, NJ: Author.

Fernandez, A. L., & Marcopulos, B. A. (2008). A comparison of normative data for the Trail Making Test from several countries: Equivalence of norms and considerations for interpretation. *Scandinavian Journal of Psychology, 49,* 239–246. doi:10.1111/j.1467-9450.2008.00637.x

Fisher, R., & Fontaine, J. R. J. (2011). Methods for investigating structural equivalence. In D. Matsumoto & F. J. van de Vijver (Eds.), *Cross-cultural research methods in psychology* (pp. 179–215). Cambridge, England: Cambridge University Press.

Foxcroft, C. D. (1997). Psychological testing in South Africa: Perspectives regarding ethical and fair practices. *European Journal of Psychological Assessment, 13,* 229–235. doi:10.1027/1015-5759.13.3.229

Funes, C. M., Rodriguez, J. H., & Lopez, S. R. (2016). Norm comparisons of the Spanish-language and English-language WAIS-III: Implications for clinical assessment and test adaptation. *Psychological Assessment, 28,* 1709–1715. doi:10.1037/pas0000302

Gasquoine, P. G., Croyle, K. L., Cavazos-Gonzalez, C., & Sandoval, O. (2007). Language of administration and neuropsychological test performance in neurologically intact Hispanic American bilingual adults. *Archives of Clinical Neuropsychology, 22,* 991–1001. doi:10.1016/j.acn.2007.08.003

Grigorenko, E. L., Geissler, P. W., Prince, R., Okatcha, F., Nokes, C., Kenny, D. A., . . . Sternberg, R. J. (2001). The organisation of Luo conceptions of intelligence: A study of implicit theories in a Kenyan village. *International Journal of Behavioral Development, 25,* 367–378. doi:10.1080/01650250042000348

Hambleton, R. K., & Swaminathan, H. (1985). *Item response theory. Principles and applications.* Boston, MA: Kluwer-Nijhoff.

Hambleton, R. K., Swaminathan, H., & Rogers, H. J. (1991). *Fundamentals of item response theory.* Newbury Park, CA: SAGE.

He, J., & van de Vijver, F. (2012). Bias and equivalence in cross-cultural research. *Online Readings in Psychology and Culture, 2*(2). https://doi.org/10.9707/2307-0919.1111

Holland, P. W., & Thayer, D. T. (1988). Differential item performance and the Mantel-Haenszel procedure. In H. Wainer, & H. I. Braun (Eds.), *Test validity* (pp. 129–145). Hillsdale, NJ: Erlbaum.

Horn, J. L. (1965). A rationale and test for the number of factors in factor analysis. *Psychometrika, 30,* 179–185. doi:10.1007/BF02289447

Hu, L. T., & Bentler, P. M. (1995). Evaluating model fit. In R. H. Hoyle (Ed.), *Structural equation modeling: Concepts, issues, and applications* (pp. 76–99). Thousand Oaks, CA: SAGE.

Hu, L. T., & Bentler, P. M. (1998). Fit indices in covariance structure modeling: Sensitivity to underparameterized model misspecification. *Psychological Methods, 3,* 424–453.

Hu, L. T., & Bentler, P. M. (1999). Cutoff criteria for fit indexes in covariance structure analysis: Conventional criteria versus new alternatives. *Structural Equation Modeling, 6,* 1–55.

Jensen, A. R. (1980). *Bias in mental testing.* New York, NY: Free Press.

Johnson, T. P. (2006). Methods and frameworks for cross-cultural measurement. *Medical Care, 44*(11 Suppl 3), S17–S20.

Johnson, T. P., Shavitt, S., & Holbrook, A. L. (2011). Survey response styles across cultures. In D. Matsumoto & F. J. van de Vijver (Eds.), *Cross-cultural research methods in psychology* (pp. 130–176). Cambridge, England: Cambridge University Press.

Kaiser, H. F. (1960). The application of electronic computers to factor analysis. *Educational and Psychological Measurement, 20,* 141–151. doi:10.1177/001316446002000116

Laher, S., & Cockcroft, K. (2014). Psychological assessment in post-apartheid South Africa: The way forward. *South African Journal of Psychology, 44,* 303–314. doi:10.1177/0081246314533634

Lorenzo-Seva, U. & ten Berge, J. M. F. (2006). Tucker's congruence coefficient as a meaningful index of factor similarity. *Methodology, 2,* 57–64. doi:10.1027/1614-1881.2.2.57

MacCallum, R. C., & Austin, J. T. (2000). Applications of structural equation modeling in psychological research. *Annual Review of Psychology, 51,* 201–226.

MacCallum, R. C., Roznowski, M., & Necowitz, L. B. (1992). Model modifications in covariance structure analysis: The problem of capitalization on chance. *Psychological Bulletin, 111,* 490–504.

Marin, G., Gamba, R. J., & Marin, B. V. (1992). Extreme response style and acquiescence among Hispanics: The role of acculturation and education. *Journal of Cross-Cultural Psychology, 23,* 498–509.

Meredith, W. (1993). Measurement invariance, factor analysis and factorial invariance. *Psychometrika, 58,* 525–543.

Meredith, W., & Teresi, J.A. (2006). An essay on measurement and factorial invariance. *Medical Care, 44*(11 Suppl. 3), S69–S77. https://doi.org/10.1097/01.mlr.0000245438.73837.89

Messick, S. (1995). Validity of psychological assessment. *American Psychologist, 50,* 741–749.

OECD. (2014). *PISA 2012 technical report.* Retrieved from https://www.oecd.org/pisa/pisaproducts/PISA-2012-technical-report-final.pdf

Pedraza, O., Graff-Radford, N. R., Smith, G. E., Ivnik, R. J., Willis, F. B., Petersen, R. C., & Lucas, J. A. (2009). Differential item functioning of the Boston Naming Test in cognitively normal African American and Caucasian older adults. *Journal of the International Neuropsychological Society, 15,* 758–768. https://doi.org/10.1017/S1355617709990361

Poortinga, Y. H. (1989). Equivalence of cross-cultural data: An overview of basic issues. *International Journal of Psychology, 24,* 737–756.

Rogers, H. J., & Swaminathan, H. (1993). Comparison of the logistic regression and Mantel–Haenszel procedures for detecting differential item functioning. *Applied Psychological Measurement, 27,* 27–51. doi:10.1177/014662169301700201

Saklofske, D. H., van de Vijver, F. J., Oakland, T., Mpofu, E., & Susuki, L. A. (2015). Intelligence and culture: History and assessment. In S. Goldstein, D. Princiotta, & J. A. Naglieri (Eds.), *Handbook of intelligence: Evolutionary theory, historical perspective, and current concepts* (pp. 341–366). New York, NY: Springer Science+Business Media.

Schmidt, F. L., Pearlman, K., & Hunter, J. E. (1980). The validity and fairness of employment and educational tests for Hispanic Americans: A review and analysis. *Personnel Psychology, 33,* 705–724. doi:10.1111/j.1744-6570.1980.tb02364.x

Sireci, S. G., Patsula, L., & Hambleton, R. K. (2005). Statistical methods for identifying flaws in the test adaptation process. In R. K. Hambleton, P. F. Merenda, & C. D. Spielberger (Eds.), *Adapting educational and psychological tests for cross-cultural assessment* (pp. 93–115). Mahwah, NJ: Erlbaum.

Sternberg, R. J., Nokes, C., Geissler, P. W., Prince, R., Okatcha, F., Bundy, D. A., & Grigorenko, E. L. (2001). The relationship between academic and practical intelligence: A case study in Kenya. *Intelligence, 29,* 401–418.

Swaminathan, H., & Rogers, H. J. (1990). Detecting differential item functioning using logistic regression procedures. *Journal of Educational Measurement, 27,* 361–370. https://doi.org/10.1111/j.1745-3984.1990.tb00754.x

Teresi, J. A. (2002). Statistical methods for examination of differential item functioning (DIF) with applications to cross-cultural measurement of functional, physical, and mental health. In J. H. Skinner, J. A. Teresi, D. Holmes, S. M. Stahl, & A. L. Stewart (Eds.), *Multicultural measurement in older populations* (pp. 23–34). New York, NY: Springer.

Teresi, J. A., Kleinman, M., & Ocepek-Welikson, K. (2000). Modern psychometric methods for detection of differential item functioning: Application to cognitive assessment measures. *Statistics in Medicine, 19,* 1651–1683.

Tucker, L. R. (1951). *A method for synthesis of factor analysis studies* (Personnel Research Section Report No. 984). Washington, DC: Department of the Army.

Tucker, L. R., & Lewis, C. (1973). A reliability coefficient for maximum likelihood factor analysis. *Psychometrika, 38,* 1–10.

Van de Vijver, F., & Leung, K. (2011). Equivalence and bias: A review of concepts, models, and data analytic procedures. In D. Matsumoto & F. J. van de Vijver (Eds.), *Cross-cultural research methods in psychology* (pp. 17–45). Cambridge, England: Cambridge University Press.

Van de Vijver, F., & Poortinga, Y. H. (2005). Conceptual and methodological issues in adapting tests. In R. K. Hambleton, P. F. Merenda, & C. D. Spielberger (Eds.), *Adapting educational and psychological tests for cross-cultural assessment* (pp. 39–63). Mahwah, NJ: Erlbaum.

Van de Vijver, F., & Poortinga, Y. H. (2016). On item pools, swimming pools, birds with webbed feet, and the professionalization of multilingual assessment. In C. S. Wells & M. Faulkner-Bond (Eds.), *Educational measurement: From foundations to future* (pp. 273-290). New York, NY: Guilford.

Van de Vijver, F., & Tanzer, N. K. (2004). Bias and equivalence in cross-cultural assessment: An overview. *Revue européenne de psychologie appliquée, 54,* 119-135. doi:10.1016/j.erap.2003.12.004

Van der Linden, W. J., & Hambleton, R. K. (Eds.). (1997). *Handbook of modern item response theory.* New York, NY: Springer-Verlag.

Vandenberg, R. J., & Lance, C. E. (2000). A review and synthesis of the measurement invariance literature: Suggestions, practices, and recommendations for organizational research. *Organizational Research Methods, 3,* 4-69.

Wheaton, B., Muthen, B., Alwin, D. F., & Summers, G. F. (1977). Assessing reliability and stability in panel models. In D. R. Heise (Ed.), *Sociology methodology 1977* (pp. 84-136). San Francisco, CA: Jossey-Bass.

Yang, S.-Y., & Sternberg, R. J. (1997). Taiwanese Chinese people's conceptions of intelligence. *Intelligence, 25,* 21-36.

Zieky, M. (2002). Ensuring the fairness of licensing tests. *CLEAR Exam Review, 12*(1), 20-26.

Index

Tables and figures are indicated by *t* and *f* following the page number

For the benefit of digital users, indexed terms that span two pages (e.g., 52–53) may, on occasion, appear on only one of those pages.

AACN. *See* American Academy of Clinical Neuropsychology
ABCA7 gene, 238–39
Academic intelligence, 254–55
Acculturation, 13*f*, 13–14, 14*f*, 218
Acquiescent response style, 256
AD. *See* Alzheimer's disease
Adaptation, 175
Adjusted GFI (AGFI), 267
Administration bias, 256–57, 270
AERA. *See* American Educational Research Association
African Americans
   AD in, 234–35, 236–38, 242, 244
   cognitive decline in, 235–36, 241–42, 243
   dementia in, 237
   depression in, 100–3, 106, 111–12
   mood disorders in, 101*t*, 102–3
   parental education, 215–16
   sociocultural factors, 242
   testing outcomes, 261
Age-related differences
   in memory, 83*f*, 83, 88–89
   positivity effects, 88–89
   in visual processing, 143
AGFI (adjusted goodness-of-fit index), 267
Aging
   effects of bilingualism on, 151–68
   effects on executive control pathways, 160–61
   effects on executive control systems, 157–59
   effects on memory systems, 156–57

population, 151, 233
risk factors for, 238–42
AHEAD (Asset and Health Dynamics Among the Oldest Old) study, 267–68
AHS (Analysis-Holism Scale), 81
Alai Mountains, 38–39
Alaska Natives, 105
All-Russian Congress on Psychoneurology, 35
Alzheimer's disease (AD), 151–52
   bilingualism and, 154–55, 156, 162
   clinicopathological associations, 236–38
   cognitive function and decline in, 235–36
   definition of, 234
   descriptive epidemiology, 234–35
   diagnosis of, 234, 242
   early-stage, 156
   future directions for research, 162
   genetics of, 238–39
   incidence of, 234–35
   neuropathology of, 152
   pathology of, 236–38
   preclinical stages, 156
   prevalence of, 234–35
   psychosocial factors, 241–42
   race-associated risk, 244
   racial disparities, 233–51
   risk factors for, 238–42
   sociocultural factors, 242
American Academy of Clinical Neuropsychology (AACN)
   position on examination language, 187
   Relevance 2050 initiative, 215, 216–17

American Educational Research
   Association (AERA), 260
   *Standards for Education and
   Psychological Testing,* 65, 253–54,
   260–61, 262
American English language
   proficiency, 222
American Indians, 105, 218, 226–27
American Psychological Association
   (APA), 70–71
   Division 45, 60–61
   *Standards for Education and
   Psychological Testing,* 65, 253–54,
   260–61, 262
Americans
   autobiographical memory, 84, 91–92
   emotional memory, 88–89, 93
   information processing, 81–82
   memory for objects and scenes, 85–86
   memory organization, 84
   memory specificity, 84
   self-reference memory, 89–90
   source/context memory, 86–87
   visual processing, 126, 127f, 130, 131f
Analysis-Holism Scale (AHS), 81
Analytical intelligence, 254–55
Analytic thought, 17
Angola, 178
Angular gyrus (AG), 205, 207f
Anthropology, 1–2
Apolipoprotein E *(APOE)* gene, 238–39
Application, 175
Arabic speakers, 216
*Archives of Clinical
   Neuropsychology,* 182–83
Argentina, 9
ASD (autism spectrum disorders), 211
Asian Americans, 104
   autobiographical memory, 91
   bilingualism, 172
   cultural considerations for, 184–85
   language diversity, 189
   mood disorders in, 104–5
   neuropsychological assessment of,
      184, 185–86
   non-English speakers, 169, 170–71
   parental education, 215–16
   self-reference memory, 91
   US population, 184
Asian and Pacific Islander languages, 169

Asian and Pacific Islanders
   cultural and historical considerations
      for, 184–85
   neuropsychological assessment
      of, 185–86
Asian Canadians, 89–90
Asian Indians, 184
Asian languages, 185–86, 189
Asians, 91, 112. *See also* East Asians
Assembly, 175
Assessment
   bicultural, 227–28
   bilingual, 225–26
   of bilingual children, 223–27
   culturally biased, 222
   of ethnic minorities, 55–80
   future directions, 227–28
   neuropsychological, 55–80, 169–99,
      187f, 223–27
   of non-English speakers, 169–99
Assessment tools, 105–11
Asset and Health Dynamics Among the
   Oldest Old (AHEAD) study, 267–68
Assimilation, 13–14, 14f
Association of Black Psychologists, 68–69
Association strength, 200–2
Ataque de nervios (attacks of nerves), 104
Attention
   caretaker feedback or attention, 125–26
   cultural differences in, 81–82, 93
   cultural influences on, 142–44
Australia, 169
Autism
   developmental changes in, 208–10, 210f
   future directions for research, 210–11
Autism spectrum disorders (ASD), 211
Autobiographical memory, 84, 91–92, 93
Aztecs, 177–78

Backgrounds, visual, 126–28
Back translation, 174–75
*Batería Neuropsicológica de Funciones
   Ejecutivas y Lóbulos Frontales*
   (BANFE, Neuropsychological Battery
   of Executive Functions and Frontal
   Lobes), 181–82
Beck Depression Inventory–second edition
   (BDI-II), 106–7, 110–11
Behavior
   cultural differences in, 139–40, 140f

cultural influences on visual
  perception, 126–38
responses associated with visual
  processing, 126–28, 127f
Belgians, 135–37
Bias, 252–53, 270
  absence of, 260
  administration, 256–57, 270
  construct, 254–55, 270
  cultural, 125–26, 141–43, 222
  definitions of, 253–54
  instrument, 256
  item, 257, 270
  method, 255–57, 270
  predictive, 253–54
  sample, 255–56
  selection, 243–44
  strategies to identify and reduce,
    262–70, 263f
  strategies to minimize risk of, 262–64,
    263f, 270–71
  strategies to mitigate cost of, 264–70
Bicultural assessment, 227–28
Biculturalism, 13–14, 14f
Bilingual assessment, 225–26
Bilingual children, 70
  culturally appropriate treatment
    for, 226–27
  neuropsychological assessment of, 223–27
Bilingualism, 62–63, 64–65, 69, 219–20
  circumstantial, 219
  clinical aspects, 173–74
  competition model, 220
  effects on cognitive reserve, 153–56,
    160–61, 161f, 163
  effects on dementia, 153
  effects on education, 220–21
  effects on executive control systems,
    157–59, 159f, 160f, 163
  effects on memory systems, 156–57
  future directions for research, 162
  and immigration, 154–55
  neural bases, 172–73
  neural mechanisms, 160–61, 161f
  in older adults, 151–68, 160f
  protective effects of, 154–56, 162–63
  sequential/successive, 171, 219–20
  simultaneous, 171, 219–20
  in US, 170, 171–74, 216
Bipolar disorder, 101t

Blacks. *See also* African Americans
  parental education, 215–16
Bolivia, 177–78
Borrowings, 179, 189
Boston Naming Test, 183
Brain
  aging, 151–68, 238–42
  cuneus, 210f, 210
  fusiform face area (FFA), 18–19,
    133–35, 134f
  gray matter volume, 156
  hippocampus, 93
  inferior frontal cortex, 205, 206f, 210f, 210
  inferior frontal gyrus (IFG), 204,
    205, 206f
  left occipitotemporal cortex, 202–3
  medial temporal lobe (MTL), 151–52,
    156, 157
  memory systems, 93
  middle temporal gyrus (MTG), 202, 203
  parahippocampal place area
    (PPA), 131–33
  ventral tegmental area (VTA), 138–39
  white matter (*see* White matter [WM]
    tracts)
Brain reserve, 152–53. *See also* Cognitive
  reserve
Brief Test of Attention, 183
Brujería (witchcraft), 177
Bruner, Jerome, 5
Buddhism, 104, 184–85
Bühler, Karl and Charlotte, 34–35
Buros Center for Testing, 180

California, 169
CALP (cognitive and academic language
  proficiency), 220–21
Calques, 179
Cambodian Americans, 184
Canadians
  cultural heterogeneity, 9
  memory organization, 83
  memory scores, 83f
  neuropsychological assessment of ethnic
    minorities, 62–63
  non-English speakers, 169
  number and mathematical
    processing, 135–37
  self-reference memory, 89–90
  source/context memory, 87f, 87–88

Cantonese–English bilinguals, 159
Cantonese–Mandarin bilinguals, 159
Caretaker feedback or attention, 125–26
Caribbean, 178
Categorical relations. *See also*
    Classification of information
  Central Asia expedition studies, 43, 43*t*
  semantic studies, 200–2, 205–6
Catholicism, 177
Center for Cross-Cultural Research, 5
Center for Epidemiological Studies
    Depression Scale (CES-D), 105,
    107–9, 110–11
Central Asia expedition studies, 36–48,
    38*f*, 41*f*, 42*f*, 43*t*
CERAD (Consortium to Establish a
    Registry for Alzheimer's Disease), 181
CES-D (Center for Epidemiological
    Studies Depression Scale), 105,
    107–9, 110–11
CFA (confirmatory factor analysis), 258–
    59, 264, 267–68
CFI (comparative fit index), 267
Change Detection Task, 126, 127*f*
Child development, 34–35
  in autism, 208–9
  neurocognitive, 200–14, 206*f*, 207*f*
  typical changes, 202–3, 204–8,
    206*f*, 207*f*
Children. *See also* Pediatric
    neuropsychology
  bilingual, 70, 223–27
  culturally appropriate treatment
    for, 226–27
  ethnic minority, 70
  neuropsychological assessment
    of, 223–27
  neuropsychology of, 215–32
  US demographics, 215–16
China, 17–18, 185–86
Chinese Americans
  bilingualism, 172
  illness beliefs, 104–5
  mood disorders in, 104–5
  in US, 184
Chinese Beck Depression Inventory, 104–5
Chinese Canadians, 135–37
Chinese language
  linguistic properties, 203–4
  orthography, 203–4

Chinese speakers
  autobiographical memory, 91–92
  bilinguals, 159
  children, 200–14, 206*f*, 207*f*
  culturally appropriate treatment
    for, 226–27
  depression in, 112
  emotional memory, 88–89
  eye movements, 128
  information organization, 82–83
  information processing, 81–82
  memory, 83*f*, 86–88, 87*f*, 89–90
  memory organization, 82–83
  neurocognitive development in, 200–14,
    206*f*, 207*f*
  number and mathematical
    processing, 135–37, 137*f*
  self-reference memory, 89–90
  semantic development in, 200–14,
    206*f*, 207*f*
  source/context memory, 86–88, 87*f*
  tests translated, normed, and validated
    with, 185–86
  in US, 170–71, 216
Ciboney Arawaks, 178
Classification of information, 82–83.
    See also Categorical relations
Cleary model, 253–54
Clinical cultural neuroscience
  introduction, 1–31
  peer-reviewed manuscripts, 5–7, 6*f*
Clinically Useful Depression Outcome
    Scale (CUDOS), 110–11
Clinical neuropsychology, 15–16
Clinical reports, 188–89
Code switching, 178–79, 189, 220
Cognition
  disparities in, 235–36
  normal, 182
  visual, 124–50
Cognitive and academic language
    proficiency (CALP), 220–21
Cognitive competence, 254–55
Cognitive decline, 235–36
Cognitive impairment, 243–44
Cognitive psychology, 32–54
Cognitive reserve (CR), 152–53, 240
  and bilingualism, 153–56, 160–62, 161*f*
  future directions for research, 162
  opportunities for building, 241

Cognitive studies
  assessment studies, 65–67, 66t
  of depression in minorities, 111–12
Cole, Michael, 5, 33–34, 44, 47–49
Collective farms (kolkhozes), 36–37, 40, 42–43
Collectivism, 17–18, 92–93
Colombian Americans, 172
Communication barriers, 69, 256–57
Comparative fit index (CFI), 267
Competition model, 220
Composite International Diagnostic Interview, 260
Computed tomography (CT), 157
Computer technology, 143–44
Confirmatory factor analysis (CFA), 258–59, 264, 267–68
Confucianism, 92, 104, 184–85
Congo, 178
Consortium to Establish a Registry for Alzheimer's Disease (CERAD), 181
Construct bias, 254–55, 270
  strategies to minimize risk of, 263f, 263–64
Construct equivalence, 258
Construct irrelevance, 254
Construct underrepresentation, 254
Construct validity, 223
Context memory, 85, 86–88, 87f
Counting, 40. See also Number processing
CR. See Cognitive reserve
Cross-cultural psychology, 8, 9, 56, 81
Cross-cultural research, 44, 252
  intelligence studies, 254–55
  strategies to identify and reduce bias in, 262–63, 263f
CT (computed tomography), 157
Cuba, 178
Cuban Americans, 103–4, 172
CUDOS (Clinically Useful Depression Outcome Scale), 110–11
Cultural bias, 125–26, 141–43, 222
Cultural competence, 71
Cultural considerations
  for Asian and Pacific Islander language speakers, 184–85
  for Spanish speakers, 176
Cultural differences, 44–45
  in attention, 81–82, 93, 142–44
  in autobiographical memory, 84, 91–92, 93
  in behavior, 139–40, 140f
  in brain functions, 93
  communication challenges that arise from, 69
  in emotional memory, 88–89, 93
  in eye movement, 128
  idioms of distress, 104–5
  in information organization, 82–84, 93
  in information processing, 81–82, 93
  in memory, 82–94, 83f, 87f
  in number and mathematical processing, 135–38, 137f
  in perception, 81–82
  in positivity effects, 88–89
  in self-reference memory, 89–91
  in self-representation, 89–91
  in visual cognition, 124–50
  in visual perception, 126–44
Cultural diversity, 215–32
Cultural-historical theory, 32–54
  relevance and contemporary implications, 46–48
  testing, 36–46
Cultural integration, 13–14, 14f
Cultural loading, 221
Culturally appropriate treatment, 226–27
Cultural neuroscience, clinical, 1–31, 6f
Cultural psychology
  central theoretical processes, 8–9
  development of, 33–36
  peer-reviewed manuscripts, 5–7, 6f
Cultural separation, 13–14, 14f
Cultural theory, 33–36
Cultural values, 176
Culture
  definition of, 7–10
  explicit dimensions, 9
  gene-culture coevolution, 10–12
  high culture, 7
  implicit dimensions, 9
  social concept of, 102
  transmission of, 10–14, 13f
Culture-related publications, 5–7, 6f
Cuneus, 210f, 210

Dairy farming, 11–12
Danes, 91–92
DASS (Depression Anxiety Stress Scales), 109
DASS-21 (DASS short version), 109

Data, reference, 188
Data collection, 256–57
Declarative memory, 156
Deduction, 43–44
Dementia, 151
    AD-related, 237, 242
    bilingualism and, 153, 155, 162
    frontotemporal, 162
    future directions for research, 162
    pathology of, 237
Dementia Rating Scale (DRS), 112
Dementia Rating Scale-2 (DRS), 183
Demographics, 215–16
Depression, 100
    cognitive studies of, 111–12
    core features of, 102
    ethnic differences in, 100–5, 101t, 106, 111–12
    in minorities, 111–12
    neuroimaging studies of, 112
    prevalence of, 100–2
Depression Anxiety Stress Scales (DASS), 109
Depression questionnaires, 106–11
Diabetes mellitus, 239–40
Diagnostic and Statistical Manual of Mental Disorders (DSM-III), 106
Diagnostic and Statistical Manual of Mental Disorders (DSM-III-R), 16
Diagnostic Interview Schedule (DIS), 106
Differential item functioning (DIF), 242, 257, 268–70, 269f
Diffusion tensor imaging (DTI), 157
DIS (Diagnostic Interview Schedule), 106
Diversity, 55
    cultural, 215–32
    language, 170–74, 189, 215–32
Documentation, 188–89
Dolor de cerebro (brain ache), 104
Dopamine, 138–39
Dopamine receptor gene DRD4, 138–39, 139f
Dopaminergic reward system, 140–41, 141f
DRS (Dementia Rating Scale), 112, 183
DSM-III *(Diagnostic and Statistical Manual of Mental Disorders)*, 106
DSM-III-R *(Diagnostic and Statistical Manual of Mental Disorders)*, 16
DTI (diffusion tensor imaging), 157

Dual inheritance theory, 11–12
Dual language system hypothesis, 219–20
Durkheim, Emile, 32–33
Dysthymia, 101t

East Asians. *See also* Asians
    autobiographical memory, 91–92, 93
    cultural orientation, 92–93
    DRD4 gene carriers, 138–39, 139f
    emotional memory, 88–89, 93
    face processing, 133–35, 134f, 136f
    information processing, 81–82
    memory, 93
    memory for items and contexts, 85
    memory for objects and scenes, 85, 86
    memory mechanisms, 92–93
    memory organization, 82–84, 93
    memory specificity, 84
    number and mathematical processing, 135–38, 137f
    self-reference memory, 89–91
    self-representation, 89
    source/context memory, 86–87
    visual cognition, 124
    visual processing, 126–33, 127f, 129–30f, 131f, 132f, 143
East Indian culture, 217–18
East–West paradigm, 17–20
Ecuador, 172, 177–78
ECVI (expected cross-validation index), 267
Education, 220–21, 240–41
    parental, 215–16, 218
Educational Testing Service (ETS), 262
EFA (exploratory factor analysis), 264, 265–66
Effort, 217–18
ELL (English Language Learning), 224
Emic approach, 1–2
Emotion(s)
    processing emotional expressions, 135, 136f
    social-emotional competence, 254–55
Emotional memory, 88–89, 93
Enculturation, 12–13, 13f
Engagement, 217–18
English language, 253
    American English language proficiency, 222
    linguistic properties, 203–4
    orthography, 203–4

INDEX

English Language Learning (ELL), 224
English language proficiency, 223
English speakers, 253
  children, 200–14
  neurocognitive development of semantics, 200–14
  number and mathematical processing, 135–37, 137f
Environment
  physical, 256–57
  socio-environmental reinforcement learning, 138–42, 141f
Equitable treatment, 260–61
Equivalence, 257–59
  construct, 258
  definitions of, 257–58
  metric or measurement unit, 259
  scalar, 259
  strategies to identify and reduce nonequivalence, 262–70
  structural, 258–59
Ethical issues, 63–64
Ethnic differences
  in Alzheimer's disease, 234–35
  in depression, 100–2
  in mood disorders, 101t, 102–5
  in visual processing of objects and backgrounds, 126–28
Ethnic groups
  subgroups, 66
  in US, 184
Ethnicity, 9, 102
Ethnic minorities, 56. *See also* Minorities
  children, 70
  cognitive assessment of, 65–67, 66t
  depression in, 100–2, 111–12
  key findings, 57–68
  mood disorders in, 100–23
  neuropsychological assessment of, 55–80
Ethnography, 2–3
Etic approach, 1–2
ETS (Educational Testing Service), 262
European Americans, 101t
European Canadians, 89–90
Event-related potentials (ERPs), 130
Evolution, 10–12
Examination language, 187–89
Executive control (EC)
  age-related declines, 158

  effects of bilingualism on, 157–59, 159f, 160f
  effects of experience on, 160–61
  future directions for research, 162
Expected cross-validation index (ECVI), 267
Experience, 160–61
Experimental beginnings, 32–54
Exploratory factor analysis (EFA), 264, 265–66
Extreme response style, 256
Eye contact, 217, 218
Eye movement
  biological factors, 128
  responses associated with visual processing, 126–28, 129–30f
Eye-tracking studies, 81–82, 128

Face processing, 133–35, 134f, 136f
Facial Action Coding System (FACS), 135, 136f
Factorial invariance, 258–59
Fairness, 259–62
Familismo, 217–18
Family, 218–19
Farmers
  collective farm workers *(kolkhoz)*, 36–37, 40, 42–43
  independent farmers *(kulaks)*, 36–37
Farsi, 62
Fatalism, 177
Feedback
  caretaker, 125–26
  language of, 226
  social, 138–39
Fergana Valley, 40–41
FFA (fusiform face area), 18–19, 133–35, 134f
Filipino Americans
  bilingualism, 172
  in US, 184
Folk psychology, 3
Fractional anisotropy, 172–73
Frame-Line Task, 81–82, 126–28, 127f
French, 62
French speakers, 216
Frontotemporal dementia, 162
Fusiform face area (FFA), 18–19, 133–35, 134f
Future directions, 162, 210–11, 227–28, 244

Galton, Francis, 32–33
Gates Foundation, 222–23
GDS (Geriatric Depression Scale), 109–10
GDS-SF (GDS short-form), 109–10
Gender differences, 217–18
Gene-culture coevolution, 10–12
Generalized Procrustes analysis, 265
Genetics, 238–39
Geriatric Depression Scale (GDS), 109–10
German, 62
German speakers, 81–82, 207–8
Gestalt psychology, 42–43
GFI (goodness-of-fit index), 267
Ghost sickness, 105
Glozman, Janna, 48–49
Goodness-of-fit index (GFI), 267
Graduate Records Exam (GRE), 59
Graphic grouping methods, 43t
Gray matter volume, 156
GRE (Graduate Records Exam), 59
Greater Antilles, 178
Greenland, 91–92

HADS (Hospital Anxiety and Depression Scale), 110–11
Hamilton Depression Rating Scale (HDRS or HAM-D), 103, 110–11
Handedness, 11–12
HDRS (Hamilton Depression Rating Scale), 103, 110–11
Head nodding, 218
Health and Retirement Study (HRS), 267–68
Heart-break syndrome, 105
Henry, John, 103
High culture, 7
Hindi, 185–86
Hinduism, 184–85
Hippocampus, 93
Hispanic Neuropsychological Society, 60–61, 68–69, 180, 216–17
Hispanics, 218–19
  bilingualism, 172
  cognitive function and decline, 235
  depression in, 112
  mood disorders in, 101t, 103–4
  neuropsychological assessment of, 62
  non-English speakers, 169, 170–71
  parental education, 215–16
  response styles, 256
  subgroups, 66
  testing outcomes, 261
Historical considerations, 2–7
  for Asian and Pacific Islander language speakers, 184–85
  for contemporary Spanish, 177–78
  cultural-historical approach, 32–54
  for Spanish speakers, 176
Hmong, 184
Holistic thought, 17–18
Hopkins Verbal Learning Test–Revised, 183
Hospital Anxiety and Depression Scale (HADS), 110–11
HRS (Health and Retirement Study), 267–68

Ichkary (illiterate women), 40, 42–43
Idioms of distress, 104–5
IDS (Inventory for Depressive Symptomatology), 110–11
Illiterate women *(ichkary)*, 40, 42–43
Illness beliefs, 104–5
Immigrants and immigration, 14
  and bilingualism, 154–55
  in US, 170–71
India, 154–55, 185–86
Indian Americans, 172
Indigenous languages, 177–78
Indigenous psychology, 8–9
Individualism, 17–18, 92–93
Indo-European languages, 169
Inference studies, 43–44
Inferior frontal cortex, 205, 206f, 210f, 210
Inferior frontal gyrus (IFG), 204, 205, 206f
Information organization, 82–83, 93
Information processing, 81–82, 93
INS (International Neuropsychological Society), 56–57
Instrument bias, 256
Integration, 13–14, 14f
Intelligence
  academic, 254–55
  analytical, 254–55
  worldly, 4–5
Intelligence studies
  cross-cultural studies, 254–55
  Western tests, 186
International Association for Cross-Cultural Psychology, 5

International Conference on Social Psychological Research in Developing Countries, 5
International Neuropsychological Society (INS), 56–57
International Test Commission, 70–71, 222–23
Standards for Education and Psychological Testing, 65, 253–54, 260–61, 262
Interpretability, 265
Interpreters, 63–64, 69, 174, 175–76
  neuropsychological assessment of bilingual children with, 224–25
  selection of, 224–25
Interviews, 105–6
Interview tools, 105–11
Inventory for Depressive Symptomatology (IDS), 110–11
Iowa Gambling Task, 181–82
IRT (item response theory), 270
Item bias, 257, 270
  strategies to minimize risk of, 263f, 264
Item response theory (IRT), 270

Janet, Pierre, 32–33
Japan, 17–18
Japanese
  emotional memory, 88
  information processing, 81–82
  memory for objects and scenes, 85
  tests translated, normed, and validated with, 185–86
  visual processing, 126, 127f
Japanese Americans, 184
Journal of Cross-Cultural Psychology, 5

Kaiser criterion, 265–66
Kamchatka peninsula studies, 48–49
Kentucky Nun Study, 152
Koffka, Kurt, 34–35, 39–43, 42f, 44–46
Köhler, Wolfgang, 39–40
Kolkhoz (collective farm) workers, 36–37, 40, 42–43
Korea, 17–18
Korean Americans, 184
Koreans, 88–89, 185–86
Kpelle tribe, 47–48
Kulaks (independent farmers), 36–37
Kyrgyzstan (Kirghizia), 4–5, 37–39, 40–41, 44–45

Lactose intolerance, 11–12
Language, 62–65
  challenges related to, 69
  cultural idioms of distress, 104–5
  of examination, 187–89
  of feedback, 226
  indigenous languages, 177–78
  verbal, 226
  written, 226
Language diversity, 185–86, 189
  in pediatric neuropsychology, 215–32
  in US, 170–74
Language proficiency
  American English language proficiency, 222
  cognitive and academic language proficiency (CALP), 220–21
  English language proficiency, 223
  and testing, 223
Latin America, 20, 169
Latinos
  culturally appropriate treatment for, 226–27
  factors impacting effort and engagement, 217–18
  mood disorders, 101t, 104
  neuropsychological assessment of, 62
  parental education, 215–16
  subgroups, 66
Lazarus, Moritz, 3
Learning
  comparable opportunity for, 261
  socio-environmental reinforcement, 138–42, 141f
Leontiev, Alexei, 3–4, 35
Leventueff, P., 40
Lewin, Kurt, 34–35, 39–40
Lewy bodies, 237
Linguistic demands, 221–22
Linguistics. *See also* Language diversity
  Chinese vs English structural differences, 203–4
List Learning, 186
Logistic regression approach, 270
Lothian Birth Cohort, 155–56
Luo people, 254–55, 258
Luoro, 254–55, 258
Luria, Alexander R., 3–5, 35–36
  Central Asia expedition studies, 36–40, 38f, 41–48, 42f, 43t
  ongoing studies, 48

Machismo, 176–77
MADRS (Montgomery-Asberg Depression Rating Scale), 110–11
Magnetic resonance imaging, functional (fMRI), 130, 256–57
Major depressive disorder, 101*t* *See also* Depression
Malaysians, 81–82, 184–85
Mandarin, 203
Mania, 101*t*
Mantel–Haenszel method, 270
Mareos (dizziness), 104
Marginalization, 13–14, 14*f*
Marx, Karl, 3–4
Marxism, 3–4
Mathematical processing
  Central Asia expedition studies, 40
  cultural differences in, 135–38, 137*f*
Mattis Dementia Rating Scale (DRS), 112, 183
MDS (multidimensional scaling), 264–65
Measurement unit equivalence, 259
Medial temporal lobe (MTL), 151–52, 156, 157
Melanesia, 2–3
Memory
  age differences in, 83*f*
  autobiographical, 84, 91–92, 93
  cultural differences in, 82–94, 83*f*, 87*f*
  declarative, 156
  effects of bilingualism on, 156–57
  emotional, 88–89, 93
  for items and contexts, 85
  mechanisms of, 92–93
  for objects and scenes, 85–86, 93
  organization of, 82–84
  quality of, 84
  quantity of, 84
  self-reference, 89–91, 93
  source/context, 85, 86–88, 87*f*
  specificity of, 84
Memory loss, early, 151–52
Method bias, 255–57, 270
  strategies to minimize risk of, 263*f*, 263–64, 270–71
Methodological challenges, 242–44
Metric equivalence, 259
Mexican Americans, 103–4

Mexicans, 91–92
Middle East, 20
Middle temporal gyrus (MTG), 202–3, 210
Migration, 154
Mindt, Rivera, 55
Mini International Neuropsychiatric Interview (MINI), 106
Mini-Mental State Exam (MMSE), 112, 234, 242
Minorities, 215–16. *See also* Ethnic minorities
Minority (neuro)psychology, 56
MMSE (Mini-Mental State Exam), 112, 234, 242
Mobile technology, 143–44
Modified Wisconsin Card Sorting Test, 183
Montana, 169
Montgomery-Asberg Depression Rating Scale (MADRS), 110–11
Mood disorders
  assessment of, 100–23
  clinical presentation of, 102–5
  ethnic differences in, 101*t*, 102–5
  in ethnic minorities, 100–23
  prevalence of, 100
  symptom profiles, 102–5
Moscow Institute of Experimental Psychology, 38–39
Moscow Institute of Psychology, 3–4
Moscow State University, 39–40
MTG (middle temporal gyrus), 202–3, 210
MTL (medial temporal lobe), 151–52, 156, 157
Multiculturalism, 71, 219–20
Multicultural Problem Solving Summit, 59
Multicultural psychology, 56
Multidimensional scaling (MDS), 264–65
Multilingualism, 154–55, 219. *See also* Bilingualism

Nahuatl, 177–78
NAN (National Academy of Neuropsychology), 56–57, 187, 216–17
Naryn River region, 38–39
National Academy of Neuropsychology (NAN), 56–57, 187, 216–17

National Council on Measurement in
  Education, 65, 253–54, 260–61, 262
National Institute on Aging (NIA), 182
Nationality, 9
Native Americans
  depression in, 112
  mood disorders in, 101t, 105
Nature vs nurture, 1–2
Neighborhood factors, 218–19
Nervios (nerves), 104
NeSBHIS (Neuropsychological Screening
  Battery for Hispanics), 181
Neurobiology, 138–42
Neurocognitive development, 200–14
Neuroimaging, 112
NEURONORMA project, 179–80, 182–83
Neurophysiology
  responses associated with face
    processing, 133–35, 134f
  responses associated with visual
    processing, 130–33, 131f, 132f
NEUROPSI: Atención y Memoria
  (NEUROPSI: Attention and Memory)
  test, 180–81
Neuropsychological assessment
  appropriate evaluators, 224
  appropriate tests, 188
  of Asian and Pacific Islander language
    speakers, 185–86
  bilingual, 225–26
  of bilingual children, 223–27
  challenges associated with, 68–70, 68t
  culturally biased, 222
  decisional algorithm for, 187f, 187
  of ethnic minorities, 55–80
  examination language, 187–89
  future directions, 227–28
  goals, 71
  indications for adjustment, 69
  with interpreters, 224–25
  interpretive approaches to, 65–67, 66t
  language, 62–65
  non-English language, 62f, 62
  of non-English speakers, 63–65, 63t,
    169–99, 187f
  normative reference data, 188
  pillars of, 222
  recommendations for, 70–72, 187–89

  of Spanish speakers, 179–83
  training for, 60–62, 60t, 61t, 71
  with translators, 224–25
Neuropsychological Assessment Survey,
  56–68, 58f
  interpretation and challenges, 65–68
  items related to assessment of ethnic
    minorities, 78–80
  limitations, 68–70
  method overview, 56–57
  response rate, 57–58
Neuropsychological Battery of Executive
  Functions and Frontal Lobes (*Batería
  Neuropsicológica de Funciones
  Ejecutivas y Lóbulos Frontales,
  BANFE*), 181–82
Neuropsychological Screening Battery for
  Hispanics (NeSBHIS), 181
Neuropsychologists
  ethnic designations, 58f, 58–59, 71
  practice time, 58–60, 59f
  professional training, 60–62, 60t, 61t, 71
  Spanish-speaking, 216–17
  survey of, 56–59, 58f
Neuropsychology
  clinical, 15–16
  cultural, 5–7, 6f, 8–9, 33–36
  cultural factors, 9
  cultural-historical approach to, 32–54
  pediatric, 215–32
Neurorehabilitation, 183
Neuroscience
  cultural, 1–31, 6f
  neural evidence for cultural influences
    on visual perception, 126–38
New Mathematics Project, 47–48
NIA (National Institute on Aging), 182
Non-English languages
  assessment in, 62f, 62–65, 63t
  in US, 170–71
Non-English language speakers
  language proficiency, 187–88
  neuropsychological assessment of,
    169–99, 187f
Nonequivalence, 262–70
Normal cognition, 182
Normative reference data, 188
Norm-referenced measures, 222

Norms, word association, 201
Nuisance factors, 254
Number processing, 135–38, 137f
Nurture vs nature, 1–2

Objects
  memory for, 85–86, 93
  visual processing of, 126–28
Older adults
  bilingual, 156–59, 160f
  emotional memory, 88–89, 93
  executive control systems, 157–59
  memory systems, 156–57
  population growth, 151
  in US, 233
  visual processing, 143
Old Order Anabaptists, 218, 226–27
Opportunity to learn, 261
Optical illusions, 44–45
Orthography, 203–4

Pakistani Americans, 184–85
Pan-Asian culture, 184–85
Parahippocampal place area (PPA), 131–33
Parental education, 215–16, 218
Paro, 254–55, 258
Patient Health Questionnaire-9 (PHQ-9), 110
Pediatric neuropsychology, 215–32. *See also* Children
Peer-reviewed manuscripts, 5–7, 6f
Perception
  cultural differences in, 81–82
  visual, 124–25
Personalismo, 217
Peru, 177–78
Peruvian Americans, 172
Philippines, 184–85
PHQ-9 (Patient Health Questionnaire-9), 110
Physical environment, 256–57
Piaget, Jean, 34–35
PISA (Programme for International Student Assessment), 262–63
Poggendorf illusion, 42–43, 44–45
Positivity effects, 88–89, 93
Postmodernism, 8
PPA (parahippocampal place area), 131–33

Predictive bias, 253–54
Present State Examination (WHO), 106
Pribram, Karl, 5
Primary Care and Obstetrics-Gynecology Studies, 110
Primary Care Evaluation of Mental Disorders questionnaire (PRIME-MD), 110
Procrustes analysis, generalized, 265
Programme for International Student Assessment (PISA), 262–63
Psychological testing, 261. *See also* Neuropsychological assessment
Psychology
  clinical neuropsychology, 15–16
  cognitive, 32–54
  cross-cultural, 8, 9, 56, 81
  cultural, 5–7, 6f, 8–9, 33–36
  folk, 3
  Gestalt, 42–43
  indigenous, 8–9
  minority (neuro)psychology, 56
  multicultural, 56
  neuropsychology, 9, 32–54, 215–32
  personal components, 45–46
Psychometric theory and methods, 254
Psychosocial distress, 104–5
Psychosocial factors, 241–42
Publications, culture-related, 5–7, 6f
PubMed, 5–7, 6f
Puerto Ricans, 103–4

QIDS (Quick Inventory of Depressive Symptomatology), 110–11
Quechua, 177–78
Questionnaires, 106–11
Quick Inventory of Depressive Symptomatology (QIDS), 110–11

Racial differences
  in AD, 233–51
  in brain aging, 238–42
  in cognitive function and decline, 235
  in cognitive risk, 243–44
  future directions for research, 244
  methodological challenges in studying, 242–44
Ratchet effect, 10

INDEX

RBANS (Repeatable Battery for the Assessment of Neuropsychological Status), 182, 186
Recording devices, 256–57
Reference data, normative, 188
Reinforcement learning, socio-environmental, 138–42, 141f
Relative chi-square ($\chi2/df$), 267
Relativism, 1–2
Relevance 2050 initiative (AACN), 215, 216–17
Religious beliefs, 177
Repeatable Battery for the Assessment of Neuropsychological Status (RBANS), 182, 186
Research
  cross-cultural, 252
  East–West paradigm, 17–20
  future directions for, 162, 210–11, 244
  peer-reviewed manuscripts, 5–7, 6f
Response styles, 256
Rey Auditory Verbal Learning Test, 186
Rey–Osterrieth Complex Figure test, 183
Rieko, 254–55, 258
Right-handedness, 11–12
Rivers, William H. R., 2–3, 4
Russian Far East studies, 48–49
Russians, 81–82

SADS (Schedule for Affective Disorders and Schizophrenia), 106
Sample bias, 255–56
Santería, 177
Scalar equivalence, 259
Scalar (strong) factorial invariance, 258–59
Scenes: memory for, 85–86, 93
Schedule for Affective Disorders and Schizophrenia (SADS), 106
Schedules for Clinical Assessment in Neuropsychiatry (WHO), 106
Schizophrenia, 102–3
Selection bias, 243–44
Self-reference effects, 89–90, 93
Self-reference memory, 89–91
Self-representation, 89–91
Semantics
  association strength, 200–2
  categorical relations, 200–2, 205–6
  future directions for research, 210–11
  neurocognitive development of, 200–14, 206f, 207f
  organization of, 202–3, 208–9, 211
  processing of, 204–8, 206f, 207f, 209–11
  typical developmental changes, 202–3, 204–8, 206f, 207f
SENAS (Spanish and English Neuropsychological Assessment Scales), 179–80, 181
Sensitivity, test, 223
Shakhimardan, 40–41, 41f, 44–45
Shamimardan, 38–39
Shen jing bing (insanity), 104–5
Singaporean East Asians, 130, 131f
Social-emotional competence, 254–55
Social feedback, 138–39
Socialization, 12–13, 13f
Socially desirable response style, 256
Social norms
  face processing differences associated with, 133–35, 134f
  violations of, 139–40, 140f
Society for Clinical Neuropsychology, 61
Society for Indian Psychologists, 60–61
Society for the Psychological Study of Culture, Ethnicity and Race, 60–61
Sociocultural factors, 217–18, 242
Sociocultural theory. *See* Cultural-historical theory
Socioeconomic status (SES), 218–19, 235
Socio-environmental reinforcement learning, 138–42, 141f
Source/context memory, 85, 86–88, 87f
South Africa, 261
South America, 177–78
South Asia, 20
Soviet Union, 36–37
Spanglish, 179
Spanish, 62, 185–86, 189, 253
  contemporary, 177–78
  historical influences on, 177–78
  in US, 170–71
Spanish and English Neuropsychological Assessment Scales (SENAS), 179–80, 181

Spanish speakers, 253
  cultural considerations for, 176
  historical considerations for, 176
  neuropsychological assessment of, 176, 179–83
  neuropsychologists, 216–17
  in US, 172, 216
Specificity, test, 223
Spiritual practices, 177
SRMR (standardized root mean squared residual), 267
Stalin, Joseph, 36–37
Standardized root mean squared residual (SRMR), 267
Standards for Education and Psychological Testing (AERA), 65, 253–54, 260–61, 262
Steinthal, Heymann, 3
Stern, Wilhelm, 34–35
Strict factorial invariance, 258–59
Strong (scalar) factorial invariance, 258–59
Stroop Color-Word Test, 181–82, 183
Structural equivalence, 258–59
Structured Clinical Interview for DSM disorders, 106
Suicidality, 106
Symbol Digit Modalities Test, 183
Symbolic posters, 40
Symbols, 40

Tagalog speakers, 170–71
Taino, 178
Taiwanese, 88
Taoism, 92, 184–85
Tashkent, Uzbekistan, 37
Task switching, 158, 160*f*
Technology, 143–44, 256–57
Test of Memory Malingering, 183
Tests and testing
  cultural loading, 221
  issues with measures, 221–22
  linguistic demands, 221–22
  norm-referenced measures, 222
  outcomes, 261
  psychological, 261
    (*see also* Neuropsychological assessment)
  test equivalence, 221

test sensitivity, 223
test specificity, 223
validity of, 222–23
Texas, 169
Thai, 184
Thailand, 186–87
Theoretical beginnings, 32–54
TLI (Tucker–Lewis index), 267
Toda people, 2–3
Tower of Hanoi test, 181–82
Trail Making Test, 183, 255–56
Training for neuropsychologists, 60–62, 60*t*, 61*t*, 71
Translation, 64–65, 69, 174–75, 220
  adaptation, 175
  application, 175
  assembly, 175
  back, 174–75
  test adaptation/translations, 222–23
Translators
  neuropsychological assessment of bilingual children with, 224–25
  selection of, 224–25
Tucker–Lewis index (TLI), 267
Tucker's phi, 266
Turks, 84

UDS (Uniform Data Set) test battery, 182
Ukrainian Psychoneurological Academy (Kharkov), 39–40
Uniform Data Set (UDS) test battery, 182
United States
  aging population, 233
  baby names, 10–11
  bilingualism, 170, 171–74
  demographics, 215–16
  ethnic groups, 184
  ethnic minorities, 55, 62–63
  fairness in testing outcomes, 261
  immigrants, 170–71
  language diversity, 170–74
  memory for objects and scenes, 86
  minority population, 215–16
  mood disorders, 100
  neuropsychological assessment of ethnic minorities, 62–63
  non-English speakers, 169
Universalism, 1–2, 15–17

Uzbekistan, 4–5, 37–41, 41f, 42–43, 44–45
Uzbek Pedagogical Academy, 39–40
Uzbek Research Institute of Samarkand, 38–39
Uzbek Socialist Soviet Republic, 39–40

Validity
  construct, 223
  test, 222–23
Values, cultural, 176
Vascular conditions, 239–40
Ventral tegmental area (VTA), 138–39
Verbal Fluency, 183
Verbal language, 226
Vico, Giambattista, 3
Vietnamese Americans, 184
Vietnamese speakers, 170–71, 216
Visual cognition, 124–50
Visual illusions
  Central Asia expedition findings, 42–43
  Central Asia expedition studies, 44–45
  Poggendorf illusion, 42–43, 44–45
Visual perception
  cultural differences in, 124–25, 138–42
  cultural influences on, 126–38, 142–44
Visual processing
  behavioral movement responses associated with, 126–28, 127f
  eye-movement responses associated with, 126–28, 129–30f
  face processing differences associated with social norms, 133–35, 134f
  neurophysiological responses associated with, 130–33, 131f, 132f
Visuospatial processing, 130, 131f
Völkerpsychologie (folk psychology), 3
Vygotsky, Lev S., 3–5, 34–35, 40
  Central Asia expedition studies, 36–39, 38f, 46–48
  cultural-historical theory, 32–54
  death of, 35, 44

WAIS-III (Wechsler Adult Intelligence Scale-III), 267–68
Washington Heights–Inwood Columbia Aging Project (WHICAP), 237–38
Weak factorial invariance, 258–59
Wechsler Adult Intelligence Scale-III (WAIS-III), 267–68
Wechsler Digit Span test, 253
Wechsler Intelligence tests, 185–86
Wechsler Memory tests, 185–86
WEIRD (Western, educated, industrialized, rich, and democratic) subject samples, 2, 16–17
Werner, Heinz, 34–35
Western, educated, industrialized, rich, and democratic (WEIRD) subject samples, 2, 16–17
Westerners
  autobiographical memory, 84, 91, 92, 93
  cultural orientation, 92–93
  DRD4 gene carriers, 138–39, 139f
  emotional memory, 88–89
  face processing, 133–35, 134f, 136f
  information processing, 81–82
  memory, 93, 94
  memory for items and contexts, 85
  memory for objects and scenes, 85–86
  memory mechanisms, 92–93
  memory organization, 82–84, 93
  memory specificity, 84
  number and mathematical processing, 135–38, 137f
  self-reference memory, 89–90, 91
  self-representation, 89
  source/context memory, 86–87
  visual cognition, 124
  visual processing, 126–33, 129–30f, 131f, 132f, 143
Western intelligence tests, 186
WHICAP (Washington Heights–Inwood Columbia Aging Project), 237–38
White matter (WM) tracts, 151–52, 156, 157
  age-related declines, 158
  effects of bilingualism on, 158, 159f
Winjo, 254–55, 258
Wisconsin Card Sorting Test, 181–82
WM. *See* White matter
Women, illiterate *(ichkary),* 40, 42–43
Word association norms, 201
World Bank toolkit, 222–23
World Health Organization (WHO), 106
Worldly intelligence, 4–5

Written language
  of feedback, 226
  orthography, 203–4
Wundt, Wilhelm, 3, 4

Yasnitsky, Anton, 44
Yordan, 38–39

Yucatan studies, 47–48

Zaporozhets, Alexander, 37
Zung Self-Rating Depression Scale
  (Zung SDS), 110–11

$\chi 2/df$ (relative chi-square), 267